Machine Learning Applied to Optical Communication Systems

Machine Learning Applied to Optical Communication Systems

Guest Editors

Jinlong Wei
Zhaopeng Xu

MDPI

Basel • Beijing • Wuhan • Barcelona • Belgrade • Novi Sad • Cluj • Manchester

Guest Editors
Jinlong Wei
Department of Optoelectronic Information and Fiber-Optic Communication
Pengcheng Laboratory
Shenzhen
China

Zhaopeng Xu
Department of Optoelectronic Information and Fiber-Optic Communication
Pengcheng Laboratory
Shenzhen
China

Editorial Office
MDPI AG
Grosspeteranlage 5
4052 Basel, Switzerland

This is a reprint of the Special Issue, published open access by the journal *Photonics* (ISSN 2304-6732), freely accessible at: www.mdpi.com/journal/photonics/special_issues/KTE30Q2DWG.

For citation purposes, cite each article independently as indicated on the article page online and as indicated below:

Lastname, A.A.; Lastname, B.B. Article Title. *Journal Name* **Year**, *Volume Number*, Page Range.

ISBN 978-3-7258-4096-0 (Hbk)
ISBN 978-3-7258-4095-3 (PDF)
https://doi.org/10.3390/books978-3-7258-4095-3

Contents

About the Editors

Jinlong Wei

Jinlong Wei is currently a Professor in Pengcheng Laboratory, Shenzhen, China, leading a research team focusing on the digital and optical signal processing of optical communication/sensing systems. He received his PhD degree from Bangor University, Bangor, UK, in 2011 and joined the University of Cambridge, UK, as a research associate (2011–2014). He was awarded an EU Marie Curie fellowship and conducted the award research in AVDA Optical Networking SE, Germany (2014–2016). He joined Huawei Technologies German Research Center, Munich, Germany, as a senior researcher and later a principal researcher (2016–2023). He has published over 200 papers in prestigious journals, including *Nature Electronics*, and conferences, as well as over 20 EU/US/CN patents. He has delivered 30 invited talks such as 2 plenary talks at international conferences, including OFC. His various pioneering work on optical access and data center networks was reported by BBC, Reuter, Yahoo, OSA, Science and Technology Daily, etc. He is a senior member of IEEE and Optica (formerly OSA) and a Marie Curie Fellow, and he was selected as one of the World's Top 2% Scientists in 2024 (career-long impact), 2022, 2021, and 2020 by Elsevier and Stanford University.

Zhaopeng Xu

Zhaopeng Xu received his B.S. degree from the School of Electronics Engineering and Computer Science, Peking University, in 2015 and his M.S. degree from the State Key Laboratory of Advanced Optical Communication Systems and Networks, Peking University, in 2018. He obtained his Ph.D. degree from the Department of Electrical and Electronic Engineering, the University of Melbourne, in 2022. Since 2023, he has been serving as a researcher at the Department of Opto-electronic Information and Fiber-Optic Communication, Pengcheng Laboratory. He has been engaged in research on optical communication, with major research interests including interference mitigation techniques in optical systems, high-speed digital signal processing, and the integration of machine learning with optical communications. He has published over 100 papers in internationally renowned journals and conferences. He has been listed among the world's top 2% scientists in 2024, and he has received a number of honors such as the Young Scientist Award at ICOCN2024, the Best Industrial Innovation Paper Award at ACP2023, and the Best Student Paper Award at ACP2020.

photonics

MDPI

Editorial

Machine Learning Applied to Optical Communication Systems

Zhaopeng Xu and Jinlong Wei *

Pengcheng Laboratory, Shenzhen 518055, China; xuzhp@pcl.ac.cn
* Correspondence: weijl01@pcl.ac.cn

1. Introduction

As the global demand for high-speed and high-capacity communication continues to surge, driven by cloud computing, artificial intelligence, 5G, virtual reality, and the Internet of Things (IoT), optical communication systems have emerged as the backbone of modern digital infrastructure [1–5]. However, the increasing complexity, performance requirements, and operational scale of these systems have begun to exceed the capabilities of traditional analytical and rule-based design methods. In this context, machine learning (ML) has become a transformative tool, enabling data-driven solutions that can adapt to dynamic conditions, extract hidden patterns, and optimize performance across the optical communication stack.

In long-haul coherent optical systems, where signal integrity is challenged by fiber non-linearities and polarization effects over hundreds of kilometers, ML techniques such as neural networks (NNs) and generative models have been deployed for non-linear compensation, channel equalization, and signal reconstruction [6–9]. For instance, deep learning models trained on simulated or real-world data can outperform conventional Volterra filters or digital backpropagation, offering both improved accuracy and flexibility [10–13]. In short-reach and direct-detection systems, including intensity-modulated direct-detection (IM/DD) links widely used in data centers, ML is increasingly used to counteract combined linear and non-linear impairments [14–17]. For example, NNs and support vector machines (SVMs) have been shown to enhance equalization, improve bit error rate (BER) performance, and adapt to non-idealities in low-cost hardware implementations, such as directly modulated lasers (DMLs), vertical cavity surface emitting lasers (VCSELs), and silicon photonic transceivers [18–21]. Passive optical networks (PONs) and optical access systems demand high bandwidth at a low cost, with minimal latency and high reliability. Here, ML is also employed for intelligent receiver implementation [22–25], enabling better system resilience and performance optimization.

Visible light communication (VLC) and optical wireless communication (OWC) are emerging as complementary technologies for indoor environments and last-meter access [26–28]. ML has proven particularly effective in these domains, supporting signal demodulation, distortion compensation, and positioning through image-based or channel-aware learning techniques. End-to-end learning using autoencoders and convolutional neural networks (CNNs) has shown strong potential to optimize the optical-to-electrical (O/E) conversion and enhance signal robustness [29–32]. In more advanced and non-traditional systems, such as chaos-based secure communication and photonic reservoir computing, ML enables reliable chaos recovery, signal decoding, and multi-channel synchronization [33–36]. These systems benefit from the adaptability and high dimensionality of learning-based methods, which can capture complex temporal and spectral features beyond the reach of conventional models. Finally, optical network management has also

 https://doi.org/10.3390/photonics12050458

embraced ML in the form of AI-enhanced network orchestration, software-defined control, and optical performance monitoring [37–40]. These applications support real-time optimization, fault localization, and predictive maintenance, aligning with the goals of intelligent and autonomous network operation [41–44].

By covering this rich spectrum of systems and applications, the papers in this Special Issue collectively demonstrate the vast potential of ML in shaping the next generation of optical communication networks. From physical-layer enhancements to system-level intelligence, ML is not only a tool for performance improvement, but also an enabler of fundamentally new architectures and functionalities.

2. An Overview of the Published Articles

This Special Issue, "Machine Learning Applied to Optical Communication Systems", brings together a diverse collection of research contributions that highlight the synergy between ML and a wide range of optical communication technologies. This Special Issue collects 14 diverse contributions, including three comprehensive review papers and 11 original research articles that exemplify the growing synergy between machine learning and optical communication systems. The selected works span various system types, including long-haul coherent transmission, short-reach interconnects, PONs, OWC, and chaos-based secure transmission systems. Each type of system poses distinct challenges ranging from non-linear impairments and chromatic dispersion to noise accumulation and hardware limitations, and ML offers promising techniques to address them. Below is a brief overview of each contribution.

For the research articles, in contribution 1, He et al. propose a modified regular perturbation (MRP) model enhanced with trainable parameters to improve the accuracy of fiber transmission modeling under dispersion and non-linearity. This hybrid physical–ML model effectively reduces fitting error, even under high launch power and dual-polarization transmission scenarios. In contribution 2, Freitas and Pires present an NN-based framework for the rapid estimation of capacity and cost in large-scale multi-fiber optical networks. The model achieves high accuracy with significant computational efficiency, serving as a practical tool for real-time network design and planning. In contribution 3, Vu et al. introduce DeepChaos+, a deep learning-based framework for chaos signal removal in wavelength division multiplexing (WDM) systems. Their model enhances detection performance while reducing the BER by about three orders of magnitude, offering a robust solution for chaos-based secure communications. In contribution 4, Srinivasan et al. propose a novel ML-driven equalization method using gradient-based optimization for VCSEL-based transceivers. The technique improves signal integrity in thermally unstable environments, a key need for data center and automotive interconnects. In contribution 5, Liem et al. present an ML-enhanced resilience mechanism for NG-EPONs to support ultra-reliable tactile Internet applications. The system utilizes ML for fault detection and software-defined networking (SDN) for proactive recovery, achieving excellent performance in reliability metrics. In contribution 6, Ji et al. apply attention-based CNNs to mitigate angle-induced distortion in camera-based visible light positioning systems. Their model significantly reduces the number of positioning errors, paving the way for practical and precise indoor navigation. In contribution 7, Osahon et al. demonstrate that a multilayer perceptron-based decision feedback equalizer (DFE) significantly outperforms the traditional methods in OWC links, particularly under high non-linear distortions. This paves the way for higher data rates in safe, short-range optical wireless links. In contribution 8, Luna-Rivera et al. address the often-neglected role of optical-to-electrical conversion in VLC systems. Using autoencoder architectures, they improve robustness and system performance, offering insights into practical VLC design for 5G and IoT networks. In contribution 9, Hung et al. propose a Mach–Zehnder interferometer (MZI)-based optical NN to regenerate the intensity-modulated signals distorted by the bandwidth limitations of Silicon micro-ring modulators. Their photonic NN matches digital NN performance, underscoring the promise of optical ML hardware for high-speed systems. In contribution 10, Zhong et al. present a photonic reservoir computing system using quantum dot spin-VCSELs for coherent optical chaos-based secure communication. They achieve robust demodulation of complex modulated signals, validating the potential of optical chaos and reservoir computing. In contribution 11, Guo et al. tackle interference from auxiliary management channels in PON systems using Gaussian mixture model (GMM)-based probabilistic shaping. The joint optimization at transmitter and receiver significantly reduces the bit error rates, improving system robustness.

For the review articles, in contribution 12, Xu et al. review NN-based equalizers for IM/DD systems, addressing their ability to mitigate non-linear impairments while highlighting the challenge of computational complexity. The paper offers a comparative analysis of network types and proposes strategies for reducing model size and power consumption, paving the way for practical deployments. In contribution 13, Shao et al. survey ML applications in short-reach systems, focusing on digital signal processing (DSP), monitoring, and control. A key contribution is their taxonomy of time series models, offering a structured view of recent advances. The review also discusses the limitations in the current methods and suggests directions for more efficient and scalable ML integration. In contribution 14, Wu et al. provide an overview of the ML techniques in self-coherent systems, emphasizing improvements in signal recovery and phase tracking. The paper discusses how ML enhances self-coherent detection performance while reducing hardware complexity, and outlines future trends in low-cost, ML-assisted self-coherent designs.

3. Conclusions

In summary, this Special Issue highlights the growing synergy between ML and optical communication systems, showcasing advancements across different optical communication applications. The eleven original research articles and three review papers collectively demonstrate how ML techniques can effectively tackle system impairments, optimize performance, and enable intelligent network management. These contributions underline the increasing role of ML as a core technology in optical systems, while also pointing toward key future challenges such as complexity reduction and real-time adaptability. ML is no longer a peripheral or auxiliary tool for optical communication; it is becoming a fundamental enabler of the next generation of optical systems. We hope that this Special Issue serves not only as a record of current advancements, but also as a roadmap for future exploration.

Acknowledgments: We would like to sincerely thank all the authors, reviewers, and editorial staff who contributed to the success of this Special Issue. We hope that this collection of works will inspire further research, foster new interdisciplinary collaborations, and accelerate the intelligent transformation of optical communication systems in the coming years.

Conflicts of Interest: The authors declare no conflicts of interest.

List of Contributions:

1. He, S.; Li, Z.; Xing, S.; Yan, A.; Zhou, Y.; Shi, J.; Shen, C.; Li, Z.; He, Z.; Chen, W.; et al. A Modified Regular Perturbation Model for the Single-Span Fiber Transmission Using Learnable Methods. *Photonics* **2024**, *11*, 1178. https://doi.org/10.3390/photonics11121178.
2. Freitas, A.; Pires, J. Using Artificial Neural Networks to Evaluate the Capacity and Cost of Multi-Fiber Optical Backbone Networks. *Photonics* **2024**, *11*, 1110. https://doi.org/10.3390/photonics11121110.

3. Vu, D.A.; Do, N.K.H.; Nguyen, H.N.T.; Dam, H.M.; Tran, T.T.T.; Nguyen, Q.X.; Truong, D.C. DeepChaos+: Signal Detection Quality Enhancement of High-Speed DP-16QAM Optical Fiber Communication Based on Chaos Masking Technique with Deep Generative Models. *Photonics* **2024**, *11*, 967. https://doi.org/10.3390/photonics11100967.
4. Srinivasan, M.; Pourafzal, A.; Giannakopoulos, S.; Andrekson, P.; Häger, C.; Wymeersch, H. Learning Gradient-Based Feed-Forward Equalizer for VCSELs. *Photonics* **2024**, *11*, 943. https://doi.org/10.3390/photonics11100943.
5. Liem, A.T.; Hwang, I.-S.; Kharga, R.; Teng, C.-H. Enhancing Tactile Internet Reliability: AI-Driven Resilience in NG-EPON Networks. *Photonics* **2024**, *11*, 903. https://doi.org/10.3390/photonics11100903.
6. Ji, W.; Hu, L.; Zhang, X.; Lou, J.; Chen, H.; Wang, Z. The Stability Optimization of Indoor Visible 3D Positioning Algorithms Based on Single-Light Imaging Using Attention Mechanism Convolutional Neural Networks. *Photonics* **2024**, *11*, 794. https://doi.org/10.3390/photonics11090794.
7. Osahon, I.N.O.; Kostakis, I.; Powell, D.; Meredith, W.; Missous, M.; Haas, H.; Tang, J.; Rajbhandari, S. Neural Network Equalisation for High-Speed Eye-Safe Optical Wireless Communication with 850 nm SM-VCSELs. *Photonics* **2024**, *11*, 772. https://doi.org/10.3390/photonics11080772.
8. Luna-Rivera, J.M.; Rabadan, J.; Rufo, J.; Gutierrez, C.A.; Guerra, V.; Perez-Jimenez, R. Impact of Optical-to-Electrical Conversion on the Design of an End-to-End Learning RGB-LED-Based Visible Light Communication System. *Photonics* **2024**, *11*, 616. https://doi.org/10.3390/photonics11070616.
9. Hung, T.-Y.; Chan, D.W.U.; Peng, C.-W.; Chow, C.-W.; Tsang, H.K. Regeneration of 200 Gbit/s PAM4 Signal Produced by Silicon Microring Modulator (SiMRM) Using Mach–Zehnder Interferometer (MZI)-Based Optical Neural Network (ONN). *Photonics* **2024**, *11*, 349. https://doi.org/10.3390/photonics11040349.
10. Zhong, D.; Wang, T.; Chen, Y.; Wu, Q.; Qiu, C.; Zeng, H.; Wang, Y.; Xi, J. Exploration of Four-Channel Coherent Optical Chaotic Secure Communication with the Rate of 400 Gb/s Using Photonic Reservoir Computing Based on Quantum Dot Spin-VCSELs. *Photonics* **2024**, *11*, 309. https://doi.org/10.3390/photonics11040309.
11. Guo, H.; Yang, C.; Chen, Z.; Li, H. Enhanced PON and AMCC Joint Transmission with GMM-Based Probability Shaping Techniques. *Photonics* **2024**, *11*, 227. https://doi.org/10.3390/photonics11030227.
12. Xu, Z.; Ji, T.; Wu, Q.; Lu, W.; Ji, H.; Yang, Y.; Qiao, G.; Tang, J.; Cheng, C.; Liu, L.; et al. Advanced Neural Network-Based Equalization in Intensity-Modulated Direct-Detection Optical Systems: Current Status and Future Trends. *Photonics* **2024**, *11*, 702. https://doi.org/10.3390/photonics11080702.
13. Shao, C.; Giacoumidis, E.; Billah, S.M.; Li, S.; Li, J.; Sahu, P.; Richter, A.; Faerber, M.; Kaefer, T. Machine Learning in Short-Reach Optical Systems: A Comprehensive Survey. *Photonics* **2024**, *11*, 613. https://doi.org/10.3390/photonics11070613.
14. Wu, Q.; Xu, Z.; Zhu, Y.; Zhang, Y.; Ji, H.; Yang, Y.; Qiao, G.; Liu, L.; Wang, S.; Liang, J.; et al. Machine Learning for Self-Coherent Detection Short-Reach Optical Communications. *Photonics* **2023**, *10*, 1001. https://doi.org/10.3390/photonics10091001.

References

1. Wei, J.; Giacoumidis, E.; Zhou, J.; Jarajreh, M.; Li, S.; Aldaya, I.; Lu, Y.; Tsokanos, A.; Richter, A. Effective Handling of Nonlinear Distortions in CO-OFDM Using Affinity Propagation Clustering. *Opt. Express* **2022**, *30*, 15697–15707. [CrossRef]
2. Ji, H.; Dong, S.; Xu, Z.; Li, J.; Unnithan, R.; Su, Y.; Shieh, W. Carrier Assisted Differential Detection with Generalized and Simplified Receiver Structure. *J. Light. Technol.* **2021**, *39*, 7159–7167. [CrossRef]
3. Wu, Q.; Xu, Z.; Zhu, Y.; Ji, H.; Yang, Y.; Qiao, G.; Liu, L.; Wang, S.; Liang, J.; Wei, J.; et al. Gradient-Descent Noise Whitening Techniques for Short-Reach IM-DD Optical Interconnects with Severe Bandwidth Limitation. *Opt. Express* **2024**, *32*, 1715–1727. [CrossRef]
4. Ji, T.; Sun, C.; Ji, H.; Xu, Z.; Shieh, W. Field Recovery at Low CSPR Using Interleaved Carrier Assisted Differential Detection. In Proceedings of the Optical Fiber Communication Conference (OFC), San Diego, CA, USA, 8–12 March 2020; p. W4A.3.

5. Tan, Z.; Yang, C.; Xu, Z.; Chen, L.; Huang, X.; Guo, H.; Zheng, Z.; Zhang, F.; Wang, Z. Experimental Demonstration for over Mbps Baseband-over-Modulation AMCC Implementation in PtP WDM-PON. In Proceedings of the 2018 Optical Fiber Communications Conference and Exposition (OFC), San Diego, CA, USA, 11–15 March 2018; p. Tu3L.3.
6. Jarajreh, M.A.; Giacoumidis, E.; Aldaya, I.; Le, S.T.; Tsokanos, A.; Ghassemlooy, Z.; Doran, N.J. Artificial Neural Network Nonlinear Equalizer for Coherent Optical OFDM. *IEEE Photonics Technol. Lett.* **2014**, *27*, 387–390. [CrossRef]
7. Silva, L.C.B.; Marciano, P.R.N.; Pontes, M.J.; Monteiro, M.E.; André, P.S.B.; Segatto, M.E.V. Integrated Photonic Neural Networks for Equalizing Optical Communication Signals: A Review. *Photonics* **2025**, *12*, 39. [CrossRef]
8. Freire, P.J.; Osadchuk, Y.; Spinnler, B.; Napoli, A.; Schairer, W.; Costa, N.; Prilepsky, J.E.; Turitsyn, S.K. Performance Versus Complexity Study of Neural Network Equalizers in Coherent Optical Systems. *J. Light. Technol.* **2021**, *39*, 6085–6096. [CrossRef]
9. Wei, J.; Yi, L.; Giacoumidis, E.; Cheng, Q.; Tao Lau, A.P. Special Issue on "Optics for AI and AI for Optics". *Appl. Sci.* **2020**, *10*, 3262. [CrossRef]
10. Liu, X.; Li, C.; Jiang, Z.; Han, L. Low-Complexity Pruned Convolutional Neural Network Based Nonlinear Equalizer in Coherent Optical Communication Systems. *Electronics* **2023**, *12*, 3120. [CrossRef]
11. Hassan, S.; Tariq, N.; Naqvi, R.A.; Rehman, A.U.; Kaabar, M.K. Performance Evaluation of Machine Learning-Based Channel Equalization Techniques: New Trends and Challenges. *J. Sens.* **2022**, *2022*, 2053086. [CrossRef]
12. He, P.; Wu, F.; Yang, M.; Yang, A.; Guo, P.; Qiao, Y.; Xin, X. A Fiber Nonlinearity Compensation Scheme with Complex-Valued Dimension-Reduced Neural Network. *IEEE Photonics J.* **2021**, *13*, 7200507. [CrossRef]
13. Deligiannidis, S.; Mesaritakis, C.; Bogris, A. Performance and Complexity Analysis of Bi-directional Recurrent Neural Network Models Versus Volterra Nonlinear Equalizers in Digital Coherent Systems. *J. Light. Technol.* **2021**, *39*, 5791–5798. [CrossRef]
14. Xu, Z.; Sun, C.; Dong, S.; Manton, J.H.; Shieh, W. Towards Low Computational Complexity for Neural Network-Based Equalization in PAM4 Short-Reach Direct Detection Systems by Multi-Symbol Prediction. In Proceedings of the Optical Fiber Communication Conference (OFC), San Francisco, CA, USA, 6–10 June 2021; p. Th1A.47.
15. Peng, C.; Chan, D.; Chow, C.; Hung, T.; Jian, Y.; Tong, Y.; Kuo, P.; Chen, G.; Liu, Y.; Yeh, C.; et al. Long Short Term Memory Neural Network (LSTMNN) and Inter-Symbol Feature Extraction for 160 Gbit/s PAM4 from Silicon Micro-Ring Transmitter. *Opt. Commun.* **2023**, *529*, 129067. [CrossRef]
16. Taniguchi, H.; Yamamoto, S.; Masuda, A.; Kisaka, Y.; Kanazawa, S. 800-Gbps PAM-4 2-km Transmission using 4-λ LAN-WDM TOSA with MLSE based on Deep Neural Network. In Proceedings of the Optical Fiber Communication Conference (OFC), San Diego, CA, USA, 6–10 March 2022.
17. Arnold, E.; Bocherer, G.; Strasser, F.; Muller, E.; Spilger, P.; Billaudelle, S.; Weis, J.; Schemmel, J.; Calabro, S.; Kuschnerov, M. Spiking Neural Network Nonlinear Demapping on Neuromorphic Hardware for IM/DD Optical Communication. *J. Light. Technol.* **2023**, *41*, 3424–3431. [CrossRef]
18. Liu, B.; Bluemm, C.; Calabro, S.; Li, B.; Schlichtmann, U. Area-Efficient Neural Network CD Equalizer for 4 × 200Gb/s PAM4 CWDM4 Systems. In Proceedings of the Optical Fiber Communication Conference (OFC), San Diego, CA, USA, 5–9 March 2023.
19. Chen, G.; Du, J.; Sun, L.; Zhang, W.; Xu, K.; Chen, X.; Reed, G.T.; He, Z. Nonlinear Distortion Mitigation by Machine Learning of SVM Classification for PAM-4 andPAM-8 Modulated Optical Interconnection. *J. Light. Technol.* **2018**, *3*, 650–657. [CrossRef]
20. Xu, Z.; Wu, Q.; Lu, W.; Ji, H.; Chen, H.; Ji, T.; Yang, Y.; Qiao, G.; Tang, J.; Cheng, C.; et al. Weight-Adaptive Joint Mixed-Precision Quantization and Pruning for Neural Network-Based Equalization in Short-Reach Direct Detection Links. *Opt. Lett.* **2024**, *49*, 3500–3503. [CrossRef]
21. Liu, X.; Li, J.; Fan, Z.; Zhao, J. Transfer-Learning Based Convolutional Neural Network for Short-Reach Optical Interconnects. In Proceedings of the Asia Communications and Photonics Conference (ACP), Shanghai, China, 24–27 October 2021.
22. Miao, X.; Bi, M.; Yu, J.; Li, L.; Hu, W. SVM-Modified-FFE Enabled Chirp Management for 10G DML-based 50Gb/s/λ PAM4 IM-DD PON. In Proceedings of the Optical Fiber Communications Conference (OFC), San Diego, CA, USA, 3–7 March 2019.
23. Kaneda, N.; Zhu, Z.; Chuang, C.; Mahadevan, A.; Farah, B.; Bergman, K.; Veen, D.; Houtsma, V. FPGA Implementation of Deep Neural Network Based Equalizers for High-Speed PON. In Proceedings of the Optical Fiber Communication Conference (OFC), San Diego, CA, USA, 8–12 March 2020.
24. Chen, Z.; Wang, W.; Zou, D.; Ni, W.; Luo, D.; Li, F. Real-Valued Neural Network Nonlinear Equalization for Long-Reach PONs Based on SSB Modulation. *IEEE Photonics Technol. Lett.* **2022**, *35*, 167–170. [CrossRef]
25. Huang, L.; Jiang, W.; Xu, Y.; Hu, W.; Yi, L. Low-Complexity Volterra-Inspired Neural Network Equalizer in 100-G Band-Limited IMDD PON System. *Opt. Lett.* **2022**, *47*, 5692–5695. [CrossRef]
26. Lu, X.; Qiao, L.; Zhou, Y.; Yu, W.; Chi, N. An I-Q-Time 3-dimensional post-equalization algorithm based on DBSCAN of machine learning in CAP VLC system. *Opt. Commun.* **2019**, *430*, 299–303. [CrossRef]
27. Haigh, P.A.; Ghassemlooy, Z.; Rajbhandari, S.; Papakonstantinou, I.; Popoola, W. Visible light communications: 170 Mb/s using an artificial neural network equalizer in a low bandwidth white light configuration. *J. Light. Technol.* **2014**, *32*, 1807–1813. [CrossRef]
28. Loureiro, P.A.; Fernandes, G.M.; Correia, S.F.; Ferreira, R.A.; Guiomar, F.P.; Monteiro, P.P. Multi-Gigabit RGB-VLC Transmission with Jointly Optimized Lighting and Communications. *J. Light. Technol.* **2024**, *42*, 4425–4432. [CrossRef]

29. Jin, R.; Wei, Y.; Zhang, J.; Shi, J.; Chi, N. Neural-Network-Based End-To-End Learning for Adaptive Optimization of Two-Dimensional Signal Generation in UVLC Systems. *Opt. Express* **2024**, *32*, 6309–6328. [CrossRef]
30. Lu, X.; Li, Y.; Chen, X.; Li, Y.; Liu, Y. Discrete wavelet transform assisted convolutional neural network equalizer for PAM VLC system. *Opt. Express* **2024**, *32*, 10429–10443. [CrossRef]
31. Xu, B.; Min, T.; Yue, C.P. Design of PAM-8 VLC Transceiver System Employing Neural Network-Based FFE and Post-Equalization. *Electronics* **2022**, *11*, 3908. [CrossRef]
32. Lu, X.; Lu, C.; Yu, W.; Qiao, L.; Liang, S.; Lau, A.P.T.; Chi, N. Memory-controlled deep LSTM neural network post-equalizer used in high-speed PAM VLC system. *Opt. Express* **2019**, *27*, 7822–7833. [CrossRef]
33. Mobini, M.; Kaddoum, G.; Herceg, M. Design of a SIMO Deep Learning-Based Chaos Shift Keying (DLCSK) Communication System. *Sensors* **2022**, *22*, 333. [CrossRef]
34. Ke, J.; Yi, L.; Yang, Z.; Yang, Y.; Zhuge, Q.; Chen, Y.; Hu, W. 32 Gb/s chaotic optical communications by deep-learning-based chaos synchronization. *Opt. Lett.* **2019**, *44*, 5776–5779. [CrossRef]
35. He, L.; Lai, R.; Shao, S.; Li, Z. Design of a Chaotic Communication System Based on Deep Learning with Two-Dimensional Reshaping. *IEEE Trans. Veh. Technol.* **2024**, *73*, 10421–10434. [CrossRef]
36. Zhang, H.; Zhang, L.; Jiang, Y.; Wu, Z. Reliable and Secure Deep Learning-Based OFDM-DCSK Transceiver Design without Delivery of Reference Chaotic Sequences. *IEEE Trans. Veh. Technol.* **2022**, *71*, 8059–8074. [CrossRef]
37. Tanimura, T.; Hoshida, T.; Kato, T.; Watanabe, S.; Morikawa, H. Convolutional Neural Network-Based Optical Performance Monitoring for Optical Transport Networks. *J. Opt. Commun. Netw.* **2019**, *11*, A52–A59. [CrossRef]
38. Wu, X.; Jargon, J.A.; Skoog, R.A.; Paraschis, L.; Willner, A.E. Applications of Artificial Neural Networks in Optical Performance Monitoring. *J. Light. Technol.* **2009**, *27*, 3580–3589.
39. Khan, F.N.; Zhou, Y.; Lau, A.P.T.; Lu, C. Modulation Format Identification in Heterogeneous Fiber-Optic Networks Using Artificial Neural Networks. *Opt. Express* **2012**, *20*, 12422–12431. [CrossRef] [PubMed]
40. Abdelli, K.; Tropschug, C.; Griesser, H.; Pachnicke, S. Fault Monitoring in Passive Optical Networks using Machine Learning Techniques. In Proceedings of the 23rd International Conference on Transparent Optical Networks, Bucharest, Romania, 2–6 July 2023; pp. 1–5.
41. Khan, F.N.; Zhong, K.; Zhou, X.; Al-Arashi, W.H.; Yu, C.; Lu, C.; Lau, A.P.T. Joint OSNR monitoring and modulation format identification in digital coherent receivers using deep neural networks. *Opt. Express* **2017**, *25*, 17767–17776. [CrossRef]
42. Wang, C.; Fu, S.; Xiao, Z.; Tang, M.; Liu, D. Long Short-Term Memory Neural Network (LSTM-NN) Enabled Accurate Optical Signal-to-Noise Ratio (OSNR) Monitoring. *J. Light. Technol.* **2019**, *37*, 4140–4146. [CrossRef]
43. Saif, W.S.; Esmail, M.A.; Ragheb, A.M.; Alshawi, T.A.; Alshebeili, S.A. Machine Learning Techniques for Optical Performance Monitoring and Modulation Format Identification: A Survey. *IEEE Commun. Surv. Tutor.* **2020**, *22*, 2839–2882. [CrossRef]
44. Musumeci, F.; Rottondi, C.E.M.; Corani, G.; Shahkarami, S.; Cugini, F.; Tornatore, M. A Tutorial on Machine Learning for Failure Management in Optical Networks. *J. Light. Technol.* **2019**, *37*, 4125–4139. [CrossRef]

photonics

MDPI

Article

A Modified Regular Perturbation Model for the Single-Span Fiber Transmission Using Learnable Methods

Shuhong He [1], Zhongya Li [1], Sizhe Xing [1], An Yan [1], Yingjun Zhou [1], Jianyang Shi [1], Chao Shen [1], Ziwei Li [1], Zhixue He [2], Wei Chen [3], Nan Chi [1,2] and Junwen Zhang [1,2,*]

1 Key Laboratory of Information Science of Electromagnetic Waves (MoE), Shanghai ERC of LEO Satellite Communication and Applications, Shanghai CIC of LEO Satellite Communication Technology, Fudan University, Shanghai 200433, China
2 PengCheng Laboratory, Shenzhen 518055, China
3 Key Laboratory of Specialty Fiber Optics and Optical Access Networks, Joint International Research Laboratory of Specialty Fiber Optics and Advanced Communication, Shanghai University, Shanghai 200444, China
* Correspondence: junwenzhang@fudan.edu.cn

Abstract: In fiber optic communication systems, the dispersion and nonlinear interaction of optical signals are critical to modeling fiber optic communication, and the regular perturbation (RP) model is a simplified modeling method composed of parallel branches, which has obvious advantages in deep learning backpropagation. In this paper, we propose a simplified single-mode fiber signal transmission model based on the RP model, which significantly improves the fitting accuracy of the model for dispersion and nonlinear interactions at the same complexity by adding trainable parameters to the standard RP model. We explain in the paper that this improvement is applicable to dual-polarization systems and still effective under the conditions of large launch power, without dispersion management, and containing amplified spontaneous emission (ASE) noise. The model uses the standard split-step Fourier method (SSFM) to generate labels and updates parameters through gradient descent method. When transmitting a dual-polarization signal with a launch power of 13 dBm, the modified regular perturbation (MRP) model proposed in the paper can reduce the fitting errors by more than 75% compared to the standard RP model after transmitting through a 120 km standard single-mode fiber.

Keywords: nonlinear Schrödinger equation; regular perturbation model; fiber nonlinearity; learnable parameters; single-span dual-polarization systems

1. Introduction

As the most widely used optical signal carrier, accurate modeling of standard single-mode fibers is an indispensable step in constructing optical communication systems. The existing digital signal processing methods are already able to effectively compensate the linear distortion of signals in fiber optic transmission, such as attenuation, dispersion, and polarization mode dispersion [1]. Therefore, nonlinear effects will become the main challenge for large launch power and long-distance fiber optic signal transmission. Due to the popularity of deep learning and end-to-end optimization methods, the accuracy of the fiber nonlinearity modeling will largely determine the effectiveness of transceiver compensation for nonlinearity [2–6].

The most widely used method for fiber optic modeling is the standard split-step Fourier method (SSFM) and its improved methods [7–9]. It obtains the numerical solution of the optical field by solving the nonlinear Schrödinger equation (NLSE) in several steps, and each step achieves an approximate analytical solution of the optical signal by considering only dispersion or nonlinearity in a series of alternating short length fibers. SSFM can achieve high accuracy in situations of high complexity and is a commonly used fiber modeling method in commercial software. However, due to the fact that the SSFM is a serial method, there are difficulties in backpropagation and risks of gradient vanishing and exploding when combined with deep learning-based transceivers [10,11]. The Gaussian noise (GN) method is a simplified method based on describing nonlinear noise in the frequency domain [12–15]. It directly adds nonlinear noise to the transmission signal by treating it as additive Gaussian noise, while ignoring that the fundamental source of nonlinear noise is the interaction between signals. Compared to the SSFM, although its computational complexity is obviously decreased, the GN model is not effective under conditions of low dispersion accumulation and large launch power [16] and is difficult to combine with advanced equalization algorithms [10]. The RP model is another commonly used simplified fiber modeling method [17], which obtains the first-order nonlinear solution of the nonlinear Schrödinger equation [18] through series matching to simplify the SSFM. Although it sacrifices some accuracy, the RP model is a parallel method that can improve the accuracy of the model by increasing the number of branches, making up for the difficulty of backpropagation in the SSFM and facilitating parameter updates in deep learning. In the past 20 years, multiple derivative algorithms have emerged for the RP method to model various waveforms, nonlinear interferences (such as SPM, IXPM, IFWM, XPM, FWM), and ASE noise interactions [19–25]. In a recent study, learnable filters were introduced and combined with machine learning to achieve high accuracy. This approach takes into account the nonlinear interactions of adjacent symbols, but the complexity increases rapidly as the number of symbols considered expand [26]. The RP model has also been applied in end-to-end learning [27] and fiber longitudinal power distribution estimation [28]. Meanwhile, due to the fact that perturbations are additive effects relative to signals, they are easily transformed into modeling and equalization, as well as combined with deep learning equalization methods [29–33]. In addition, the RP model can flexibly introduce high-order solutions to tackling the problem of insufficient accuracy of the first-order perturbation models for large-power signal input [34], but this will significantly increase the complexity of the algorithm. In response to this phenomenon, researchers have proposed various simplification schemes, among which [35,36] reduce the number of triplets required for fitting by quantifying perturbation parameters, while other studies [23,37] convert the multiplication operation of signals into signals rotation, reducing the number of multiplication required in the algorithm. Recent studies have shown that the RP model can flexibly allocate the proportion of pre-equalization and post-equalization at the transmitter and the receiver [38] or be applied to super-channel systems [39]. Additionally, analyzing the capacity of wavelength division multiplexing (WDM) using the RP model is also an interesting topic [40].

This paper proposes a modified RP algorithm that provides higher accuracy, more accurate signal-to-noise ratio, and a larger launch power range without increasing the complexity of traditional RP models. The low-complexity MRP model proposed in this paper sets trainable parameters for the step size and nonlinear coefficients of each branch of the RP model. Waveform-wise data generated by standard SSFM is used as labels, and supervised learning is used to optimize each trainable parameter. The trained MRP model can achieve high-precision simulation of fiber dispersion and nonlinear interactions, reducing the fitting errors by more than 75% compared to the standard RP model after transmitting through a 120 km standard single-mode fiber. In addition, this paper analyzed the complexity of the MRP model and walked through different complexity, fiber length, and launch power conditions to verify the robustness of the model. The results show that the model has high accuracy and a wide range of applicability. In this paper, we only considered the scenario of the single-span fiber optic transmission, but we emphasize that the basic structure of this model will not change under multi-span systems. We will discuss and demonstrate the detail of the performance of this model under multi-span systems in our future research.

2. MRP Models and Algorithm Description

2.1. Principle of MRP Method

When the polarization state of a bipolarized signal changes rapidly enough due to the birefringence phenomenon of the fiber, the transmission of the bipolarized optical signal in the fiber can be represented by a generalized Manakov expression of the NLSE [41,42]:

$$\frac{\partial E_{x,y}}{\partial z}=-\frac{\alpha}{2}E_{x,y}-\frac{\beta_2}{2}j\frac{\partial^2 E_{x,y}}{\partial t^2}+\frac{\beta_3}{6}\frac{\partial^3 E_{x,y}}{\partial t^3}+j\frac{8}{9}\gamma\left(\left|E_{x,y}\right|^2+\left|E_{y,x}\right|^2\right)E_{x,y} \tag{1}$$

where $E_{x,y}$ is the optical field of x or y polarization signal, β_2 and β_3 are second-order and third-order dispersion coefficients, respectively, α is the fiber loss, $\gamma = 2\pi n_2/\lambda A_{eff}$ is the nonlinear parameter, and z is the fiber length.

Here, we use normalization to represent the optical field in fibers, using:

$$E(z,t)=U(z,t)e^{-\int_0^z \frac{\alpha(z')}{2}dz'} \tag{2}$$

Then, we assume that the attenuation of the fiber is a constant α, and the inline amplifier after each span of fiber completely compensates for the fiber link loss, i.e., $\alpha(z)=\alpha\left(z-L_{sp}\lfloor z/L_{sp}\rfloor\right)$. Where L_{sp} is the single span length of the optical fiber, $\lfloor\cdot\rfloor$ is the remainder operator. Therefore, substitute Equation (2) into Equation (1), we update the representation of the Manakov formula.

$$\frac{\partial U_{x,y}}{\partial z}=-\frac{\beta_2}{2}j\frac{\partial^2 U_{x,y}}{\partial t^2}+\frac{\beta_3}{6}\frac{\partial^3 U_{x,y}}{\partial t^3}+j\frac{8}{9}\gamma e^{-\alpha z}\left(\left|U_{x,y}\right|^2+\left|U_{y,x}\right|^2\right)U_{x,y} \tag{3}$$

Assuming that the solution of the equation is a non-singular problem, i.e., small perturbations do not alter the properties of the equation solution, consider the solution in the conventional perturbation form [18,43]:

$$U(z,t)=U^{(0)}(z,t)+\gamma U^{(1)}(z,t)+\gamma^2U^{(2)}(z,t)+o(\gamma^2) \tag{4}$$

In order to avoid high computational complexity, we do not consider nonlinear perturbation terms above second order. Instead, we substitute the first two terms on the right side of Equation (4) into Equation (3) and use series matching to derive the differential equations corresponding to the zeroth-order solution and the first-order solution, respectively:

$$\frac{\partial U^{(0)}_{x,y}}{\partial z}=-\frac{\beta_2}{2}j\frac{\partial^2 U^{(0)}_{x,y}}{\partial t^2}+\frac{\beta_3}{6}\frac{\partial^3 U^{(0)}_{x,y}}{\partial t^3}+j\frac{8}{9}\gamma e^{-\alpha z}\left(\left|U^{(0)}_{x,y}\right|^2+\left|U^{(0)}_{y,x}\right|^2\right)U^{(0)}_{x,y} \tag{5}$$

$$\frac{\partial U^{(1)}_{x,y}}{\partial z}=-\frac{\beta_2}{2}j\frac{\partial^2 U^{(1)}_{x,y}}{\partial t^2}+\frac{\beta_3}{6}\frac{\partial^3 U^{(1)}_{x,y}}{\partial t^3}+j\frac{8}{9}\gamma e^{-\alpha z}\left(\left|U^{(0)}_{x,y}\right|^2+\left|U^{(0)}_{y,x}\right|^2\right)U^{(0)}_{x,y} \tag{6}$$

Using Fourier transform to obtain the frequency domain of differential equations, and then using inverse Fourier transform to give the time domain forms of the zeroth-order solution and the first-order solution:

$$U^{(0)}_{x,y}(z,t)=\mathcal{F}^{-1}\left[e^{j(\frac{\beta_2}{2}\omega^2+\frac{\beta_3}{6}\omega^3)z}\mathcal{F}\left[U^{(0)}_{x,y}(0,t)\right]\right] \tag{7}$$

$$U^{(1)}_{x,y}(z,t)=j\frac{8}{9}\mathcal{F}^{-1}\left[\int_0^z dz'e^{j(\frac{\beta_2}{2}\omega^2+\frac{\beta_3}{6}\omega^3)(z-z')}\cdot\mathcal{F}\left[U^{(0)}_{x,y,tri}(z',t)\right]e^{-\alpha(z'-L_{sp}\lfloor\frac{z'}{L_{sp}}\rfloor)}\right] \tag{8}$$

$$U^{(0)}_{x,y,tri}(z',t)=\left(U^{(0)}_{x,y}(z',t)U^{(0)*}_{x,y}(z',t)+U^{(0)}_{y,x}(z',t)U^{(0)*}_{y,x}(z',t)\right)U^{(0)}_{x,y}(z',t)$$

Here, $\mathcal{F}$ and $\mathcal{F}^{-1}$ represent Fourier transform and inverse Fourier transform, respectively. On this basis, we assume $z = L$, where L is the total length of the fiber, and approximate the definite integral in (8) by step. We derive the first-order perturbation solution of NLSE as follows:

$$
\begin{aligned}
&MRP_L\left[U_{x,y}(0,t))\right] = U_{x,y}(L,t) = U_{x,y}^{(0)}(L,t) + \gamma U_{x,y}^{(1)}(L,t) \\
&= \mathcal{F}^{-1}\left[k_0 \cdot e^{j(\frac{\beta_2}{2}\omega^2 + \frac{\beta_3}{6}\omega^3)L} \mathcal{F}\left[U_{x,y}^{(0)}(0,t)\right]\right] \\
&+ j\frac{8}{9}\gamma\mathcal{F}^{-1}\sum_{n=1}^{N_{step}}\left[k_n \cdot e^{j(\frac{\beta_2}{2}\omega^2 + \frac{\beta_3}{6}\omega^3)(L - \frac{\delta_n}{2} - \sum_{i=1}^{n-1}\delta_i)} \cdot \mathcal{F}[U_{x,y,tri}^{(0)}(\frac{\delta_n}{2} + \sum_{i=1}^{n-1}\delta_i, t)] \cdot \delta_{i,eff} \cdot e^{-\alpha f(\frac{\delta_n}{2} + \sum_{i=1}^{n}\delta_i)}\right] \\
&\delta_{i,eff} = \frac{1 - e^{-\alpha\delta_i}}{\alpha}, \ \ f(l) = l - L_{sp}\left\lfloor \frac{l}{L_{sp}} \right\rfloor
\end{aligned}
\tag{9}
$$

Here, N_{step} represents the total number of terms converted from integration to summation, which is also the total number of branches of the RP model, δ_i is the additional step size considered for each summation term, $\delta_{i,eff}$ is the effective length of the optical fiber, and $f(l)$ is the attenuation term of the optical fiber. Note that on the basis of the RP model, we have added a learnable parameter k_n (n = 0~N_{step}). Additionally, we have adjusted the step size, δ_i, to become a learnable real parameter, and we refer to this improved model as the MRP model. For k_n, when n = 0, it corresponds to the zeroth-order equation solution, so k_0 is a learnable parameter of complex value. k_0 can control the power ratio between the linear and nonlinear parts in the MRP model. When n = 1~N_{step}, it corresponds to the solution of the first-order equation, where k_n is a learnable parameter of complex value that can control the magnitude of the nonlinear amplitude and phase shift in each branch. Figure 1 shows the mathematical model of a single-span fiber optic communication system considered in our work.

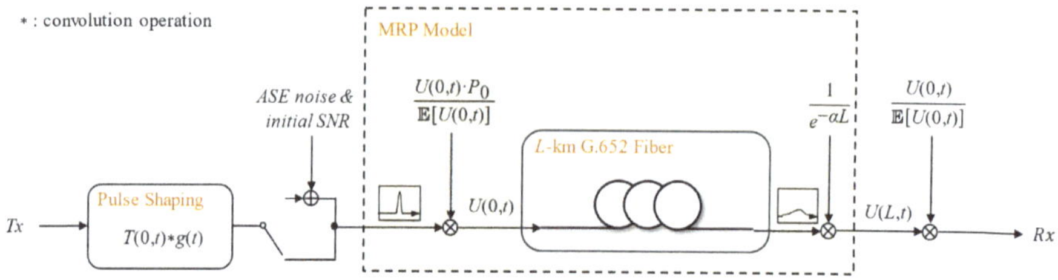

Figure 1. Conceptual diagram of the single-span fiber communication system.

For general RP models, taking $\delta_i = L/N_{step}$, k_n = 1 + 0j (n = 0~N_{step}). Therefore, readers can understand our intention to enhance the RP branches by incorporating the learnable flexible step size, learnable linear parameters, and learnable nonlinear parameters in the RP branches to improve the accuracy of the integration approximation of ordinary RP models. This method maintains the same computational complexity when the step size remains consistent. Its specific structure is shown in Figure 2.

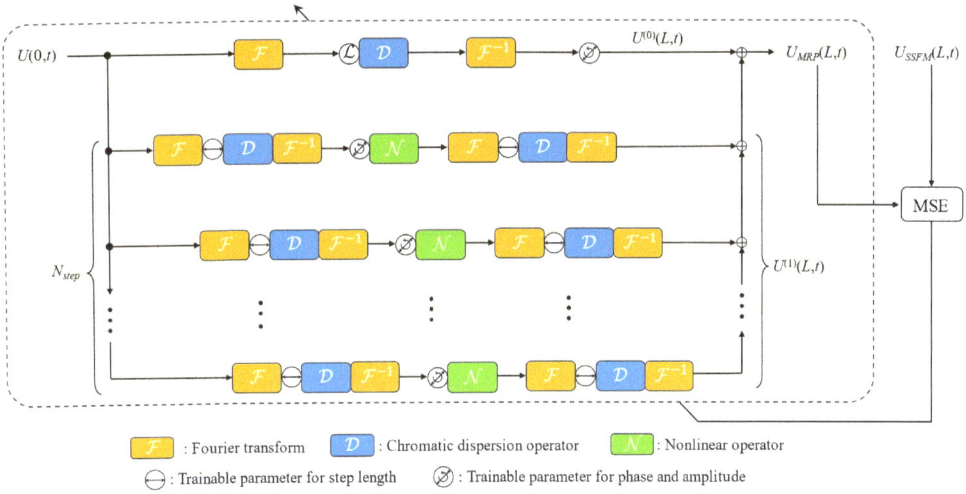

Figure 2. Conceptual diagram of MRP model structure and training method.

2.2. Complexity Analysis of the MRP Model

The main analytical steps of the derivation of the complexity of the model are reported in Appendix A. The total computational complexity of the MRP model is:

$$\begin{aligned} N_{\times}^{(real)}(S) &= 4S\log_2 S + 4S + N_{step}[8S\log_2 S + 20S] \\ N_{+}^{(real)}(S) &= 6S\log_2 S + 2S + N_{step}[12S\log_2 S + 13S] \end{aligned} \tag{10}$$

Assume that the sequence processed by the model consists of two sampling points per symbol, we can calculate the computational complexity of each symbol corresponding to different branch numbers and sequence lengths, S, according to Equation (10), as shown in Figure 3.

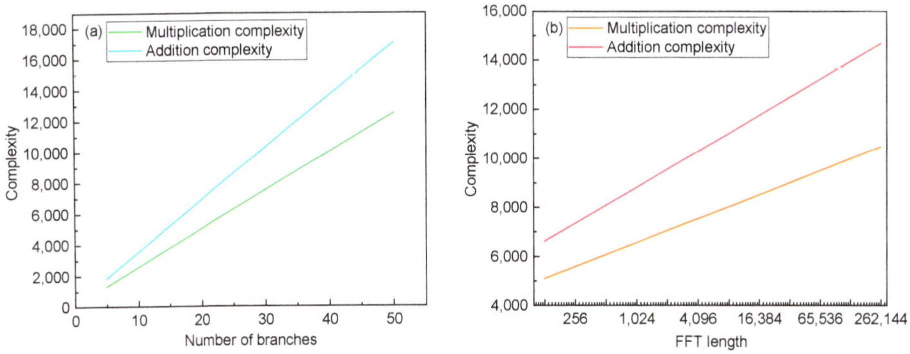

Figure 3. Complexity of MRP models under different conditions: (**a**) N_{step} and (**b**) FFT length.

It is evident that the complexity of the MRP model is directly proportional to both the number of branches and the length of the input sequence. It is worth noting that FFT is the main source of algorithm complexity. For sequences with a length of S = 4096, the number of multiplication and addition used by radix-2 FFT accounts for 84% and 61% of the total number of multiplication and addition in the MRP model, respectively. In addition, we remind readers that using the calculation method of base radix-4 FFT can further reduce the complexity of the algorithm. Under the same sequence length, S, radix-4 FFT can reduce the number of multiplications to 3/4 as opposed to radix-2 FFT. In addition, when calculating both x-polarization and y-polarization simultaneously, the results of mode squared can be reused to reduce computational complexity.

2.3. Optimization and Training Procedure

In this part, we will mainly explain the optimization methods for various parameters of the MRP model. We use supervised learning to update the learnable parameters in the model. Our model processes both x-pol and y-pol optical signal data, so the model can be simply expressed as:

$$\left[U_x(z,t), U_y(z,t)\right] = MRP\left[U_x(0,t), U_y(0,t)\right] \tag{11}$$

We use a randomly uniformly distributed signal of length, S_0, which is filtered by a pulse shaping filter and is upsampled to two samples per symbol to generate the input sequence (i.e., sampling length of $S = 2 \times S_0$). The label uses the SSFM to generate sequences of the same length. The SSFM can be represented by:

$$\begin{aligned} &SSFM_\delta\left[U_{x,y}(0,t))\right] = U_{x,y}(\delta,t) \\ &= \mathcal{F}^{-1}\left[e^{j(\frac{\beta_2}{2}\omega^2+\frac{\beta_3}{6}\omega^3)\frac{\delta}{2}}\mathcal{F}\left[e^{j\frac{8}{9}\gamma\delta_{eff}\cdot[-\alpha f(\frac{\delta}{2})]\{|U_{x,y}(\frac{\delta}{2},t)|^2+|U_{y,x}(\frac{\delta}{2},t)|^2\}}U_{x,y}^{(0)}(\tfrac{\delta}{2},t)\right]\right] \\ &SSFM_L\left[U_{x,y}(0,t))\right] = U_{x,y}(L,t) = \underbrace{SSFM_\delta\left[SSFM_\delta \ldots\ldots \left[SSFM_\delta\left[U_{x,y}(0,t))\right]\right]\right]}_{iterate\ \times L/\delta} \end{aligned} \tag{12}$$

The meanings of each symbol here are the same as those of the MRP model, with the subscripts δ and L is the distance of optical signal transmission. The $U_{x,y}^{(0)}$ here is an operator rather than a simple function, which will not cause ambiguity when defining the MRP model, but special clarification is needed here. After generating labels, due to the obvious power increment of the RP model and its derivative models, we use a simple mean normalization to align the power of the model's outputs and the labels.

$$Y_{x,y}(L,sT) = \frac{|U_{x,y}(L,sT)|}{\mathrm{mean}(|U_{x,y}(L,sT)|)}, \quad s = 1,2,\cdots,S \tag{13}$$

Here, Y is the MRP output sequence or the SSFM label and T is the sampling period.

To minimize the error between the labels and the outputs of the MRP model, we update the learnable parameters in the model using the batch gradient descent method. The loss function is defined as the mean squared error (MSE) loss:

$$L = \min_{\delta,k}\frac{1}{2B}\left[\sum_{s=1}^{S}\left|\hat{Y}_x - Y_x\right|^2 + \left|\hat{Y}_y - Y_y\right|^2\right] \tag{14}$$

Here, B represents the amount of the data input to the model to calculate the gradient at a time, i.e., batch size. In addition, we choose the Adam algorithm for parameter updates to minimize the loss function, and the specific learning rate and batch size will be explained in the next section.

Due to the form of the loss function, we do not include ASE noise in either the generated labels or the training inputs, as the addition of ASE noise can affect the model's convergence, especially at lower launch power levels. However, in the comparison of results in the next section, we compared the accuracy of the model in both cases without and with ASE noise.

For learnable parameters, δ, representing step size distribution, we made a slight modification on the basis of the RP model and used the step size of the logarithmic distribution to δ. The results in [44,45] indicate that using a logarithmic step equalization model for channel estimation has better accuracy, as it increases the step calculation density for the majority of power in the fiber optic channel, that is, the proportion of large-power components in the total computational complexity. This has a certain effect on improving the accuracy of channel modeling from integral form to summation form. The formula for the logarithmic step distribution is expressed as follows:

$$\begin{gathered}\delta_n = -\frac{1}{\mu\alpha}\ln\left(\frac{1-n\kappa}{1-(n-1)\kappa}\right),\ n = 1,2,\cdots,N_{step} \\ \kappa = \left(1-e^{-2\alpha L}\right)/N_{step}\end{gathered} \tag{15}$$

Based on the results of [45], we take the initial value $\mu = 0.4$, but this is not a strict constraint. Adjusting the initial value of μ slightly does not have a significant impact on the final convergence value of the loss function. Moreover, in the next section of result analysis, the RP model used for comparison also adopts this distribution.

3. Results and Discussion

In this chapter, we will provide a detailed explanation of various hyperparameter settings related to the model and reveal the performance of the MRP model in various aspects. We will introduce the RP model and the SSFM model as comparison objects, and their mathematical expressions have been detailed in Equations (9) and (12), respectively.

3.1. Fiber and DSP Setup

We consider a standard single-mode fiber with a span of $L = 120$ km, transmitting a dual-polarized 50-GBaud signal. The wavelength of lasers is 1552.6 nm. Table 1 shows the fiber's parameters, which will be applied to all models mentioned in this chapter, including SSFM, RP and MRP models. We intentionally chose a larger group velocity coefficient, β_2, to demonstrate that even when the symbol length affected by nonlinear inter-symbol interference is longer, the model's performance remains strong. We also note that using other reasonable dispersion values does not significantly impact the improvement in model performance. For the signal output by the model, we process the waveform-wise signal through a standard DSP process to remove linear distortion in the fiber optic, its specific structure is shown in Figure 4.

Table 1. Fiber parameters.

Nonlinear Parameter γ	$1.2\ \mathrm{W}^{-1}\cdot\mathrm{km}^{-1}$
Fiber attenuation α	$0.2\ \mathrm{dB}\cdot\mathrm{km}^{-1}$
Second-order dispersion β_2	$-27.77 \times 10^{-27}\ \mathrm{s}^2\cdot\mathrm{m}^{-1}$
Third-order dispersion β_3	$17.68 \times 10^{-41}\ \mathrm{s}^3\cdot\mathrm{m}^{-1}$

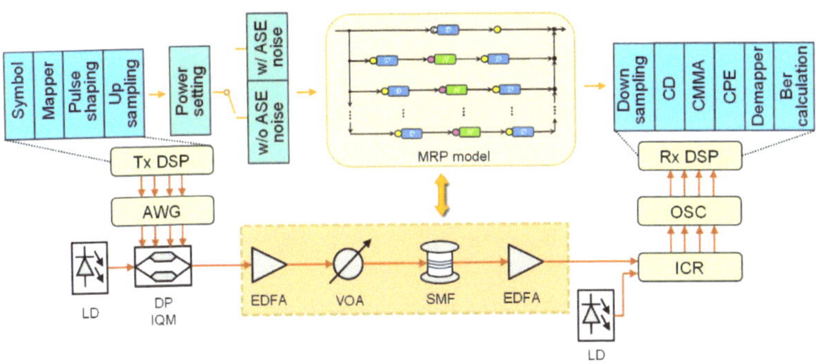

Figure 4. The setup of the system simulated by the MRP model.

We quantify the accuracy of the model by evaluating and comparing the symbol-wise data solved by the MRP model and the SSFM. Normalized root mean square error

(RMSE) and error vector magnitude (EVM) are used here to evaluate the model, and their definitions are as follows:

$$RMSE = \sqrt{\frac{1}{2}\left(\frac{\mathbb{E}\left\{\left|R_x - \hat{R}_x\right|^2\right\}}{\mathbb{E}\left\{\left|R_x\right|^2\right\}} + \frac{\mathbb{E}\left\{\left|R_y - \hat{R}_y\right|^2\right\}}{\mathbb{E}\left\{\left|R_y\right|^2\right\}}\right)} \tag{16}$$

$$EVM = \frac{1}{2}\left(\sqrt{\frac{\mathbb{E}\left\{\left|\hat{R}_x - T_x\right|^2\right\}}{\mathbb{E}\left\{\left|T_x\right|^2\right\}}} + \sqrt{\frac{\mathbb{E}\left\{\left|\hat{R}_y - T_y\right|^2\right\}}{\mathbb{E}\left\{\left|T_y\right|^2\right\}}}\right) \tag{17}$$

where R are the symbol-wise data passed through DSP at the receiving end, and T are the initial data sent by the transmitting end. Due to the model being a stationary process, the expectation, $\mathbb{E}$, are transformed into a sum of the points in the sequence.

3.2. Training Setup

The training setup details are provided in Table 2. The initial data we use is the 16-QAM signal generated from uniformly distributed random numbers following Gray mapping, and the length of both the input sequences and the label sequences is S = 8192. Before entering the model, the signal is upsampled and filtered by a shaping filter with a roll-off factor of 0.01. Due to the small amounts of trainable parameters, we can train the model even when the amount of training data is limited. We use 512 different sequences to generate labels and participate in training, with 87.5% of the sequences as the training set and 12.5% as the test set to train and save the best model. And we tested the trained model using a sequence with a length of 32,768. We believe that 1500 epochs are sufficient to generate the optimal model, and in actual testing, the model is often accurate enough at 500 epochs. It is worth noting that when the launch power is low, we use a smaller initial learning rate, α_0. As the launch power increases, α_0 should also be appropriately increased.

Table 2. Training parameters.

Parameter Initialization	$k_n = 1 + 0j$ δ_i = RHS of Equation (15)
Loss function	MSE
Optimizer	Adam
Learning Rate	$\alpha_{epoch} = \alpha_0 \times 0.9^{\lfloor \frac{epoch}{20} \rfloor}$
Epoch number	1500
Batch size	8

3.3. The Discussion of the MRP Model

Firstly, we will analyze the impact of different branches numbers, N_{step}, on the accuracy of the model, which guides us to further validate the required step size standards and also determines the computational complexity of the model. We fixed a large launch power P_0 = 13 dBm and traversed the performance of different N_{step} at a fiber length of L = 120 km, as shown in Figure 5. We note that, unless otherwise stated, the pulse shaping filter uses a root raised cosine (RRC) filter with the roll-off factor of 0.01.

As shown in Figure 5, it can be observed that there is a limit to the fitting accuracy for both the RP model and the MRP model. Increasing the number of branches beyond this limit will not further improve system performance. N_{step} directly affects the structure and the complexity of the model; therefore, we chose N_{step} = 10, 20, and 30 as the branches of the MRP model and the RP model for comparison.

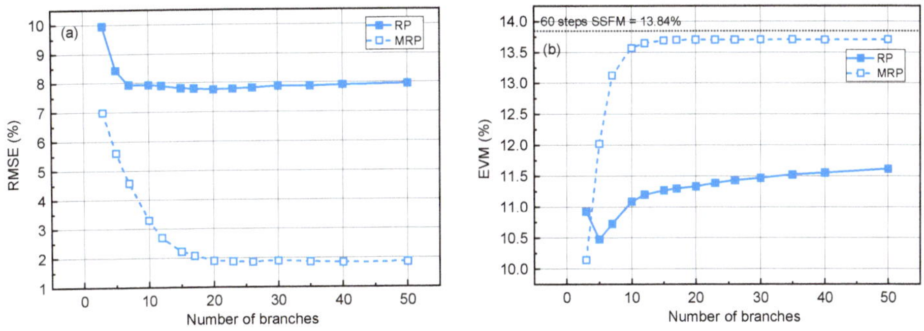

Figure 5. RMSE (**a**) and EVM (**b**) of the MRP and RP models under different branch numbers.

Figures 6 and 7 show the constellation and waveform differences between the MRP model and the RP model compared to the SSFM. It can be clearly observed that compared to the RP model, the constellation points and the EVM of the MRP model are significantly closer to the SSFM, especially for the outer ring constellation points. From the perspective of waveforms, the MRP model outperforms the RP model at almost each sampling point.

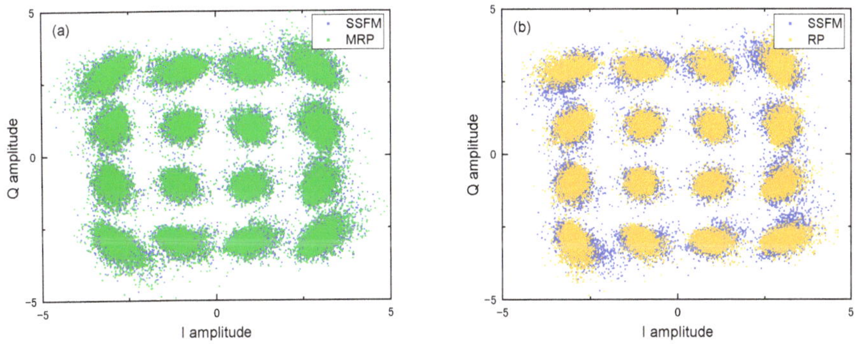

Figure 6. Constellation diagrams comparison between SSFM and (**a**) MRP or (**b**) RP models.

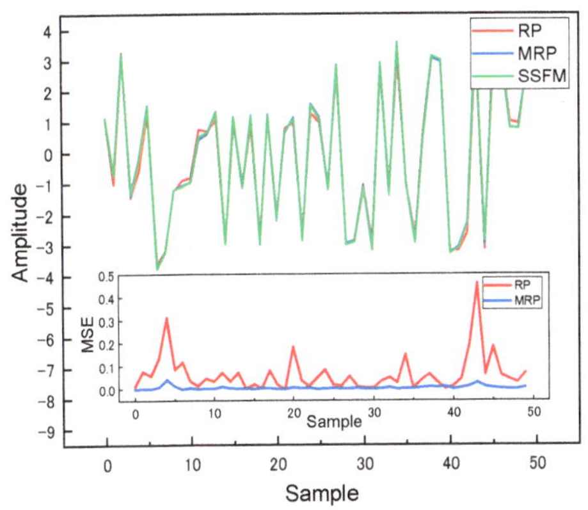

Figure 7. Comparison of waveforms and MSE between SSFM and RP or MRP models.

Figure 8 shows the relationship between RMSE and launch power for RP and MRP with different numbers of branches. Here, a 60-step SSFM is used as the standard quantity for calculating RMSE. It is obvious that the fitting error between the MRP model and SSFM is smaller in the large launch power range than standard RP. Taking the results at 14 dBm as an example, the RMSE of the 30-step RP model is 11.82%, while the RMSE of the 30-step MRP model is only 2.77%, reducing the error by more than 75%, and the computational complexity of the two models is completely the same. As we have already gone through the standard DSP process and the same normalization process before calculating RMSE, the error here is almost entirely composed of nonlinear differences. If factors such as phase rotation are considered, the performance gap between the RP model and the MRP model will further increase. In practical systems, both the signal and amplifier are not ideal. Due to the presence of ASE noise and arbitrary waveform generator AWG sampling noise, the signal at the transmitting end is modeled as the sum of ideal signal and Gaussian noise. Nonlinear effects will occur between the noise and signal, resulting in the Gordon–Mollenauer Effect [46]. To verify the accuracy of the model's estimation of non-ideal signals, we set the initial signal-to-noise ratio of the signal to 37 dB, and we consider an additive Gaussian noise of a constant power of 0.01 mW before entering the fiber to simulate sampling noise and ASE noise separately. Note that we use multiple independent Gaussian noise sources to ensure that the sum of the noise remains Gaussian distributed. Figure 9 shows the accuracy of the model in estimating the received signals' EVM without and with ASE noise. When the launch power is large, the MRP model is significantly better than the RP model. In addition, we found that both the RP model and MRP model tend to underestimate the EVM of the signal, and the underestimation degree of the MRP model is significantly lighter than that of the RP model. At a launch power of 13 dBm with ASE noise, the EVM difference between the 20-step MRP and SSFM is only 0.13%, which is much lower than the 2.51% corresponding to the 20-step RP model. Moreover, this difference will further expand with the increase in fiber launch power.

It is worth noting that the joint analysis of Figures 8 and 9 shows that as the launch power increases, the growth trend of the EVM of the RP model slows down significantly, while the MRP model can continue to maintain its approximation with the SSFM, but the RMSE of both increases exponentially. We speculate that this result may be due to two reasons: firstly, when the launch power is too large, the nonlinear power also increases significantly, and a limited number of branches can no longer effectively adapt to nonlinear damage. Secondly, as the launch power increases, the proportion of high-order nonlinear interactions in the signal also increases, and the existing model structures are not effectively adapted to this part of nonlinearity. Therefore, increasing the number of branches, N_{step}, has a limited effect on improving RMSE performance. We hope to address this issue in our future work by adding branches with high-order nonlinearity.

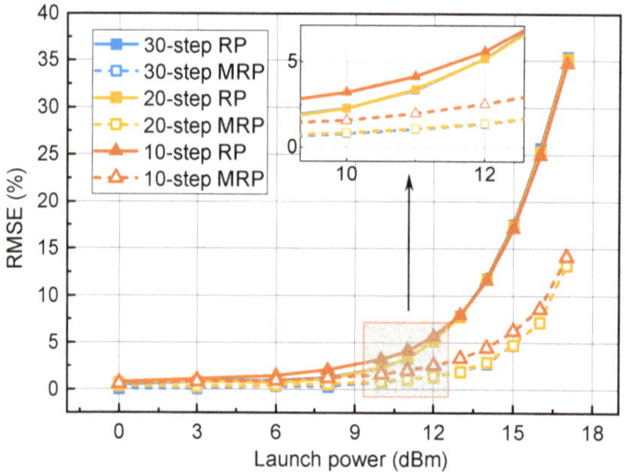

Figure 8. RMSE of the MRP and RP models under a different number of branches.

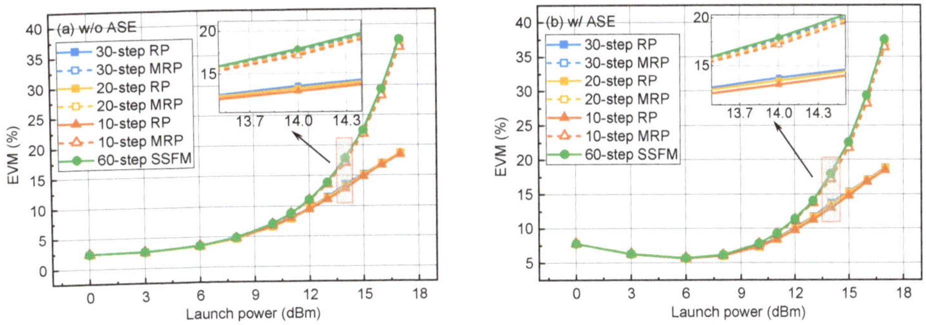

Figure 9. EVM of the MRP and RP models (**a**) w/o and (**b**) w/ ASE noise under a different number of branches, N_{step} (initial SNR = 37 dB, AWGN power = 0.01 mW).

To further analyze the performance of the model, we compared the results under different distances and waveform conditions, as shown in Figures 10 and 11. Here, we use a 2 km/step SSFM model as the training label and the object for calculating RMSE and EVM. Consistent with expectations, the fitting accuracy of RMSE and EVM of MRP model is obviously better than that of RP model. As the distance increases, the errors between the RP model and MRP model with SSFM tend to stabilize, that is because the fiber attenuation makes the contribution of the fiber beyond 80 km to the signals' nonlinearity nearly negligible. For filters with different roll-off factor, the RMSE and EVM of the MRP model are also significantly better than those of the RP model. Figure 12 shows the number of steps and corresponding computational complexity required to achieve the same accuracy for MRP and SSFM when the training label is generated by 600-step SSFM. The results indicate that the MRP model with 20 branches performs as well as 200-step SSFM. Since the trainable parameters introduced by the MRP model do not increase the computational complexity of the RP model. This means that, when achieving the same degree of nonlinear fitting, the number of real multiplications and real additions required by the MRP model are only 10.23% and 10.35% of those required by SSFM, respectively.

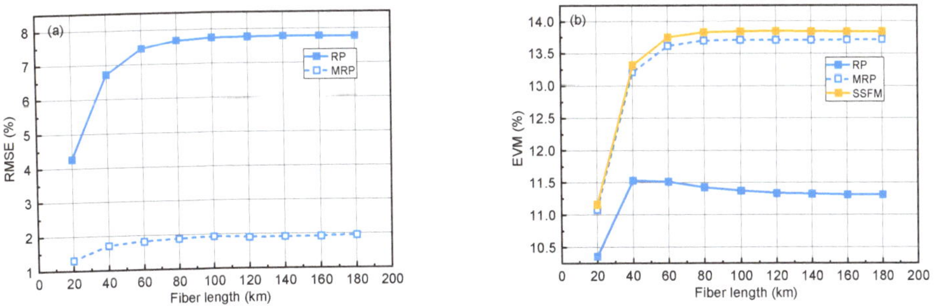

Figure 10. (**a**) RMSE and (**b**) EVM of the MRP and RP models under different fiber lengths (launch power = 13 dBm, MRP branches number N_{step} = 20).

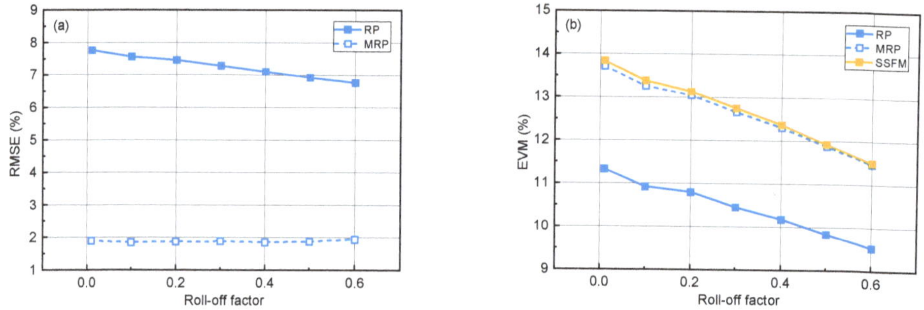

Figure 11. (**a**) RMSE and (**b**) EVM of the MRP and RP models under different roll-off factors (fiber length = 120 km, launch power = 13 dBm, branches number N_{step} = 20).

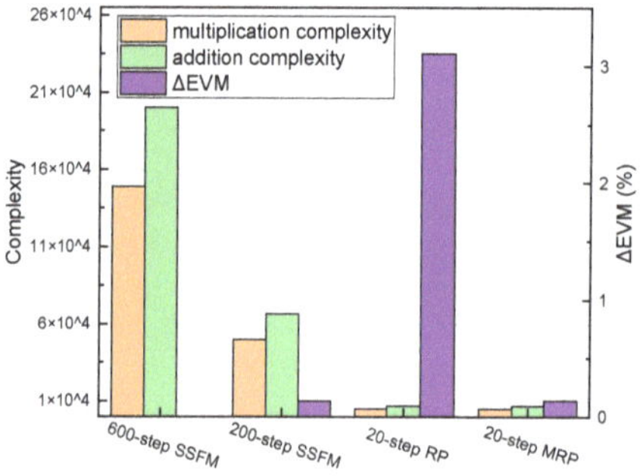

Figure 12. Computational complexity and ΔEVM of the SSFM and MRP or RP models (fiber length = 120 km, launch power = 13 dBm, sequence length S = 8192).

4. Conclusions

In this paper, we propose a modified RP model by adding trainable parameters to the standard RP model branches, using the SSFM to generate labels, and optimizing parameters through traditional backpropagation. While maintaining the same complexity as the RP model, the accuracy of the MRP model is greatly improved. When the input signal is a dual-polarization signal with a launch power of 13 dBm, the fitting error of the RP model is reduced by more than 75% after transmission through a 120 km standard single-mode fiber. We have also noted that under all launch power conditions in our paper, the accuracy of the MRP model is significantly higher than that of the RP model. In general, the paper validates the MRP model's performance under different launch powers, numbers of branches, fiber lengths, and pulse shaping filters, all of which show significant performance improvements. Furthermore, we point out that, compared to SSFM, the MRP model significantly reduces complexity while achieving the same degree of nonlinear fitting.

Author Contributions: Conceptualization, S.H.; methodology, S.H.; validation, Z.L. (Zhongya Li); visualization, A.Y.; formal analysis, S.H.; investigation, S.H., Z.L. (Zhongya Li), S.X. and A.Y.; resources, J.Z. and N.C.; data curation, S.H., Z.L. (Zhongya Li) and S.X.; writing—original draft preparation, S.H.; writing—review and editing, J.Z., N.C., Y.Z., J.S., C.S., Z.L. (Ziwei Li), Z.H. and W.C.; supervision, J.Z.; project administration, J.Z.; funding acquisition, J.Z. and N.C. All authors have read and agreed to the published version of the manuscript.

Funding: This work was supported in part by the National Key Research and Development Program of China (2023YFB2905700), in part by the National Natural Science Foundation of China under Grants (62171137, 62235005 and 61925104), in part by Key Research and Development Program of Jiangsu Province, China (No. SBE2022020259), Key Project of Jiangsu Province Industrial Prospect and Key Core Technology (BE2022055-4).

Institutional Review Board Statement: Not applicable.

Informed Consent Statement: Not applicable.

Data Availability Statement: The data that support the findings of this study are available from the authors upon reasonable request.

Conflicts of Interest: The authors declare no conflicts of interest.

Appendix A

Here, we analyze the complexity of this model, assuming that the signal sampling length is $S = 2 \times S_{symbol}$ and the number of branches in the MRP model is N_{step}. For the linear part of MRP, i.e., the zeroth-order solution term, $U_{x,y}^{(0)}$, the model needs to perform one fast Fourier transform (FFT) and one inverse fast Fourier transform (IFFT) on the sequence in order to perform dispersion operations without nonlinearity. For the dispersion operations of S-point radix-2 FFT and sequence length, S, the required complex multiplication and complex addition quantities are as follows:

$$\begin{gathered} N_{\times,fft}^{(complex)}(S) = \tfrac{S}{2}\cdot\log_2 S\,,\ \ N_{+,fft}^{(complex)}(S) = S\cdot\log_2 S \\ N_{\times,D}^{(complex)}(S) = S \end{gathered} \quad \text{(A1)}$$

where $N_{\times,fft}^{(complex)}(S)$ and $N_{+,fft}^{(complex)}(S)$, respectively, represent the number of complex multiplications and complex additions required for a fast Fourier transform. $N_{\times,D}^{(complex)}(S)$ represents the complex multiplication required to add dispersion to the signal. Therefore, the complexity of the zeroth-order solution for a sequence of length, S, is $2\times N_{\times,fft}^{(complex)}(S) + N_{\times,D}^{(complex)}(S)$ complex multiplication and $2\times N_{+,fft}^{(complex)}(S)$ complex addition. Due to the operation rules of complex numbers, one complex multiplication is equal to four real multiplication and two real addition, and one complex addition is equal to two real addition. Therefore, the computational complexity of the zeroth-order solution is as follows:

$$N_{\times}^{(0,real)}(S) = 4S\cdot(\log_2 S + 1)\ \ ,\ \ N_{+}^{(0,real)}(S) = 6S\cdot\log_2 S + 2S \quad \text{(A2)}$$

The number 0 or 1 in the subscript above the letter $N_{\times,fft}^{(0,\text{real})}$ here represents the order of the solution. For the first-order nonlinear solutions, each branch requires two FFT operations, two IFFT operations, two dispersion operations, and one nonlinear operation. The complexity of single FFT and dispersion operations is the same as Equation (A1), while nonlinear operations include $2\times S$ complex multiplication and one complex modulo squared operation. The complex modulo squared operation of dual polarization is equivalent to four real multiplication and three real addition operations. Therefore, for the first-order part of an MRP model with N_{step} branches, the computational complexity is as follows:

$$\begin{aligned} N_{\times}^{(1,real)}(S) &= N_{step}\cdot\left\{4\cdot\left[\underbrace{4N_{\times,fft}^{(complex)}(S)}_{fft\ \&\ ifft} + \underbrace{2N_{\times,D}^{(complex)}(S)}_{dispersion} + \underbrace{2S}_{nonlinear}\right] + \underbrace{4S}_{modulus}\right\} \\ &= N_{step}\cdot[8S\log_s S + 20S] \\ N_{+}^{(1,real)}(S) &= N_{step}\cdot\left\{\begin{array}{l} 2\cdot\left[\underbrace{4N_{\times,fft}^{(complex)}(S)}_{fft\ \&\ ifft} + \underbrace{2N_{\times,D}^{(complex)}(S)}_{dispersion} + \underbrace{2S}_{nonlinear}\right] \\ +2\cdot\left[\underbrace{4N_{+,fft}^{(complex)}(S)}_{fft\ \&\ ifft}\right] + \underbrace{3S}_{modulus} + \underbrace{2S}_{branch} \end{array}\right\} \\ &= N_{step}[12S\log_2 S + 13S] \end{aligned} \quad \text{(A3)}$$

where the last term $2S$ of $N_{+}^{(1,\mathrm{real})}$ in Equation (A3) represents the addition operation between the branches of the MRP model. In addition, the reason why there are only two complex multiplication operations for nonlinear operations is that the nonlinear coefficients can be calculated before signal transmission to directly multiply the signal and its modulus square. Therefore, the total computational complexity of the MRP model is show as Equation (13).

References

1. Faruk, M.S.; Savory, S.J. Digital Signal Processing for Coherent Transceivers Employing Multilevel Formats. *J. Light. Technol.* **2017**, *35*, 1125–1141. [CrossRef]
2. Zhang, S.; Yaman, F.; Nakamura, K.; Inoue, T.; Kamalov, V.; Jovanovski, L.; Vusirikala, V.; Mateo, E.; Inada, Y.; Wang, T. Field and Lab Experimental Demonstration of Nonlinear Impairment Compensation Using Neural Networks. *Nat. Commun.* **2019**, *10*, 3033. [CrossRef]
3. Kamalov, V.; Jovanovski, L.; Vusirikala, V.; Zhang, S.; Yaman, F.; Nakamura, K.; Inoue, T.; Mateo, E.; Inada, Y. Evolution from 8QAM Live Traffic to PS 64-QAM with Neural-Network Based Nonlinearity Compensation on 11000 Km Open Subsea Cable. In Proceedings of the 2018 Optical Fiber Communications Conference and Exposition (OFC), San Diego, CA, USA, 11–15 March 2018; IEEE: New York, NY, USA, 2018.
4. Niu, Z.; Yang, H.; Zhao, H.; Dai, C.; Hu, W.; Yi, L. End-to-End Deep Learning for Long-Haul Fiber Transmission Using Differentiable Surrogate Channel. *J. Light. Technol.* **2022**, *40*, 2807–2822. [CrossRef]
5. Li, S.; Hager, C.; Garcia, N.; Wymeersch, H. Achievable Information Rates for Nonlinear Fiber Communication via End-to-End Autoencoder Learning. In Proceedings of the 2018 European Conference on Optical Communication (ECOC), Rome, Italy, 23–27 September 2018; IEEE: New York, NY, USA, 2018.
6. Karanov, B.; Chagnon, M.; Thouin, F.; Eriksson, T.A.; Buelow, H.; Lavery, D.; Bayvel, P.; Schmalen, L. End-to-End Deep Learning of Optical Fiber Communications. *J. Light. Technol.* **2018**, *36*, 4843–4855. [CrossRef]
7. Agrawal, G.P. *Nonlinear Fiber Optics*, 2nd ed.; Academic: New York, NY, USA, 1995.
8. Secondini, M.; Marsella, D.; Forestieri, E. Enhanced Split-Step Fourier Method for Digital Backpropagation. In Proceedings of the 2014 European Conference on Optical Communication (ECOC), Cannes, France, 21–25 September 2014; IEEE: New York, NY, USA, 2014.
9. Hager, C.; Pfister, H.D. Physics-Based Deep Learning for Fiber-Optic Communication Systems. *IEEE J. Sel. Areas Commun.* **2021**, *39*, 280–294. [CrossRef]
10. Neskorniuk, V.; Carnio, A.; Marsella, D.; Turitsyn, S.K.; Prilepsky, J.E.; Aref, V. Memory-Aware End-to-End Learning of Channel Distortions in Optical Coherent Communications. *Opt. Express* **2023**, *31*, 1–20. [CrossRef]
11. Freire, P.J.; Napoli, A.; Spinnler, B.; Costa, N.; Turitsyn, S.K.; Prilepsky, J.E. Neural Networks-Based Equalizers for Coherent Optical Transmission: Caveats and Pitfalls. *IEEE J. Sel. Top. Quantum Electron.* **2022**, *28*, 7600223. [CrossRef]
12. Poggiolini, P.; Carena, A.; Curri, V.; Bosco, G.; Forghieri, F. Analytical Modeling of Nonlinear Propagation in Uncompensated Optical Transmission Links. *IEEE Photonics Technol. Lett.* **2011**, *23*, 742–744. [CrossRef]
13. Poggiolini, P. The GN Model of Non-Linear Propagation in Uncompensated Coherent Optical Systems. *J. Light. Technol.* **2012**, *30*, 3857–3879. [CrossRef]
14. Carena, A.; Bosco, G.; Curri, V.; Jiang, Y.; Poggiolini, P.; Forghieri, F. EGN Model of Non-Linear Fiber Propagation. *Opt. Express* **2014**, *22*, 16335–16362. [CrossRef] [PubMed]
15. Carena, A.; Curri, V.; Bosco, G.; Poggiolini, P.; Forghieri, F. Modeling of the Impact of Nonlinear Propagation Effects in Uncompensated Optical Coherent Transmission Links. *J. Light. Technol.* **2012**, *30*, 1524–1539. [CrossRef]
16. Dar, R.; Feder, M.; Mecozzi, A.; Shtaif, M. Properties of Nonlinear Noise in Long, Dispersion-Uncompensated Fiber Links. *Opt. Express* **2013**, *21*, 25685–25699. [CrossRef] [PubMed]
17. Vannucci, A.; Serena, P.; Bononi, A. The RP Method: A New Tool for the Iterative Solution of the Nonlinear Schrodinger Equation. *J. Light. Technol.* **2002**, *20*, 1102–1112. [CrossRef]
18. Surjan, P.; Angyan, J. Perturbation-Theory for Non-Linear Time-Independent Schrodinger-Equations. *Phys. Rev. A* **1983**, *28*, 45–48. [CrossRef]
19. Garcia-Gomez, F.J.; Kramer, G. Mismatched Models to Lower Bound the Capacity of Dual-Polarization Optical Fiber Channels. *J. Light. Technol.* **2021**, *39*, 3390–3399. [CrossRef]
20. Mecozzi, A.; Essiambre, R.-J. Nonlinear Shannon Limit in Pseudolinear Coherent Systems. *J. Light. Technol.* **2012**, *30*, 2011–2024. [CrossRef]
21. Zhao, Y.; Dou, L.; Tao, Z.; Yan, M.; Oda, S.; Tanimura, T.; Hoshida, T.; Rasmussen, J.C. Improved Analytical Model for Intra-Channel Nonlinear Distortion by Relaxing the Lossless Assumption. In Proceedings of the 39th European Conference and Exhibition on Optical Communication (ECOC 2013), London, UK, 22–26 September 2013; pp. 1–3.
22. Zhao, Y.; Dou, L.; Tao, Z.; Xu, Y.; Hoshida, T.; Rasmussen, J.C. Nonlinear Noise Waveform Estimation for Arbitrary Signal Based on Nyquist Nonlinear Model. In Proceedings of the 2014 The European Conference on Optical Communication (ECOC), Cannes, France, 21–25 September 2014; pp. 1–3.

23. Tao, Z.; Zhao, Y.; Fan, Y.; Dou, L.; Hoshida, T.; Rasmussen, J.C. Analytical Intrachannel Nonlinear Models to Predict the Nonlinear Noise Waveform. *J. Light. Technol.* **2015**, *33*, 2111–2119. [CrossRef]
24. Fan, Y.; Dou, L.; Tao, Z.; Lei, L.; Oda, S.; Hoshida, T.; Rasmussen, J.C. Modulation Format Dependent Phase Noise Caused by Intra-Channel Nonlinearity. In Proceedings of the 2012 38th European Conference and Exhibition on Optical Communications, Amsterdam, The Netherlands, 16–20 September 2012; pp. 1–3.
25. Liang, X.; Kumar, S. Multi-Stage Perturbation Theory for Compensating Intra-Channel Nonlinear Impairments in Fiber-Optic Links. *Opt. Express* **2014**, *22*, 29733–29745. [CrossRef]
26. Barreiro, A.; Liga, G.; Alvarado, A. Data-Driven Enhancement of the Time-Domain First-Order Regular Perturbation Model. *J. Light. Technol.* **2023**, *41*, 2691–2706. [CrossRef]
27. Neskorniuk, V.; Carnio, A.; Bajaj, V.; Marsella, D.; Turitsyn, S.K.; Prilepsky, J.E.; Aref, V. End-to-End Deep Learning of Long-Haul Coherent Optical Fiber Communications via Regular Perturbation Model. In Proceedings of the 2021 European Conference on Optical Communication (ECOC), Bordeaux, France, 13–16 September 2021; IEEE: New York, NY, USA, 2021.
28. Sasai, T.; Yamazaki, E.; Nakamura, M.; Kisaka, Y. Closed-Form Expressions for Fiber-Nonlinearity-Based Longitudinal Power Profile Estimation Methods. In Proceedings of the 2022 European Conference on Optical Communication (ECOC), Basel, Switzerland, 18–22 September 2022; IEEE: New York, NY, USA, 2022.
29. Mecozzi, A.; Clausen, C.B.; Shtaif, M. Analysis of Intrachannel Nonlinear Effects in Highly Dispersed Optical Pulse Transmission. *IEEE Photonics Technol. Lett.* **2000**, *12*, 392–394. [CrossRef]
30. Tao, Z.; Dou, L.; Yan, W.; Li, L.; Hoshida, T.; Rasmussen, J.C. Multiplier-Free Intrachannel Nonlinearity Compensating Algorithm Operating at Symbol Rate. *J. Light. Technol.* **2011**, *29*, 2570–2576. [CrossRef]
31. Soman, S.K.O.; Amari, A.; Dobre, O.A.; Venkatesan, R. Second-Order Perturbation Theory-Based Digital Predistortion for Fiber Nonlinearity Compensation. *J. Light. Technol.* **2021**, *39*, 5474–5485. [CrossRef]
32. Liu, X.; Li, C.; Jiang, Z.; Han, L. Low-Complexity Pruned Convolutional Neural Network Based Nonlinear Equalizer in Coherent Optical Communication Systems. *Electronics* **2023**, *12*, 3120. [CrossRef]
33. He, P.; Wu, F.; Yang, M.; Yang, A.; Guo, P.; Qiao, Y.; Xin, X. A Fiber Nonlinearity Compensation Scheme With Complex-Valued Dimension-Reduced Neural Network. *IEEE Photonics J.* **2021**, *13*, 7200507. [CrossRef]
34. Kumar, S.; Yang, D. Second-Order Theory for Self-Phase Modulation and Cross-Phase Modulation in Optical Fibers. *J. Light. Technol.* **2005**, *23*, 2073–2080. [CrossRef]
35. Redyuk, A.; Averyanov, E.; Sidelnikov, O.; Fedoruk, M.; Turitsyn, S. Compensation of Nonlinear Impairments Using Inverse Perturbation Theory With Reduced Complexity. *J. Light. Technol.* **2020**, *38*, 1250–1257. [CrossRef]
36. Peng, W.-R.; Li, Z.; Zhu, F.; Bai, Y. Training-Based Determination of Perturbation Coefficients for Fiber Nonlinearity Mitigation. In Proceedings of the 2015 Optical Fiber Communications Conference and Exhibition (OFC), Los Angeles, CA, USA, 22–26 March 2015; IEEE: New York, NY, USA, 2015.
37. Peng, W.-R.; Li, Z.; Zhu, F.; Bai, Y. Extending Perturbative Nonlinearity Mitigation to PDM-16QAM. In Proceedings of the 2014 European Conference on Optical Communication (ECOC), Cannes, France, 21–25 September 2014; IEEE: New York, NY, USA, 2014.
38. Tharranetharan, S.; Soman, S.K.O.; Lampe, L. Joint Fiber Nonlinearity Mitigation and Compensation for Digital Sub-Carrier Multiplexing System. *IEEE Photonics J.* **2024**, *16*, 7201517. [CrossRef]
39. He, Z.; Mirani, A.; Karlsson, M.; Schroder, J. Perturbation-Based Joint SPM and XPM Compensation for Superchannel System. *IEEE Photonics Technol. Lett.* **2024**, *36*, 1349–1352. [CrossRef]
40. Ramachandran, V.; Liga, G.; Barreiro, A.; Alvarado, A. Capacity Region Bounds for Optical WDM Channels Based on First-Order Regular Perturbation. *J. Light. Technol.* **2023**, *41*, 31–40. [CrossRef]
41. Mateo, E.F.; Zhou, X.; Li, G. Improved Digital Backward Propagation for the Compensation of Inter-Channel Nonlinear Effects in Polarization-Multiplexed WDM Systems. *Opt. Express* **2011**, *19*, 570–583. [CrossRef]
42. Mateo, E.F.; Yaman, F.; Li, G. Efficient Compensation of Inter-Channel Nonlinear Effects via Digital Backward Propagation in WDM Optical Transmission. *Opt. Express* **2010**, *18*, 15144–15154. [CrossRef] [PubMed]
43. Fillion, N.; Corless, R.M. Perturbation Theory. *arXiv* **2022**, arXiv:2212.07380.
44. Sinkin, O.V.; Holzlöhner, R.; Zweck, J.; Menyuk, C.R. Optimization of the Split-Step Fourier Method in Modeling Optical-Fiber Communications Systems. *J. Light. Technol.* **2003**, *21*, 61–68. [CrossRef]
45. Zhang, J.; Li, X.; Dong, Z. Digital Nonlinear Compensation Based on the Modified Logarithmic Step Size. *J. Light. Technol.* **2013**, *31*, 3546–3555. [CrossRef]
46. Minzioni, P.; Pusino, V.; Cristiani, I.; Marazzi, L.; Martinelli, M.; Degiorgio, V. Study of the Gordon-Mollenauer Effect and of the Optical-Phase-Conjugation Compensation Method in Phase-Modulated Optical Communication Systems. *IEEE Photonics J.* **2010**, *2*, 284–291. [CrossRef]

photonics

MDPI

Article

Using Artificial Neural Networks to Evaluate the Capacity and Cost of Multi-Fiber Optical Backbone Networks

Alexandre Freitas [1] and João Pires [2,*]

[1] Department of Electrical and Computer Engineering, Instituto Superior Técnico, Universidade de Lisboa, Avenida Rovisco Pais 1, 1049-001 Lisboa, Portugal; alexandre.costa.freitas@tecnico.ulisboa.pt
[2] Department of Electrical and Computer Engineering and Instituto de Telecomunicações, Instituto Superior Técnico, Universidade de Lisboa, Avenida Rovisco Pais 1, 1049-001 Lisboa, Portugal
* Correspondence: jpires@lx.it.pt

Abstract: A possible solution to address the enormous increase in traffic demands faced by network operators is to rely on multi-fiber optical backbone networks. These networks use multiple optical fibers between adjacent nodes, and, when properly designed, they are capable of handling petabits of data per second (Pbit/s). In this paper, an artificial neural network (ANN) model is investigated to estimate both the capacity and cost of a multi-fiber optical network. Furthermore, a fiber assignment algorithm is also proposed to complement the network design, enabling the generation of datasets for training and testing of the developed ANN model. The model consists of three layers, including one hidden layer with 50 hidden units. The results show that for a large network, such as one with 100 nodes, the model can estimate performance metrics with an average relative error of less than 0.4% for capacity and 4% for cost, while achieving a computation time nearly 800 times faster than the heuristic approach used in network simulation. Additionally, the network capacity is around 5 Pbit/s.

Keywords: multi-fiber optical networks; artificial neural networks; machine learning; network capacity and cost; fiber assignment

1. Introduction

In recent years, data traffic has increased significantly, a trend expected to continue due to the growth of applications and services that require high bandwidth and generate large amounts of data. Examples include video streaming services, cloud computing, machine-to-machine applications, online gaming, and the adoption of emerging technologies like 5G and beyond and advanced artificial intelligence applications [1]. This evolving scenario places special requirements on the backbones of network operators, which could experience traffic flows between their nodes reaching tens of Tb/s in the medium term, and even up to hundreds of Tb/s in the long term [1]. This situation presents a significant challenge for the design of future optical networks, particularly their backbone segments.

Optical networks are communication infrastructures, owned by telecommunication operators (telcos) or internet companies (e.g., Google, Microsoft, Meta), that utilize light for transmission, processing, and routing information and rely on optical fibers as their transmission medium. A fundamental technology in the field of optical networking is Wavelength Division Multiplexing (WDM). WDM allows the simultaneous transmission of multiple optical signals (also designated as optical channels) on the same optical fiber, with each channel using a different wavelength. The number of optical channels that can be transmitted over an optical fiber is limited to about 100 when using the traditional C-band, restricting the maximum WDM transmission capacity to well below 100 Tb/s for significant distances [2]. To greatly increase the number of optical channels to cope with the enormous growth in bandwidth demand, one can rely on space division multiplexing (SDM) techniques. This approach can be implemented using a multi-fiber (MF) solution, i.e., multiple standard single-mode fiber pairs per link instead of just one, as it is typical, or, alternatively, advanced fibers such as multicore fibers or few-mode fibers, with both solutions still operating in the C-band [2]. By relying on these solutions, it is feasible to design petabit-class optical networks, which are networks capable of handling data at speeds reaching or exceeding one petabit per second (Pb/s) [3].

For designing MF networks, it is crucial to define, in addition to the traditional routing and wavelength assignment solutions, a strategy for allocating fibers to the network, specifically, a fiber assignment strategy. In [4], two approaches were proposed to optimize network capacity by adding extra fibers. In the first approach, fibers were added to links supporting the maximum number of traffic demands, while in the second, fibers were added to links exhibiting the highest number of adjacent demands. Furthermore, in [5], the idea is to add extra fibers to links that are responsible for blocking traffic demands due to spectrum exhaustion, with the goal of minimizing the number of fibers added.

Network capacity is a key performance metric in optical networks. This capacity can be defined as the maximum amount of data that the entire network can handle per unit of time, and it is closely related to the concept of channel capacity introduced by Claude Shannon in 1948 [6]. The estimation of network capacity is a challenging task because it depends not only on physical layer aspects related to optical fibers and other optical devices but also on networking aspects such as physical and logical topology, routing, as well as wavelength and modulation assignment. Consequently, it suffers from the hurdle of long computation times, especially when dealing with large-scale networks. Although the problem of predicting optical network capacity has been the focus of many studies, (see [7–10]), to the best of the authors' knowledge, none of the published research has relied on machine learning (ML) techniques for this purpose, despite these techniques being widely used in the context of optical networks to address other problems [11–13]. The closest study is reported in [14], where a routing and wavelength assignment (RWA) problem is treated using ML techniques by transforming it into a multi-classification problem, which is then solved using logistic regression and deep neural network techniques. However, the network capacity estimation problem, although also involving RWA calculations, is more general than this. Furthermore, the complexity of the problem for MF networks is even higher due to the necessity of using fiber assignment techniques.

In this paper, we investigate the utilization of an ML solution, specifically an artificial neural network (ANN) model [12], to estimate both the capacity and cost metrics of an MF-based optical network capable of handling Pb/s of data, with the cost being defined as the total length of optical fiber required in the network. The goal is to determine whether it is possible to significantly speed up the computations of these two metrics in comparison with heuristic methods, while still achieving accurate results.

To generate the large sets of synthetic data needed to train and test the model, we used a tool previously developed by the authors [10]. This tool not only generates random network topologies that aim to mimic real optical backbone networks but also performs routing and fiber assignment operations on these networks using heuristics developed specifically for this purpose, including the fiber assignment algorithm that is described in this work, which is a crucial component of our methodology.

The rest of the paper is organized as follows: Section 2 reviews important aspects of network modeling and random network generation and explains how both network capacity and cost can be computed. It also describes the fiber assignment algorithm proposed here for allocating fibers in MF networks. Section 3 details the ANN model introduced in this work. Section 4 presents some simulation results and, finally, Section 5 summarizes and concludes the paper.

2. Network Aspects and Dataset Generation

2.1. Network Modeling

In an abstract way, an optical network can be described as an undirected weighted graph $G(V, E)$, with $V = \{v_1, \ldots, v_N\}$ denoting a set of nodes and $E = \{e_1, \ldots, e_K\}$ denoting a set of links, where $N = |V|$ is the number of nodes and $K = |E|$ is the number of links. In transparent optical networks, all node functionalities take place in the optical domain, and the nodes are built using reconfigurable optical add–drop multiplexers (ROADMs), which are responsible for switching optical channels between different fibers, among other functions. Interconnection between these elements and client equipment is achieved using transponders, which are devices responsible for mapping client signals into optical channels. Meanwhile, an optical link represents a physical interconnection between two nodes, implemented using a pair of optical fibers, along with optical amplifiers spaced appropriately to compensate for fiber losses. Note that in the case of MF networks, multiple pairs of fibers are used instead. Furthermore, each optical fiber supports WDM signals, meaning it carries a specific number of optical channels. Each link $(v_i, v_j) \in E$ is characterized by three attributes: $l_{i,j}$, the link length in kilometers between the nodes v_i and v_j; $nf_{i,j}$, the number of optical fiber pairs in the link; and $u_{i,j}$, the link capacity measured in terms of the number of optical channels denoted as N_{ch}. In this work, we assume that fiber transmission takes place in the extended C-band, which has a bandwidth of 4800 GHz, enabling the support of $N_{ch,max} = 75$ channels, with a channel spacing of 64 GHz corresponding to a baud rate of 64 Gbaud.

In addition to N and K, other important parameters of the graph G are the node degree $\delta(G)$, the network diameter $d(G)$, and the algebraic connectivity $a(G)$. $\delta(G)$ defines the number of links connected to a given node, $d(G)$ is the length of the longest shortest path between any two nodes, and $a(G)$ is the second smallest eigenvalue of the graph's Laplacian matrix [15].

In the context of ANNs, it is necessary to have very large datasets for training and testing purposes. To achieve this, it is useful to be able to generate numerous network topologies, which can be performed through random graphs designed to adequately describe the characteristics of real-world optical networks. In [10], we described a tool that we developed to generate random networks appropriate for describing optical backbone networks. The tool is based on a modified Waxman model and can generate networks that are resilient to single-link failures. In a simplified way, this model works by dividing a two-dimensional (2D) square plane with area $A = L^2$ (L is the side length of the plane) into a set of regions. In these regions, N nodes are randomly placed, and then the nodes are interconnected with links according to the Waxman probability, which is characterized by the α and β parameters, both in the range $[0, 1]$.

2.2. Routing, Fiber Assignment, Capacity, and Cost

Network capacity refers to the maximum amount of data that the network can theoretically handle per unit of time, typically measured in bits per second (bit/s). This metric depends on many network parameters, including the physical topology defined by the graph $G(V, E)$ and the logical topology, which describes the way how the information flows between all the network nodes. The logical topology is defined by the traffic matrix $T = [t_{s,d}]$, where each entry $t_{s,d}$ represents a traffic demand, or in other terms, the volume of traffic flowing from a source node s to a destination node d, with $s, d \in V$. For each traffic demand $t_{s,d}$, it is necessary to find a path in the graph $G(V, E)$, between node s and node d. This is the role of the routing process. The routing process can be implemented using rigorous mathematical techniques, such as integer linear programing (ILP), or heuristics, as an alternative [16]. As ILPs become computationally infeasible for large-scale networks, we have to rely on heuristics in this work, as the analysis of such networks is paramount.

When the number of channels N_{ch} per fiber is limited, as in this work, the routing process is known as constrained routing and can lead to traffic demand blocking whenever no channels (wavelengths) are available on one or more links of the path. To overcome such a limitation, one can add more pairs of optical fibers per link as needed, as it is the case for MF networks. This leads to a new process referred to as unconstrained routing plus fiber assignment. This process can be implemented using the heuristics proposed in [10]. In a simplified way, the heuristic method first computes the shortest paths between each pair of nodes in $G(V, E)$ using the Dijkstra algorithm, with distance as the metric. Traffic demands between node pairs are then prioritized according to a specific sorting strategy and routed along their designated paths. Each path is assigned a wavelength using a first-fit strategy, thereby forming an optical channel. Finally, an optical fiber is allocated to each channel using Algorithm 1, which is described below.

To generate the datasets required to train and test the ANN, we have applied the referred heuristics to the randomly generated networks using the modified Waxman model assuming a uniform traffic demand between all the network node pairs, which can be defined as

$$t_{s,d} = \begin{cases} 1 & s \neq d \\ 0 & s = d \end{cases}. \tag{1}$$

Taking into account the traffic matrix $T = [t_{s,d}]$ of size $N \times N$, and assuming that $u_{i,j} = \infty$, we can apply unconstrained routing to each network graph $G(V, E)$ to compute the list of established paths $\mathrm{P} = [\pi_{s,d}]$, with the path $\pi_{s,d}$ having the length $l(\pi_{s,d}) = \sum_{i,j} l_{i,j}$. Additionally, we compute the link wavelength matrix, $W = [w_{i,j}]$, which is also a $N \times N$ matrix, where $w_{i,j}$ is the list of all the wavelengths λ_k present in the link (i, j), i.e., $w_{i,j} = [\lambda_k]$. As referred to before, fiber assignment is a central process in MF networks. To implement this process, we propose Algorithm 1, which allows us to obtain the fiber matrix $NF = [nf_{i,j}]$, representing the number of fibers per link, taking into account that the maximum number of wavelengths per link is $N_{\lambda,max} = N_{ch,max}$.

Algorithm 1: Fiber Assignment

Input: graph $G(V, E)$, wavelength matrix $W = [w_{i,j}]$, number of wavelengths $N_{\lambda,max}$.
Output: fiber matrix $NF = [nf_{i,j}]$.
Initialize NF, with $nf_{i,j} = 0, \forall (i, j) \in E$.
for each pair of nodes (i, j) in W **do**
    if G has an edge (i, j) **then**
        if there are no wavelengths used in (i, j), i.e., $w_{i,j} = 0$ **then**
            $nf_{i,j} \leftarrow 1$: At least one fiber is required
        else
            normalized wavelengths $\leftarrow$ $w_{i,j}$ mapped into the range 1 to $N_{\lambda,max}$
            num_fibres $\leftarrow$ maximum number of wavelengths repetitions in normalized wavelengths
            $nf_{i,j} \leftarrow$ num_fibres
        end if
    else
        $nf_{i,j} \leftarrow 0$: Case there is no edge (*i*,*j*)
    end if
end for
return NF

Note that with the unconstrained routing, the number of wavelengths in each link is not limited, so the value assigned to a given λ_k can be any natural number, in contrast to constraint routing, where it is bounded by $N_{\lambda,max}$. In the algorithm, to determine the number of fibers needed in each link, the maximum number of "repeated wavelengths" in that link must be determined. A wavelength is considered a "repeated wavelength" when its value modulo $N_{\lambda,max}$ (where the modulo operation returns the remainder after division) is equal to that of another wavelength also present in that link. For instance, if $N_{\lambda,max}$ is 75, then wavelengths 1 and 76 are "repeated" because 76 modulo 75 equals 1. This implies that both wavelengths would occupy the same channel in a link, hence they are "repeated". This concept is crucial in determining the number of fibers needed for a link, ensuring that each "repeated" wavelength has its own fiber. Finding the maximum count of "repeated wavelengths" will ensure that there are enough fibers to accommodate all the wavelengths, thus assuring that there are no channels with the same wavelength on the same fiber.

By knowing the length of the path $\pi_{s,d}$, $l(\pi_{s,d})$, it is possible to compute its maximum capacity value, $C(\pi_{s,d})$, also denoted as the Shannon capacity, measured in bits per second. This calculation uses the optical reach values of the path (see Table 2 of [10]), where optical reach is defined as the maximum length of the path for which a certain value of the capacity can be achieved assuming a baud rate of 64 Gbaud. Furthermore, after obtaining the capacity of all the established paths, one arrives at the network capacity, which is given by

$$C_{net} = \sum_{s,d} C(\pi_{s,d}). \tag{2}$$

Another important metric is the network's cost. The overall cost of an optical network is the sum of the costs of all nodes and links, with node costs primarily driven by transponders and link costs by optical amplifiers. It is reasonable to assume that, in optical backbone networks, link costs are the dominant contributors to the network costs. As a result, these costs are predominantly determined by fiber length, since this parameter defines the number of optical amplifiers required [17]. In this case, the network cost is given by

$$\Lambda_{net} = \sum_{i,j} l_{i,j} \times nf_{i,j}. \tag{3}$$

3. Neural Network Design

An artificial neural network is a network of units, also called neurons, which are organized in multiple layers, including an input layer, a variable number of hidden layers, and an output layer. These layers operate in a fully connected way, meaning that each neuron of a given layer is connected to all the neurons of the next layer. Each neuron has a variable weight per input, denoted as $\omega_{m,i}$, with m defining the neuron position in a layer and i its input, which are summed together along with a bias term b_m. The result of this operation is then passed through an activation function to obtain the output of that neuron. The activation function used in this study for the hidden layers is the ReLU (Rectified Linear Unit) function, which is given by

$$g(x) = \max(0,\ x) \tag{4}$$

while for the output layer we have the linear activation function, that is,

$$g(x) = x. \tag{5}$$

Note that both activation functions are commonly used in regression problems, such as the one we are considering here [18].

The training of neural networks consists of determining the values for all $\Omega = [\omega_{m,i}]$ matrices and bias vectors $B = [b_m]$ that minimize a given loss function with a given iterative method (optimizer algorithm). For the training process, it is necessary to randomly generate a large number of datasets using the procedures described previously. Each dataset includes an array of inputs $X = [x_1,\ x_2 \ldots x_n]$, called features, and an array of outputs $Y = [y_1,\ y_2]$, obtained by network simulation, called labels. The features include the number of nodes, the number of links, the network diameter, the algebraic connectivity, and quantities such as the maximum, minimum, average, and variance of both link length and node degree. Furthermore, the labels include the network capacity $y_1 = C_{net}$, given by (2), and the network cost $y_2 = \Lambda_{net}$, given by (3).

In the training process, each dataset is split into a training set (the data used to determine the model's parameters), a validation set (used to make an unbiased evaluation of the model's performance during training), and a test set (used to assess the model's performance after the training is complete). Before the data are split into these three sets, they need to be pre-processed and shuffled. Data pre-processing consists of preparing the data to make them more suitable for the training process.

The loss function is used to measure the difference between the value predicted by the ANN and the actual value obtained by simulation. In other words, it measures the error associated with the model's predictions. For regression problems, the mean squared error (MSE) is commonly used as the loss function [18]. MSE can be expressed as follows:

$$MSE = \frac{1}{M}\sum_{i=1}^{M}(\hat{y}_i - y_i)^2 \tag{6}$$

where M is the number of data values being considered, $\hat{y}_i$ are the estimated values, and y_i are the actual values.

The optimizer algorithm is the method that determines how the weight matrices and bias vectors are updated during the training process. Common optimizers include the Stochastic Gradient Descent (SGD) and the Adaptive Momentum Estimation (Adam), with the former being used in this work. The updating of the network parameters requires the computation of the gradient of the loss function, a task performed by the backpropagation algorithm [19]. An important parameter related to the optimizer is the learning rate. This parameter determines the magnitude of the updates applied to the weights and biases during each iteration. Another important parameter is the batch size, which refers to the size of subsets into which the training data are divided. The dropout regularization can also be used to prevent overfitting, which occurs when a model learns the training data too closely but fails to make accurate predictions on the testing data.

A key aspect of training an ANN is optimizing the hyperparameters. Hyperparameters are the variables that configure how the model learns from the data. This includes the number of hidden layers, the number of units in each hidden layer, the learning rate, the batch size, and dropout regularization. During training, various hyperparameter combinations are tested to achieve the best performance on the validation set. This operation is called hyperparameter tuning.

In this work, the tuning operation is performed using the R^2 score metric, which is defined as

$$R^2 = 1 - \frac{\sum_{i=0}^{M}(y_i - \hat{y}_i)^2}{\sum_{i=0}^{M}(y_i - \overline{y}_i)^2} \tag{7}$$

where y_i is the actual value, $\hat{y}_i$ is the predicted value, $\overline{y}$ is the mean of the actual values, and M is the number of data values being considered. The R^2 score will take values between 0 and 1, where a value of 1 indicates that the model fits the data perfectly and 0 that it does not fit the data at all. That means that the closer the values are to 1, the better the model is performing [20].

From the hyperparameter tuning process, the ANN's structure was defined (see Figure 1). The model that achieved the best performance on the validation set has one hidden layer with 50 hidden units ($m = 50$), considering the number of features equal to 12 ($n = 12$). The learning rate was optimized to 0.1, the batch size was set to 64, and no dropout regularization was needed.

This model structure and learning rate resulted in relatively high R^2 scores: 0.9994 for y_1 and 0.9962 for y_2. Additionally, it has a relatively low number of trained parameters (the total number of weights and biases), with 752 parameters, which represents a good balance between model complexity and performance. To build and optimize the ANN model, we used the PyTorch 2.2 framework [21].

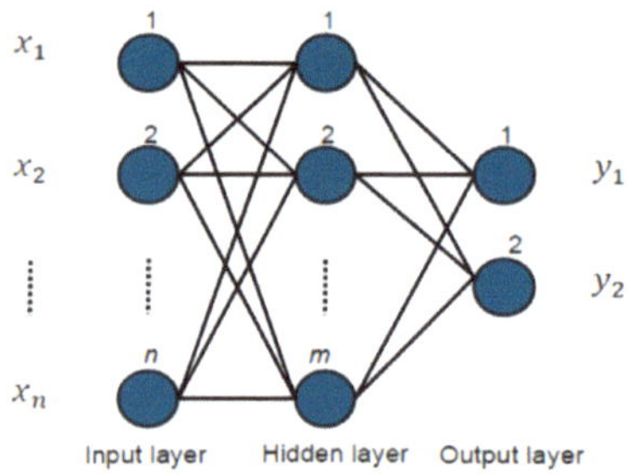

Figure 1. Model of the ANN network with 1 hidden layer.

4. Simulation Results and Discussion

To train the ANN model, a set of 8480 networks was used. These networks were generated with the tool described in [10] considering a 2D square plane with side lengths varying from 1000 km to 5000 km in increments of 1000 km, the number of regions in the plane set to 4, the number of nodes varying from 5 to 100, the number of links varying from 5 to 231, and an average node degree varying from 2 to around 5. The Waxman parameters chosen were $\alpha = \beta = 0.4$. Furthermore, the maximum number of channels per links was set to $N_{ch,max} = 75$.

Once the model is trained, the final step is to evaluate its performance through testing. For this purpose, a dataset of 1440 random networks was generated under the same conditions as those used to train the model. The network simulation took around 1 h and 16 min for the entire dataset, while the prediction time for the ANN model was just 11 milliseconds.

The mean relative errors for this test dataset, as defined by (8), are as follows: 2.47% for the network capacity (y_1) and 5.29% for the total fiber cost (network cost) (y_2) predictions. Figure 2 shows the scatter plot of the relative errors against the number of nodes for both outputs. Each blue dot represents the relative error (RE) for each individual network in the set, given by

$$RE = \frac{y_i - \hat{y}_i}{y_i} \quad (8)$$

with y_i being the value determined from the simulation solution and $\hat{y}_i$ the prediction made with the ANN model.

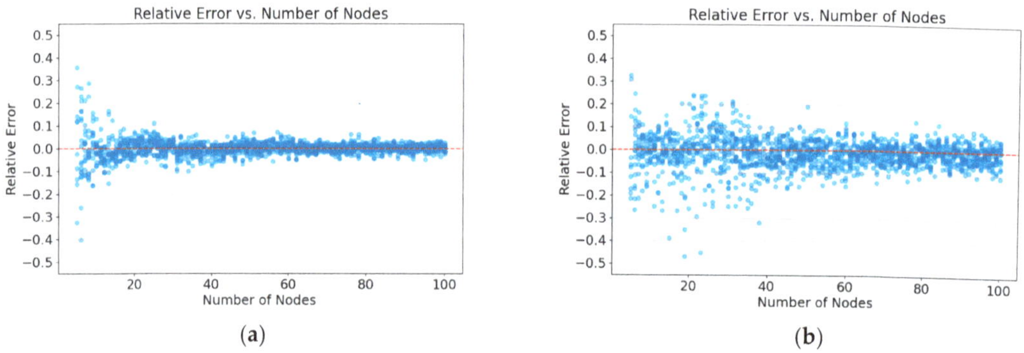

Figure 2. Relative errors as a function of the number of nodes for both outputs of the ANN model (N ranging from 5 to 100): (**a**) total network capacity; (**b**) total fiber cost.

It was also shown that for the total network capacity (Figure 2a), 89.45% of the examples have a relative error below 5%, and 96.67% of the examples have a relative error below 10%. In the case of the total fiber cost (Figure 2b), 87.02% of the examples have a relative error below 10%, and 94.24% of the examples have a relative error below 15%. It can be seen that the model tends to perform better on networks with a higher number of nodes, while its performance is more irregular on networks with fewer nodes. A possible explanation for this behavior is that smaller networks might exhibit more variability in their features as well as in the relationships between features and labels, which makes it more challenging for the model to learn stable patterns that are crucial for making accurate predictions. This irregular performance of smaller-scale networks is particularly evident for the label "network cost" when the number of nodes is 10 or fewer. On the other hand, larger networks could be more homogeneous, exhibiting more uniform and consistent patterns that the model can learn and predict more effectively.

In order to analyze how the ANN model behaves with testing datasets that have a number of nodes outside the training range, we generated 3920 additional networks under the same conditions as the previous sets, but with the number of nodes ranging from 5 to 200. Generating this set took 55 h and 31 min, while the ANN model predicted the corresponding set in only 79 milliseconds. Figure 3 shows a scatter plot comparing the relative errors as a function of the number of nodes for this set of networks. The plots in Figure 3 show that the results are identical to those of Figure 2 when the number of nodes ranges from 5 to 100. However, outside this range, the model's performance becomes unreliable, although it still performs quite well for up to about 115 nodes. The cause of this irregular behavior differs from that observed in small-scale networks and is due to the model being trained on a specific data range (number of nodes ranging from 5 to 100), and extrapolating beyond this range can lead to less reliable predictions.

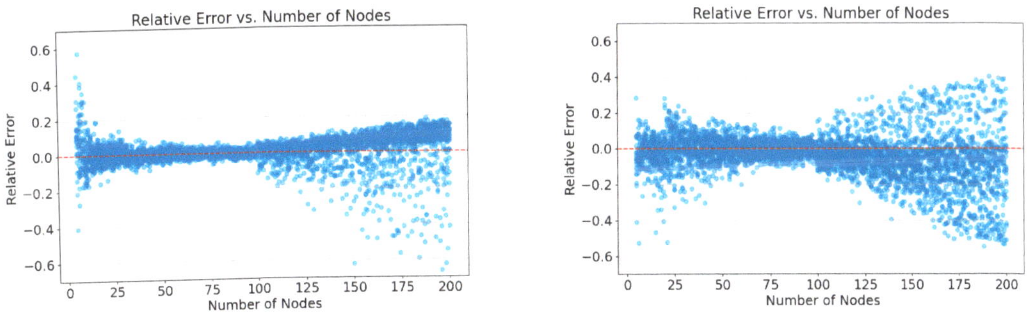

(a) (b)

Figure 3. Relative errors as a function of the number of nodes for both outputs of the ANN model (N ranging from 5 to 200): (**a**) total network capacity; (**b**) total fiber cost.

The capability of a model to perform well in the presence of unseen data, which it was not trained on, is known as Out-of-Distribution (OOD) generalization [22]. This is a challenge that conventional supervised learning methods (such as ANNs) often find difficult to handle effectively as these types of models fundamentally assume that the training and test datasets originate from the same distribution. Note that addressing the OOD generalization problem is an active area of research in the field of ML [22].

Table 1 compares the results predicted by the ANN model for the total network capacity ($\hat{y}_1$) and total fiber cost ($\hat{y}_2$) with the corresponding results obtained by applying a heuristic approach to different random networks, using the tool described in [10], as well as Algorithm 1 for the fiber assignment task. These results show that the ANN models tend to have a good performance in the generated networks within this range of nodes, with the relative errors generally being low. Furthermore, the prediction times with the ANN are always significantly faster than the computation times obtained with the network simulation tool. For example, for a network with 100 nodes, the prediction time is about

17.1 milliseconds, whereas the computation time is about 13.5 s. This means that the ANN model is roughly 800 times faster than the heuristic approach, while achieving low relative errors of about 0.4% for network capacity and about 4% for network cost.

Table 1. Accuracy of ANN prediction: y_1: capacity; y_2: cost.

N	y_1 [Tb/s]	$\hat{y}_1$ [Tb/s]	RE (%)	y_2 [10^3 km]	$\hat{y}_2$ [10^3 km]	RE (%)
10	48.0	45.4	5.44	24.47	24.08	1.59
20	303.2	317.9	4.86	14.71	14.00	4.85
30	708.0	705.2	0.39	27.05	26.62	1.57
40	803.0	820.1	2.13	122.27	123.21	0.77
50	1244.2	1214.4	2.40	231.63	257.36	11.1
60	1837.2	1937.3	5.45	267.02	247.97	7.14
70	3432.8	3393.9	1.13	128.44	131.79	2.61
80	3185.0	3197.8	0.40	626.00	597.00	4.63
90	5898.6	5864.0	0.59	189.51	194.54	2.66
100	5394.6	5376.5	0.34	718.09	690.02	3.91

Remarkably, the network capacity for a number of nodes greater than or equal to 50 nodes exceeds 1 Pb/s, reaching about 5 Pb/s for 100 nodes. However, this comes at the cost of significantly increasing the required optical fiber length in the network, as this work is based on the MF paradigm, where additional fibers are added whenever a link reaches its maximum supported number of optical wavelengths.

A key point in the analysis is understanding how the ANN model performs on real optical network topologies, despite being trained on synthetic data generated from random networks. To address this point, Table 2 provides results for four real reference networks: COST239 ($N = 11, K = 26, \bar{l} = 462.6$ km) [23], DTAG ($N = 14,\ K = 23,\ \bar{l} = 236.5$ km) [9], NSFNET ($N = 14,\ K = 21,\ \bar{l} = 1211.3$ km) [23], and UBN ($N = 24,\ K = 43,\ \bar{l} = 993.2$ km) [23], with $\bar{l}$ being the average link length.

Table 2. Accuracy of DNN predictions in reference networks. y_1: capacity; y_2: cost.

Network	y_1 [Tb/s]	$\hat{y}_1$ [Tb/s]	RE (%)	y_2 [10^3 km]	$\hat{y}_2$ [10^3 km]	RE (%)
COST239	81.2	82.8	2.01	24.06	23.07	4.11
DTAG	147.4	145.9	1.01	10.88	10.95	0.69
NSFNET	98.0	104.1	6.18	45.39	38.63	14.87
UBN	272.8	269.9	1.10	85.42	101.42	18.73

The results show that the ANN model predicts both outputs with low relative errors in the majority of cases and achieves computation times approximately 10 times faster than the heuristic method. However, there are instances where higher relative errors have been observed, with two cases exceeding a 10% relative error for the network cost: the NSFNET and UBN cases. Interestingly, these two cases correspond to the networks with larger average link lengths. An explanatory hypothesis for this behavior is that these networks exhibit significant variability in their features, making it more difficult for the ANN model to accurately capture the relationships between features and labels, a trend similar to the one shown in Figure 2b for networks with fewer than 40 nodes.

5. Conclusions

In this paper, the problem of estimating the capacity and cost of multi-fiber optical networks was addressed, using for this purpose an ANN model. These networks, by using multiple fiber pairs per link, can achieve very high network capacities, even in the order of petabits per second.

To generate the datasets required to train and test the ANN, we applied an appropriate heuristic that relies on a fiber assignment algorithm which was also proposed in the context of this work.

The implemented model was an ANN with 12 inputs (parameters related to the physical topology of the optical network), 2 outputs, and 1 hidden layer. The outputs correspond to two metrics: the network capacity, measured in Tbit/s, and the network cost, quantified by the total length of optical fiber deployed in the network, measured in km.

The ANN was trained with a number of nodes varying between 5 and 100, and it was extensively tested within the same range. The results showed good performance with a mean relative error of 2.47% and 5.29% for the first and second metric, respectively. The ANN model also showed significantly faster performance compared to the heuristic method, with the ANN predictions never taking more than a few tens of milliseconds, while the network simulation could take up to tens of seconds to reach the results in larger networks.

Remarkably, the network capacity for 50 or more nodes exceeds 1 Pb/s, reaching about 5 Pb/s for 100 nodes. However, this comes at the cost of a significant increase in the length of the total optical fiber required in the network.

Author Contributions: Conceptualization, A.F. and J.P.; methodology, A.F. and J.P.; software, A.F.; validation, A.F.; formal analysis, A.F. and J.P.; investigation, A.F. and J.P.; writing—original draft preparation, J.P.; writing—review and editing, J.P.; visualization, A.F.; supervision, J.P. All authors have read and agreed to the published version of the manuscript.

Funding: This research received no external funding.

Institutional Review Board Statement: Not applicable.

Informed Consent Statement: Not applicable.

Data Availability Statement: Data are contained within the article.

Conflicts of Interest: The authors declare no conflicts of interest.

References

1. Ruiz, M.; Hernández, J.A.; Quagliotti, M.; Salas, E.H.; Riccardi, E.; Rafel, A.; Velasco, L.; de Dios, O.G. Network traffic analysis under emerging beyond-5G scenarios for multi-band optical technology adoption. *J. Opt. Commun. Netw.* **2023**, *15*, F36–F47. [CrossRef]
2. Winzer, P.J. The future of communications is massively parallel. *J. Opt. Commun. Netw.* **2023**, *15*, 783–787. [CrossRef]
3. Furukawa, H.; Luís, R.S. Petabit-class optical networks based on spatial-division multiplexing technologies. In Proceedings of the 2020 International Conference on Optical Network Design and Modelling (ONDM), Barcelona, Spain, 18–21 May 2020. [CrossRef]
4. Parker, M.C.; Wright, P.; Lord, A. Multiple fiber, flexgrid elastic optical network design using MaxEnt optimization. *J. Opt. Commun. Netw.* **2015**, *7*, B194–B201. [CrossRef]
5. Lopez, V.; Zhu, B.; Moniz, D.; Costa, N.; Pedro, J.; Xu, X. Optimized design and challenges for C&L band optical line systems. *J. Lightw. Technol.* **2020**, *38*, 1080–1091. [CrossRef]
6. Shannon, C.E. A mathematical theory of communication. *Bell Syst. Tech. J.* **1948**, *27*, 379–423. [CrossRef]
7. Vincent, R.J.; Ives, D.J.; Savory, S.J. Scalable capacity estimation for nonlinear elastic all-optical core networks. *J. Lightw. Technol.* **2019**, *37*, 5380–5391. [CrossRef]
8. Matzner, R.; Semrau, D.; Luo, R.; Zervas, G.; Bayvel, P. Making intelligent topology design choices: Understanding structural and physical property performance implications in optical networks. *J. Opt. Commun. Netw.* **2021**, *13*, D53–D67. [CrossRef]
9. Ives, D.J.; Bayvel, P.; Savory, S.J. Routing, modulation, spectrum and launch power assignment to maximize the traffic throughput of a nonlinear optical mesh network. *Photon. Netw. Commun.* **2015**, *29*, 244–256. [CrossRef]
10. Freitas, A.; Pires, J. Investigating the impact of topology and physical impairments on the capacity of an optical backbone network. *Photonics* **2024**, *11*, 342. [CrossRef]
11. Musumeci, F.; Rottondi, C.; Nag, A.; Macaluso, I.; Zibar, D.; Ruffini, M.; Tornatore, M. An overview on application of machine learning techniques in optical networks. *IEEE Commun. Surv. Tuts.* **2019**, *21*, 1383–1408. [CrossRef]
12. Gu, R.; Yang, Z.; Ji, Y. Machine learning for intelligent optical networks: A comprehensive survey. *arXiv* **2020**, arXiv:2003.05290v1. [CrossRef]
13. Morais, R.M.; Pedro, J. Machine learning models for estimating quality of transmission in DWDM Networks. *J. Opt. Commun. Netw.* **2018**, *10*, D84–D99. [CrossRef]

14. Martín, I.; Troia, S.; Hernández, J.A.; Rodríguez, A.; Musumeci, F.; Maier, G.; Alvizu, R.; de Dios, O.G. Machine learning-based routing and wavelength assignment in software-defined optical networks. *IEEE Trans. Netw. Serv. Manag.* **2019**, *16*, 871–883. [CrossRef]
15. Châtelain, B.; Bélanger, M.P.; Tremblay, C.; Gagnon, F.; Plant, D.V. Topological wavelength usage estimation in transparent wide area networks. *J. Opt. Commun. Netw.* **2009**, *1*, 196–203. [CrossRef]
16. Santos, J.; Pedro, J.; Monteiro, P.; Pires, J. Optimized routing and buffer design for optical transport networks based on virtual concatenation. *J. Opt. Commun. Netw.* **2011**, *3*, 725–738. [CrossRef]
17. Çetinkaya, E.K.; Alenazi, M.J.F.; Cheng, Y.; Peck, A.M.; Sterbenz, P.G. A comparative analysis of geometric graph models for modelling backbone networks. *Opt. Switch. Netw.* **2014**, *14*, 95–106. [CrossRef]
18. Chugh, S.; Ghosh, S.; Gulistan, A.; Rahman, B. Machine learning regression approach to the nanophotonic waveguide analyses. *J. Lightw. Technol.* **2019**, *37*, 6080–6089. [CrossRef]
19. Le Cun, Y. Theoretical framework for back-propagation. In Proceedings of the 1988 Connectionist Models Summer School; Morgan Kaufman: Pittsburg, PA, USA, 1988; pp. 21–28.
20. Coefficient of Determination, R-Squared. Available online: https://www.ncl.ac.uk/webtemplate/ask-assets/external/maths-resources/statistics/regression-and-correlation/coefficient-of-determination-r-squared.html (accessed on 7 July 2024).
21. Pytorch. Available online: www.archive.org (accessed on 20 April 2024).
22. Liu, J.; Shen, Z.; He, Z.Y.; Zhang, X.; Xu, R.; Yu, H.; Cui, P. Towards out-of-distribution generalization: A survey. *arXiv* **2023**, arXiv:2108.13624. [CrossRef]
23. Pires, J.J.O. On the capacity of optical backbone networks. *Network* **2024**, *4*, 114–132. [CrossRef]

photonics

MDPI

Article

DeepChaos+: Signal Detection Quality Enhancement of High-Speed DP-16QAM Optical Fiber Communication Based on Chaos Masking Technique with Deep Generative Models

Dao Anh Vu [1], Nguyen Khoi Hoang Do [2], Huyen Ngoc Thi Nguyen [2], Hieu Minh Dam [3], Thuy Thanh Thi Tran [1], Quyen Xuan Nguyen [4] and Dung Cao Truong [1,*]

1 Department of Electronic Engineering, Post and Telecommunications Institute of Technology, Hanoi 152638, Vietnam; daova@ptit.edu.vn (D.A.V.); thuyttt@ptit.edu.vn (T.T.T.T.)
2 Department of Information Technology, Post and Telecommunications Institute of Technology, Hanoi 152638, Vietnam; nguyendhk@ptit.edu.vn (N.K.H.D.)
3 Department of Computer Science, Swinburne University of Technology, Hanoi 100000, Vietnam
4 School of Electrical and Electronic Engineering, Hanoi University of Science and Technology, Hanoi 100000, Vietnam; quyen.nguyenxuan@hust.edu.vn
* Correspondence: dungtc@ptit.edu.vn

Abstract: In long-haul WDM (wavelength division multiplexing) optical communication systems utilizing the DP-16QAM modulation scheme, traditional methods for removing chaos have exhibited poor performance, resulting in a high bit error rate of 10^{-2} between the original signal and the removed chaos signal. To address this issue, we propose DeepChaos+, a machine learning-based approach for chaos removal in WDM transmission systems. Our framework comprises two key points: (1) DeepChaos+ automatically generates a dataset that accurately reflects the features of the original signals in the communication system, which eliminates the need for time-consuming data simulation, streamlining the process significantly; (2) it allows for the training of a lightweight model that provides fast prediction times while maintaining high accuracy. This allows for both efficient and reliable signal reconstruction. Through extensive experiments, we demonstrate that DeepChaos+ achieves accurate reconstruction of the original signal with a significantly reduced bit error rate of approximately 10^{-5}. Additionally, DeepChaos+ exhibits high efficiency in terms of processing time, facilitating fast and reliable signal reconstruction. Our results underscore the effectiveness of DeepChaos+ in removing chaos from WDM transmission systems. By enhancing the reliability and efficiency of chaotic secure channels in optical fiber communication systems, DeepChaos+ holds the potential to improve data transmission in high-speed networks.

Keywords: wavelength division multiplexing; optical fiber communication; machine learning; variational auto encoder; knowledge distillation

1. Introduction

In the field of communication, for several decades recently, the wavelength division multiplexing (WDM) technique has been widely applied in optical transmission systems for both high-speed backbone and access systems, up to 400 Gbps per wavelength, aiming to utilize the huge bandwidth of optical fiber [1–4] in order to respond to the huge requirement of broadband services, e.g., broadband mobile access service, data mining, cloud computing, augmented reality, and virtual reality experiments. To enhance the capacity of communication channels, we can combine the WDM technique with several different methods, such as spatial division multiplexing [5], multi-carrier [6], super-channel [7], advanced multilevel modulation format [8], etc. This is due to the linear independence of such mentioned formats from the wavelength. Lots of optical fiber WDM communication systems use advanced modulation schemes such as quadrature amplitude modulation

(QAM) [9,10] or phase shift keying (PSK) [11] techniques in combination with a dual polarization scheme to increase the significant spectral efficiency by mean of maximization of wavelength bandwidth. For example, if an optical fiber transmission migrates from the mono polarization-QPSK format to the dual polarization-QPSK format, the capacity of the channel will double, while the bit error rate of the optical link will be negligibly affected [11].

In high-security communication systems, such as military communication, banking communication, and government communication systems [12], advanced and complicated encryption systems have been used to encode communication bit streams, such as FPGA-based hardware [13], RSA [14,15], DES [16], alternatively interleaved AES [16–18], etc. However, these methods have some drawbacks, such as: (i) requiring complex electronic circuits for encoding and decoding, which are costly; (ii) increasing information processing delays; and (iii) being vulnerable to brute-force algorithms due to the development of computer-supported algorithms [19,20]. Therefore, an economical approach increasingly used in physical layer security is the use of chaotic techniques [21–23] due to the superior characteristics of chaotic features, such as random pseudo-noise and spread spectrum [24]. On the other hand, chaos is also a deterministic phenomenon, so data can be decrypted if the synchronization process can be controlled. Information data are mixed with chaotic sequences through mechanisms such as scrambling, modulation, and encryption in order to enhance security so that eavesdroppers cannot successfully detect the encrypted information [17,25,26]. Nowadays, the chaos phenomenon is applied in various applications, such as wireless and radio communication [27], free-space optical communication [25], short-range chaotic optical communication [28,29], visible light communication (VLC) [30,31], underwater communication [32–34], automatic control [35], sensors [36], etc. Some related works have proposed utilizing the chaos effect arising from the dynamic properties of semiconductor laser systems for high-intensity applications in intensity-modulated direct detection (IM-DD) systems. However, generating chaotic laser beams with significant amplitude variations for WDM IM-DD systems remains challenging [37,38]. In addition, chaotic techniques have not been widely applied in fiber optic communication systems due to the lack of research on the chaos phenomenon in wideband optical fiber communication systems, such as WDM systems.

One of the most significant technological advancements in recent years is artificial intelligence (AI). In particular, deep learning (DL) models [39] have brought new dimensions to many fields, such as human–machine interaction [40], robotics [41], natural language processing [42,43], etc. Recently, deep learning models have been applied in the field of information by effectively representing them through deep autoencoders for optical information signals to reduce nonlinearity and nonlinear balance. Specifically, in chaotic optical communication systems, the Informer model is utilized to improve the BER performance quality of chaos shift keying (CSK) modulation communication systems due to a deep understanding of the dynamic behaviors of chaos through data-driven analysis [44].

In this study, we introduce DeepChaos+, a novel framework designed to address the chaos problem to enhance the performance of high-speed DP-16QAM optical fiber communication systems. This framework tackles the chaos challenge in WDM systems, which previously caused high bit error rates, by using deep learning models to reduce the bit error rate to approximately 10^{-5}. Furthermore, with the use of advanced machine learning techniques, namely, data augmentation and knowledge distillation, reduction in both training time and inference time is achieved. This offers a reliable and efficient solution for enhancing the quality of optical fiber communications, which is critical for the advancement of high-speed networks.

2. Background and Problem Setting

2.1. Long-Haul WDM Optical Communication System Using DP-16QAM Modulation Scheme

The conceptual diagram of the chaotic optical communication with two wavelength multiplexed channels using DP-16QAM for each channel is exhibited and described in

Figure 1. In this diagram, one channel is for chaotic optical communication (COC), and another is for conventional fiber optic communication (CFOC). Each channel, defined by the wavelength of its carrier wave, is coupled into the same optical fiber in DP-16QAM data format [45,46]. The chaos cryptography technique encrypts some crucial information at a given WDM channel λ_c. In this paper, the chaos sequence is created by a logistic map using a retrieval rule as follows [17]:

$$z_{n+1} = 4z_n(1 - z_n), \tag{1}$$

where $n = 1, 2, 3\ldots$ is a positive integer, and $z_0 = [0, 1]$ is a starting real number between 0 and 1. Therefore, it is easy to see that z_n always satisfies $0 < z_n < 1$. The chaotic function has a probability distribution density as follows:

$$p(z) = \frac{1}{\pi\sqrt{z(1-z)}} \quad \text{for } 0 < z < 1. \tag{2}$$

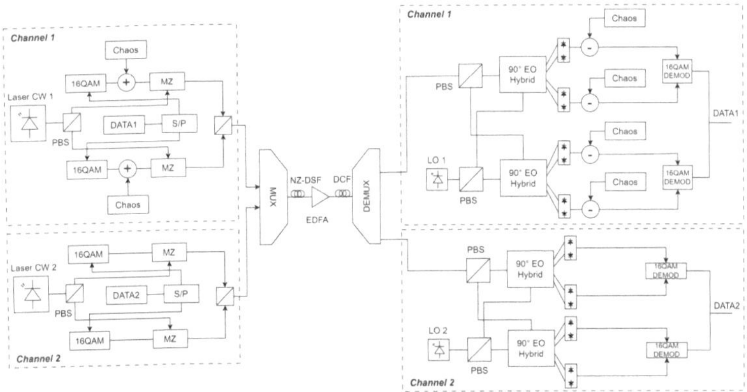

Figure 1. Conceptual Conceptional diagram of the COC and CFOC channels in the long-haul WDM optical communication system using the $DP - 16QAM$ modulation scheme.

In this proposed system, both the laser diode transmitter (LDT) and the laser diode receiver (LDR) are constructed from a single-mode semiconductor laser with an external reflector and the same configuration. Both the COC channel and the CFOC channel are (de)multiplexed by means of a wavelength (de)multiplexer in the C-band of the third telecom window. The transmitter laser (LDT) of the COC channel emits a chaotic carrier, and an optical isolator (ISO) is used to ensure unidirectional transmission. The original message is superimposed on a chaotic carrier by the chaos masking sequence (CMS). As seen in Figure 1, the chaotic signal is implemented by simply adding the CMS signal to the output of the conventional 16QAM modulated signal. On the receiver side, the chaotic signal is decoded by simply subtracting the received COC signal and a CMS signal that is synchronized to the form of the transmitter side.

Information propagating along the long-haul optical fiber link is greatly affected by fiber loss, dispersion, and nonlinear effects. We place an erbium-doped fiber amplifier (EDFA) for the fiber loss compensation and a dispersion compensating fiber (DCF) for the dispersion compensation.

The EDFA amplifier has the gain coefficient G, determined by the formula $G = \alpha L$. Here, α is the fiber loss coefficient, L is the total length of the transmission link, and L_{DCF} is the length of the DCF fiber. On the receiver, the optical signal is de-multiplexed by the wavelength de-multiplexer (DEMUX) and photodetector after propagating over a long-haul section.

The dynamic behaviors of a couple of transmitters and receivers that are set up in a COC system can be described by well-known Lang–Kobayashi rate equations, with optical feedback and injection terms [29,45] as follows:

$$\begin{aligned}\frac{dE_{T,R}(t)}{dt} &= \frac{1}{2}(1+i\psi)\left[\frac{G[N_{T,R}(t)-N_0]}{1+\varepsilon|E_{T,R}(t)|^2}\right]E_{T,R}(t),\\ &+k_{T,R}E_{T,R}(t-\tau)\exp(-i\omega\tau)+k_{\mathrm{irj}}E_{\mathrm{ext}}(t),\end{aligned} \tag{3}$$

where E and N correspond the slowly varying complex electric field amplitudes and the carrier density in the laser cavity, respectively; T and R stand for transmitter and receiver; ω is the angle frequency of the free operation laser; τ is the round-trip time; and E_{est} is the transmission link. Then, the COC signal is decoded, and the DP-16QAM signal is also demodulated to recover the baseband signals of the external electric field amplitude at the input of the receiver. For the proposed COC and CFOC parallel transmission system, we consider a two-channel WDM system (each subscript denotes the channel number). The light propagation through the fiber is described in terms of the well-known nonlinear Schrödinger (NLS) equation [46]:

$$\begin{aligned}\frac{dN_{T,R}(t)}{dt} &= \frac{I_{T,R}}{qV}-\frac{1}{\tau_n}N_{T,R}(t)\\ &-\frac{G[N_{T,R}(t)-N_0]}{1+\varepsilon|E_{T,R}(t)|^2}|E_{T,R}(t)|^2.\end{aligned} \tag{4}$$

Here, E_j and E_k are slowly varying complex electric field amplitudes of the j-th and k-th channels; equally α is the fiber loss coefficient; β_2 is the second-order dispersion coefficient of optical fiber; and γ is the nonlinear coefficient. In this implementation, we use non-zero dispersion shifted fiber (NZ-DSF) following the ITU-T G.655 recommendation, and these typical parameters are determined as $\alpha = 0.2$ dB/km, $\beta_2 = 5.1$ ps$^2 \cdot$ km^{-1}/km, and $\gamma = 1.5$ W$^{-1} \cdot$ km^{-1}. Other hyperparameters are listed in Table 1.

Table 1. Hyperparameters for the DeepChaos+ framework include the VAE model, the student model, and the Informer Aggregation model.

Hyperparameter	Value
Learning rate for the VAE model	0.0003
Learning rate for the student model	0.001
Optimizer	Adam [47]
Total epochs per update	8
Update time step	600
Mini-batch size	128
Aggregation model for VAE and student models	Informer (Attention and Convolution)
Activation function for the VAE model	ELU
VAE–student coefficient λ	0.6
Gradient norm	0.5

2.2. Problem Definition

In our objective to employ a deep learning-based approach for eliminating the chaos introduced at the transmitter side from the received signal in a long-haul WDM optical communication system utilizing the DP-16QAM modulation scheme, our main goal is to minimize the bit error rate (BER) as a critical performance metric in communication systems [48]. To achieve this, we aim to find a mapping function $\mathcal{F}_\theta$ parameterized by θ. The mapping function $\mathcal{F}_\theta$ should be capable of removing the chaos from the received signal, denoted as $\mathcal{X}^\dagger$, and recovering the original signal, denoted as $\mathcal{X}$. We define the set of original signals in the system as $\mathcal{X} \in \{0,1\}^{(1\times d)}$, where d represents the length of the original sampled signals. The chaos adding function is denoted as I, such that $\mathcal{X}^\dagger = I(\mathcal{X})$.

The mapping function $\mathcal{F}_\theta : \mathcal{X}^\dagger \rightarrow \tilde{\mathcal{X}}$ can be optimized by finding the optimal parameters θ^* for $\mathcal{F}_\theta$ that minimize the expected BER, expressed as:

$$\theta^* := \arg\min_\theta \mathcal{B}(\tilde{\mathcal{X}}, \mathcal{X}). \tag{5}$$

Here, $\mathcal{B}(\tilde{X}, \mathcal{X})$ represents the BER between the reconstructed signal $\tilde{\mathcal{X}}$ from the received signal with added chaos ($\mathcal{X}^\dagger$) and the original signal ($\mathcal{X}$). By optimizing the θ parameters of the mapping function $\mathcal{F}_\theta$, we aim to train a model that can effectively eliminate the chaos from the received signal. This deep learning-based approach offers the advantage of faster noise removal compared to traditional methods. Additionally, the trained model can generalize well to handle unseen signals with similar properties, providing robust chaos elimination in a wide range of scenarios.

3. Related Work

Recently, the prominent advantages of deep neural networks as well as the advancement of algorithms and deep learning models have become very attractive for viable applications thanks to their ability to automatically learn feature representations from input data without the need for human intervention. Deep neural networks and deep learning models have been able to automatically extract important features from input signals effectively, thereby improving the quality and accuracy of digital signal processing in fiber optic information systems and chaotic-modulated optical information systems. For example, the digital back-propagation through DNNs has been applied to eliminate the nonlinear effect limit in order to enhance the quality of digital signal processing in amplified fiber optic communication systems, as demonstrated by the work of Q. Fan et al. [49]. Similarly, deep learning models have been effectively used to address the challenges of dispersion and nonlinearity compensation in high-speed wavelength division multiplexing (WDM) fiber optic communication systems employing multilevel modulation channels like 64QAM, as demonstrated in [50–55]

For chaotic communication systems, recently, a multi-carrier chaos shift keying (DL-IM-MCDCSK) system utilizing deep learning (DL) and index mapping (IM) techniques to mitigate the information leakage risk associated with conventional MC-DCSK systems has been proposed [56]. The proposed system operates without a reference signal and utilizes a two-dimensional reshaping (TDR) index mapping structure to equalize the chaotic signals in both frequency and time domains. The offline-trained DNN classifier can significantly improve the bit error rate (BER) performance during information recovery without requiring conventional maximum likelihood estimation (MLE). In addition, a chaos synchronization that does not require hardware implementations [57] or reference chaotic sequences [58] can be achieved via deep learning models to provide high-level physical layer security for optical communications. For another actual implementation, very recently, a high-speed chaotic receiver with up to 32 Gb/s messages hidden in a wideband chaotic optical carrier has been experimentally demonstrated over a 20 km fiber link, showing a significant simplification while still guaranteeing security [59]. These promising potentials prove that both deep neural networks and deep learning models are effective and viable performance quality enhancements for solving signal detection and signal processing problems in chaotic secure communication systems as well as in high bit rate optical fiber communication systems.

4. Our Solution: DeepChaos+

This section presents DeepChaos+, a framework designed to enhance the performance of high-speed DP-16QAM optical fiber communication systems. We first provide an overview of the framework and then introduce its end-to-end learning objective.

4.1. Overview Process of DeepChaos+

Let us define $\mathcal{E}(\tilde{\mathcal{X}})$ as the set of incorrect predictions (error bits) made by the model $\mathcal{F}_\theta$ given the true label $\mathcal{X}$ (transmitted bits). In our system, we employ the bit error rate (BER) metric, similar to the approach proposed by Dao et al. [60], defined as follows:

$$\mathcal{B}(\tilde{\mathcal{X}}, \mathcal{X}) = \frac{|\mathcal{E}(\tilde{\mathcal{X}})|}{|\mathcal{X}|}. \tag{6}$$

Generating a large dataset $\mathcal{X}$ to train the model $\mathcal{F}_\theta$ is time-consuming, and balancing the training $\mathcal{X}^{train} \in \mathcal{X}$ and testing sets $\mathcal{X}^{test} \in \mathcal{X} \setminus \mathcal{X}^{train}$ is challenging. Too much training data may cause overfitting, while too little can lead to underfitting, both increasing the BER $\mathcal{B}(\tilde{\mathcal{X}}, \mathcal{X})$.

DeepChaos+ (as depicted in Figure 2) combines a Variational Autoencoder (VAE) $\mathcal{F}\theta$ and a lightweight Informer Network $f^{student}\omega$ to optimize communication system performance while using limited original signal data $\mathcal{X}^{train}$. The VAE is first trained on $\mathcal{X}^{train}$ to generate synthetic data $\mathcal{X}^g$, which is combined with $\mathcal{X}^{train}$ to form an augmented dataset $\mathcal{D}$. This process continues iteratively, refining the VAE until it can minimize the BER on $\mathcal{X}^{test}$. Optionally, chaos can be added to $\mathcal{X}^g$ to better represent original signal characteristics.

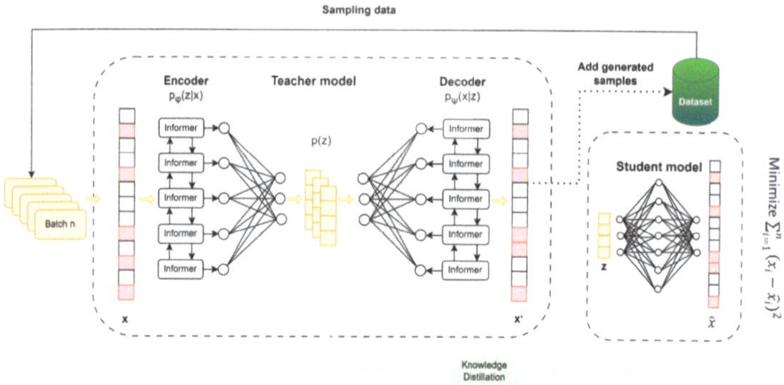

Figure 2. Overview of the DeepChaos+ framework. The framework introduces two key models: the Variational Autoencoder (VAE) and the lightweight Informer Network. The VAE is trained to generate interpolated data from the set $\mathcal{X}$. The generated data are then combined with the dataset $\mathcal{D}$ and used to iteratively retrain the VAE. The lightweight Informer Network, with fewer parameters but functionality equivalent to the VAE's decoder, is trained to predict a set $\tilde{\mathcal{X}}$ that minimizes the bit error rate $\mathcal{B}(\tilde{\mathcal{X}}, \mathcal{X})$. Knowledge Distillation is employed to ensure the Informer achieves similar performance to the decoder while enabling faster inference time.

In parallel, the Informer Network $f^{student}\omega$ is trained using the augmented dataset $\mathcal{D}$, with fewer parameters than the VAE's decoder but with similar functionality. Knowledge Distillation is used to transfer knowledge from the VAE to $f^{student}\omega$ by training it to mimic the output of the VAE's decoder, enabling it to achieve comparable performance while enabling faster inference.

4.2. End-to-End Learning Objective

Mathematically, we further decompose the Variational Autoencoder (VAE) model $\mathcal{F}_\theta$ into two models: the encoder denoted as $\mathcal{G}_\psi$ and the decoder denoted as $\mathcal{M}_\phi$. Formally, we have:

$$\mathcal{F}_\theta = \mathcal{G}_\psi \circ \mathcal{M}_\phi, \tag{7}$$

$$\mathcal{X}^{train} = \mathcal{F}_\theta(\mathcal{X}^{train}) = \mathcal{M}_\phi\Big(\mathcal{G}_\psi(\mathcal{X}^{train})\Big) = \mathcal{M}_\phi(\mathcal{Z}). \tag{8}$$

The encoder $\mathcal{G}\psi$ maps $\mathcal{X}^{train}$ to a latent space $\mathcal{Z}$, and the decoder $\mathcal{M}\phi$ reconstructs $\mathcal{X}^{train}$ from $\mathcal{Z}$. The VAE assumes a latent variable $\mathcal{Z} \in \mathbb{R}^{1\times v}$, with v as the latent space dimension. This latent variable captures the features of the original signal and follows a latent distribution $p_\phi(\mathcal{Z})$. The complete generative process can be described by:

$$p_\phi(\mathcal{Z} \mid \mathcal{X}) = \frac{p_\phi(\mathcal{X} \mid \mathcal{Z})p_\phi(\mathcal{Z})}{p_\phi(\mathcal{X})}. \tag{9}$$

To approximate the intractable posterior distribution $p_\phi(\mathcal{Z} \mid \mathcal{X})$, the model $\mathcal{G}_psi$ learns a simpler distribution $q_\psi(\mathcal{Z} \mid \mathcal{X})$. The objective is to have $p_\phi(\mathcal{Z} \mid \mathcal{X}) \approx q_\psi(\mathcal{Z} \mid \mathcal{X})$, which is achieved by minimizing the KL divergence $D_{KL}(q_\psi \| p_\phi)$. This is equivalent to maximizing the evidence lower bound objective (ELBO):

$$\mathcal{L}^{\text{ELBO}} = \mathbb{E}_{q_\psi}\left[\log p_\phi(\mathcal{X} \mid \mathcal{Z})\right] - \mathbb{E}_{q_\psi}\left[\log \frac{q_\psi(\mathcal{Z} \mid \mathcal{X})}{p_\phi(\mathcal{Z})}\right]. \tag{10}$$

The ELBO includes the expected reconstruction error $\log p_\phi(\mathcal{X} \mid \mathcal{Z})$ learned by the decoder model. The DeepChaos+ framework introduces a student model $f_\omega^{\text{student}}$ trained on augmented data $\mathcal{D}$ generated by the VAE (teacher model). The student model is trained using a mean squared error (MSE) loss, defined as:

$$\mathcal{L}^{\text{MSE}} = \frac{1}{N}\sum_{i=1}^{N}\left|\tilde{\mathcal{X}}^{(i)} - \hat{\mathcal{X}}_\zeta^{(i)}\right|_2^2, \tag{11}$$

where N is the number of samples in the dataset $\mathcal{D}$. The overall objective function for training the student model combines the ELBO loss and the MSE loss:

$$\mathcal{L}^{\text{total}} = \lambda\mathcal{L}^{\text{ELBO}} + (1-\lambda).\mathcal{L}^{\text{MSE}} \tag{12}$$

Here, λ is a hyperparameter that balances the trade-off between the ELBO and MSE losses. By optimizing this combined loss function, DeepChaos+ effectively learns from the synthetic data, enhancing the student model's ability to predict the original signals. This approach addresses the challenge of limited original signal data by generating synthetic data that capture the essential features of the original signals, leading to more accurate predictions by the student model.

5. Experiment

In this section, we present a comprehensive evaluation of the performance of our proposed DeepChaos+ framework in removing chaos from simulated original signals in long-haul WDM optical communication systems utilizing the DP-16QAM modulation scheme. We conduct experiments under various settings to assess the effectiveness of DeepChaos+. We begin by describing the experiment setup, which includes the selection of hyperparameters, dataset description, and comparison methods. Hyperparameters such as the learning rate, batch size, and network architecture are carefully chosen to ensure the optimal performance of DeepChaos+. Additionally, we compare the performance of DeepChaos+ with traditional methods to establish its superiority.

5.1. Experiment Setup

Our objective is to evaluate the bit error rate (BER), as defined in Equation (6), and the time efficiency of DeepChaos+. Regarding BER, we aim to demonstrate that DeepChaos+ achieves competitive performance even with a limited amount of training data. To accomplish this, we divided the dataset $\mathcal{X}$ into different proportions for training and testing: 20%, 40%, 60%, and 80%. For the second term, we analyze the training time and inference time of DeepChaos+ and compare it with other machine learning-based methods.

Hardware Configuration. In order to run our framework efficiently, the following hardware requirements should be met. A GPU is recommended for faster training and inference, with a minimum of an NVIDIA GTX 1060 (6GB VRAM), though more powerful options like the NVIDIA RTX 3090 or A100 are ideal for larger datasets. A minimum of 8GB of RAM is required, but 16GB or more is recommended for on larger datasets. Additionally, a multi-core CPU (quad-core or higher) is beneficial for data preprocessing and managing overall system performance. Our experiments is run on a system equipped with Intel Core i7 Processor, a NVIDIA GTX 4090i (24GB VRAM) and 64GB of RAM.

Comparison Methods And Metrics. We compare DeepChaos+ with several comparison methods, including BiLSTM [61], Informer [62], GRU-D [63], and the chaos-solving module proposed within the system. For convenience, we denote the chaos-solving module in our system as TDiS. These machine learning-based methods were chosen as they represent state-of-the-art approaches in the field of time series analysis and chaos-based modeling. The comparison is based on two metrics: bit error rate and inference time (in seconds).

Dataset. The dataset used in our experiment was collected by simulating the COC (chaos on carrier) and CFOC (chaos frequency on carrier) channels in a long-haul WDM (wavelength division multiplexing) optical communication system using the DP-16QAM modulation scheme. We generated a sequence of 1,000,000 bits and passed them through the communication system. At the transmitter side, the generated sequence was combined with chaos, resulting in a chaotic-modulated signal. At the receiver side, the received signal was subjected to noise removal, resulting in a de-noised signal. It is important to note that this de-noised signal is used to evaluate the performance of the proposed TDiS (chaos-solving module) method.

Hyperparameter Settings. The detailed hyperparameters used for training the model are provided in Table 1. As mentioned earlier, we employ a Knowledge Distillation technique to reduce the model size while preserving accuracy for faster inference. The student model, which is obtained through Knowledge Distillation, is partitioned into different sizes: tiny, with a total of 240,712 parameters; small, with 461,283 parameters; medium, with 920,784 parameters; and large, with 1,911,365 parameters. These variations in model size allow us to evaluate the trade-off between model complexity and performance, enabling us to select the most suitable configuration based on our specific requirements. The learning rates for the VAE (0.0003) and student (0.001) models ensure stable convergence, and the Adam optimizer is selected for its adaptive learning rate benefits. A total of 8 epochs per update and an update time step of 600 allow the model to learn effectively over time, while the mini-batch size of 128 balances speed and memory efficiency. The aggregation model uses the Informer, incorporating attention and convolution for capturing long-range dependencies, and ELU is chosen as the activation function for the VAE to prevent vanishing gradients. The VAE–student coefficient λ is set to 0.6 for optimal knowledge transfer, and a gradient norm of 0.5 prevents gradient explosion during training.

5.2. Training Efficiency Analysis

In Figure 3 (left), it is evident that DeepChaos+ (medium size) achieves fast convergence to a BER of approximately 2×10^{-5} in $\mathcal{X}^{test}$. However, the other learning models, such as GRU-D and BiLSTM, also exhibit fast convergence, but the only reach a BER of around 2×10^{-3}. Despite utilizing 60% of the original signal set $\mathcal{X}$ and having larger model sizes compared to DeepChaos+, they do not attain the same level of accuracy. This highlights the superiority of DeepChaos+ in effectively capturing the underlying patterns and optimizing the BER, even with a smaller proportion of training data. The key factor behind this phenomenon is the ability of DeepChaos+ to generate additional data variants that effectively represent the original data using only 60% of the original data. It selectively learns from the best-performing generated data, resulting in the lowest BER for the $\mathcal{X}^{test}$ set. Over time, DeepChaos+ autonomously generates data that accurately capture the underlying features of the remaining 40% of the data, further enhancing its performance and ultimately achieving the lowest BER among the three methods.

On the other hand, in Figure 3 (right), an interesting observation is made when experimenting with different sizes of the student model for the remaining 40% subset. It is noted that DeepChaos+ did not converge to 100% accuracy on the test set with the tiny and small sizes. This suggests that these smaller-sized models would have a higher bit error rate (BER). However, starting from the medium-sized model and above, DeepChaos+ achieved nearly 100% fit with a recorded BER of 2.3×10^{-4}. This observation emphasizes the significance of model size in attaining higher accuracy. Larger-sized models, such as medium and above, possess the ability to effectively capture and represent the underlying dynamics of the data, leading to better convergence and lower BER. Additionally, it is worth noting the training time (red line in the figure), which significantly increases with each model size. In our experiments, the medium-sized student model proved to be the most suitable choice, as it offered a balance between fast training time and achieving accuracy comparable to the large-sized model.

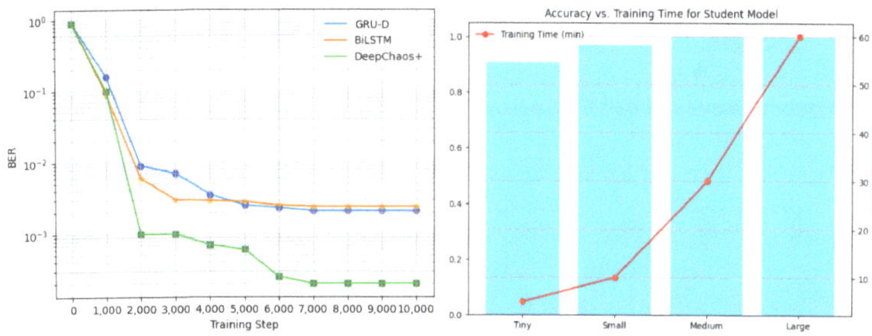

Figure 3. The training performance of DeepChaos+ in the 60% dataset is shown in the **left** figure, while the learning performance of the student model is depicted for different sizes in the **right** figure. The red line in the right figure represents the training time, indicating that, as the size of the student model increases, the training time also lengthens.

5.3. Inference Efficiency Analysis

We have implemented a prediction model using batch processing with a batch size of 128 data points at a time. This ensures that the model predicts 128 data points in parallel, thereby increasing the prediction speed. It is important to note that setting the batch size hyperparameter depends on the GPU configuration. Note that if we use a dataset consisting of 20% of the original sampling signal for training, the remaining 80% of the data points need to be predicted. The more data we use for training, the fewer test points we have for prediction. The total time required for predicting the entire dataset can be calculated by multiplying the batch size by the number of batches. Since the model predicts 128 data points simultaneously (due to the batch size of 128), the model only needs to predict a certain number of batches. The overall prediction time of the model is then divided by the total number of data points to obtain the average prediction time per data point. By organizing and optimizing the prediction process in batches, we can leverage parallel processing and enhance the prediction speed, especially when dealing with large datasets [64].

Based on Table 2, we can analyze the effectiveness of DeepChaos+ compared to other methods. Note that DeepChaos+$_{tiny}$, DeepChaos+$_{small}$, DeepChaos+$_{medium}$, and DeepChaos+$_{large}$ are referred to as student models in this context. Looking at the runtime performance, both BiLSTM and GRU-D exhibit increasing runtime values as the percentage of the original data increases. However, the DeepChaos+ models consistently show significantly lower runtimes across different data percentages. Even the largest student model, DeepChaos+$_{large}$, has remarkably lower runtimes compared to BiLSTM and GRU-D. As the model size increases from tiny to small, medium, and large, the DeepChaos+ models have slower execution times as the number of parameters increases.

Table 2. Comparing the inference time of DeepChaos+ with other state-of-the-art (SOTA) models across different training set and testing set ratios of 20%, 40%, 60%, and 80% for training.

Model	20%	40%	60%	80%	Each Data Point (Average)
BiLSTM	3.2984 s	2.4752 s	1.5684 s	0.6956 s	0.0054 s
GRU-D	5.6874 s	3.315 s	2.1064 s	1.1896 s	0.0092 s
DeepChaos+$_{tiny}$	0.051 s	0.03388 s	0.02464 s	0.01248 s	0.00007 s
DeepChaos+$_{small}$	0.102 s	0.06776 s	0.04928 s	0.02496 s	0.00016 s
DeepChaos+$_{medium}$	0.153 s	0.10164 s	0.07392 s	0.03744 s	0.00025 s
DeepChaos+$_{large}$	0.204 s	0.13552 s	0.09856 s	0.04992 s	0.00034 s
DeepChaos+	2.5764 s	1.0248 s	0.5596 s	0.2804 s	0.0021 s

5.4. Quantitative Analysis

We compared the effectiveness of DeepChaos+ on test sets of 20%, 40%, 60%, and 80%, as shown in Figure 4 (left). For all test sets, DeepChaos+ was able to generate data and learn until achieving near 100% accuracy on the 40%, 60%, and 80% sets. However, on the 20% set, the model only achieved about 84% accuracy due to the initial lack of data, which was insufficient for effective inference and generation of samples. The remaining competing methods showed significantly lower effectiveness, as they could not infer features like DeepChaos+, especially on the 40% dataset, where DeepChaos+ achieved an accuracy of nearly 95%. Methods like BiLSTM only reached around 67%, and GRU-D reached 72%. The traditional approach also achieved accuracy similar to that of DeepChaos+ across all four datasets. However, when it comes to BER (discussed in the following section), this method yields a lower performance compared to DeepChaos+.

Regarding the BER of the test sets (Figure 4, right), we only show the results for the 60% and 80% sets since DeepChaos+ has not yet reached 100% accuracy on the other two sets, resulting in higher BER. DeepChaos+ clearly has the best performance, as seen in the 60% set where its BER falls within the range of 2.3×10^{-4}. The remaining methods have BER ranging from approximately 1.9×10^{-3} to 3×10^{-3}. For the 80% set, DeepChaos+ demonstrates superior performance, with a BER of around 1.5×10^{-4}. This can be explained by the fact that, as DeepChaos+ has more datasets, the interpolation for generating data becomes more accurate. However, when reaching a certain threshold, the other methods also start to generalize and narrow the gap slightly compared to DeepChaos+.

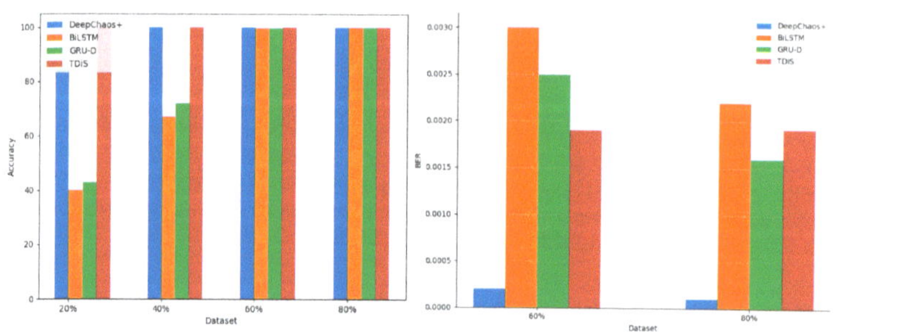

Figure 4. The figure on the **left** illustrates the performance of DeepChaos+ on the testing set of training datasets of 20%, 40%, 60%, and 80%. The figure on the **right** displays the BER (bit error rate) of DeepChaos compared to the other methods, particularly on the 60% and 80% datasets.

5.5. Discussion

While the experimental results presented effectively demonstrate the prowess of the *DeepChaos+* framework in reducing the bit error rate in WDM optical fiber communication systems, it is important to acknowledge the limitations associated with the use of simulated datasets. The simulated environment, although carefully designed to mimic real-world conditions, may not fully capture the inherent complexities of actual optical communication systems. To address this limitation, future research should focus on validating the DeepChaos+ framework using actual experimental data obtained from real-world optical communication systems. This approach will assess the robustness and generalizability of our model under more complicated conditions. Additionally, incorporating real-world data will provide deeper insights into the practical applications of DeepChaos+ for reliably improving the performance of optical fiber communication systems, ensuring that the proposed solutions can be effectively implemented in practical, large-scale deployments.

6. Conclusions

In this study, we addressed the challenge of removing chaos in long-haul WDM optical communication systems utilizing the DP-16QAM modulation scheme. DeepChaos+ introduced two key components to enhance the performance of chaos removal. Through extensive experiments, we demonstrated the effectiveness of DeepChaos+ in accurately reconstructing the original signal with a significantly reduced bit error rate. The achieved bit error rate of approximately 10^{-5} highlighted the superiority of DeepChaos+ compared to traditional methods. Additionally, DeepChaos+ exhibited high efficiency in terms of processing time, enabling fast signal reconstruction. By enhancing the reliability and efficiency of chaotic secure channels in optical fiber communication systems, DeepChaos+ has the potential to significantly improve data transmission in high-speed networks.

Author Contributions: Conceptualization, D.A.V., Q.X.N. and D.C.T.; methodology, D.A.V., D.C.T. and N.K.H.D.; data collection, H.N.T.N., T.T.T.T. and D.A.V.; software, T.T.T.T., H.N.T.N., H.M.D. and N.K.H.D.; validation, D.A.V., T.T.T.T., N.K.H.D., Q.X.N. and D.C.T.; formal analysis, H.N.T.N., H.M.D. and N.K.H.D.; investigation, D.A.V., T.T.T.T., H.N.T.N. and H.M.D.; resources, D.C.T. and Q.X.N.; data curation, T.T.T.T., N.K.H.D., H.N.T.N. and H.M.D.; writing—original draft preparation, D.A.V., T.T.T.T., H.M.D. and H.N.T.N.; writing—review and editing, H.M.D., H.N.T.N. and D.C.T.; visualization, T.T.T.T. and H.N.T.N.; supervision, D.C.T. and Q.X.N.; project administration, D.C.T.; funding acquisition, D.A.V. and D.C.T. All authors have read and agreed to the published version of the manuscript.

Funding: This research received no external funding.

Data Availability Statement: The simulation dataset used in our experiments is publicly available at: https://drive.google.com/drive/folders/1PwM4Z79csBFITiqdQcsq_gdbEwMzloxd.

Conflicts of Interest: The authors declare no conflicts of interest.

References

1. Igarashi, K.; Tsuritani, T.; Morita, I.; Tsuchida, Y.; Maeda, K.; Tadakuma, M.; Saito, T.; Watanabe, K.; Imamura, K.; Sugizaki, R.; et al. Super-Nyquist-WDM transmission over 7326-km seven-core fiber with capacity-distance product of 1.03 Exabit/s·km. *Opt. Express* **2014**, *22*, 1220–1228. [CrossRef] [PubMed]
2. Zheng, Z.; Qian, Z.; Shou, G.; Hu, Y. OCDMA over WDM System Based on Chebyshev-Map Chaotic Spread Spectrum in Passive Optical Network. In Proceedings of the 2009 WRI World Congress on Computer Science and Information Engineering, Los Angeles, CA, USA, 31 March–2 April 2009; Volume 1, pp. 208–211. [CrossRef]
3. Guo, P.; Hou, W.; Guo, L.; Ning, Z.; Obaidat, M.S.; Liu, W. WDM-MDM Silicon-Based Optical Switching for Data Center Networks. In Proceedings of the ICC 2019, 2019 IEEE International Conference on Communications (ICC), Shanghai, China, 20–24 May 2019; pp. 1–6. [CrossRef]
4. Zhang, J.; Yu, J.; Chi, N. Generation and transmission of 512-Gb/s quad-carrier digital super-Nyquist spectral shaped signal. *Opt. Express* **2013**, *21*, 31212–31217. [CrossRef] [PubMed]
5. Yang, L.; Zhou, T.; Jia, H.; Yang, S.; Ding, J.; Fu, X.; Zhang, L. General architectures for on-chip optical space and mode switching. *Optica* **2018**, *5*, 180–187. [CrossRef]
6. Bosco, G.; Curri, V.; Carena, A.; Poggiolini, P.; Forghieri, F. On the Performance of Nyquist-WDM Terabit Superchannels Based on PM-BPSK, PM-QPSK, PM-8QAM or PM-16QAM Subcarriers. *J. Light. Technol.* **2011**, *29*, 53–61. [CrossRef]

7. Zhou, Y.R.; Smith, K.; Payne, R.; Lord, A.; Raddatz, L.; Bertolini, M.; Van De Velde, T.; Colombo, C.; Korkmaz, E.; Fontana, M.; et al. 1.4Tb real-time alien superchannel transport demonstration over 410 km installed fiber link using software reconfigurable DP-16QAM/QPSK. In Proceedings of the OFC 2014, San Francisco, CA, USA, 9–13 March 2014; pp. 1–3. [CrossRef]
8. Tokle, T.; Serbay, M.; Jensen, J.; Geng, Y.; Rosenkranz, W.; Jeppesen, P. Investigation of Multilevel Phase and Amplitude Modulation Formats in Combination With Polarization Multiplexing up to 240 Gb/s. *IEEE Photonics Technol. Lett.* **2006**, *18*, 2090–2092. [CrossRef]
9. Buchali, F.; Steiner, F.; Böcherer, G.; Schmalen, L.; Schulte, P.; Idler, W. Rate Adaptation and Reach Increase by Probabilistically Shaped 64-QAM: An Experimental Demonstration. *J. Light. Technol.* **2016**, *34*, 1599–1609. [CrossRef]
10. Yu, J.; Dong, Z.; Chien, H.C.; Shao, Y.; Chi, N. 7Tb/s (7 × 1.284 Tb/s/ch) Signal Transmission over 320 km Using PDM-64QAM Modulation. *IEEE Photonics Technol. Lett.* **2012**, *24*, 264–266. [CrossRef]
11. Charlet, G.; Renaudier, J.; Mardoyan, H.; Tran, P.; Pardo, O.B.; Verluise, F.; Achouche, M.; Boutin, A.; Blache, F.; Dupuy, J.Y.; et al. Transmission of 16.4-bit/s Capacity Over 2550 km Using PDM QPSK Modulation Format and Coherent Receiver. *J. Light. Technol.* **2009**, *27*, 153–157. [CrossRef]
12. Wollschlaeger, M.; Sauter, T.; Jasperneite, J. The Future of Industrial Communication: Automation Networks in the Era of the Internet of Things and Industry 4.0. *IEEE Ind. Electron. Mag.* **2017**, *11*, 17–27. [CrossRef]
13. Said, S.; Azzaz, M.S.; Djeddou, M.; Benssalah, M. An FPGA real-time implementation of the Chen's chaotic system for securing chaotic communications. *Int. J. Nonlinear Sci.* **2009**, *7*, 467–474.
14. Kocarev, L.; Sterjev, M.; Fekete, A.; Vattay, G. Public-key encryption with chaos. *Chaos* **2005**, *14*, 1078–1082. [CrossRef] [PubMed]
15. Çavuşoğlu, Ü.; Akgul, A.; Zengin, A.; Pehlivan, I. The design and implementation of hybrid RSA algorithm using a novel chaos based RNG. *Chaos Solitons Fractals* **2017**, *104*, 655–667. [CrossRef]
16. Kun, Y.; Han, Z.; Zhaohui, L. An Improved AES Algorithm Based on Chaos. In Proceedings of the 2009 International Conference on Multimedia Information Networking and Security, Hubei, China, 18–20 November 2009; Volume 2, pp. 326–329. [CrossRef]
17. Qiu, J.; Zhang, L.; Li, D.; Liu, X. High security chaotic multiple access scheme for visible light communication systems with advanced encryption standard interleaving. *Opt. Eng.* **2016**, *55*, 066121. [CrossRef]
18. Çavuşoğlu, Ü.; Kacar, S.; Zengin, A.; Pehlivan, I. A novel hybrid encryption algorithm based on chaos and S-AES algorithm. *Nonlinear Dyn.* **2018**, *92*, 1745–1759. [CrossRef]
19. Fu, C.; Lin, B.B.; Miao, Y.S.; Liu, X.; Chen, J.J. A novel chaos-based bit-level permutation scheme for digital image encryption. *Opt. Commun.* **2011**, *284*, 5415–5423. [CrossRef]
20. Enayatifar, R.; Abdullah, H.; Isnin, I. Chaos-based image encryption using a hybrid genetic algorithm and a DNA sequence. *Opt. Lasers Eng.* **2014**, *56*, 83–93. [CrossRef]
21. Hameed Al-Moliki, Y.; Alresheedi, M.; Al-Harthi, Y. Physical-Layer Security against Known/Chosen Plaintext Attacks for OFDM-based VLC system. *IEEE Commun. Lett.* **2017**, *21*, 2606–2609. [CrossRef]
22. Xiang, S.Y.; Pan, W.; Luo, B.; Yan, L.S.; Zou, X.H.; Li, N.; Zhu, H.N. Wideband Unpredictability-Enhanced Chaotic Semiconductor Lasers with Dual-Chaotic Optical Injections. *IEEE J. Quantum Electron.* **2012**, *48*, 1069–1076. [CrossRef]
23. Li, N.; Susanto, H.; Cemlyn, B.; Henning, I.; Adams, M. Secure communication systems based on chaos in optically pumped spin-VCSELs. *Opt. Lett.* **2017**, *42*, 3494–3497. [CrossRef]
24. Illing, L. Digital communication using chaos and nonlinear dynamics. *Nonlinear Anal. Theory Methods Appl.* **2009**, *71*, e2958–e2964. [CrossRef]
25. Annovazzi-Lodi, V.; Aromataris, G.; Benedetti, M.; Merlo, S. Secure Chaotic Transmission on a Free-Space Optics Data Link. *Quantum Electron. IEEE J.* **2008**, *44*, 1089–1095. [CrossRef]
26. Xiang, S.Y.; Pan, W.; Luo, B.; Yan, L.S.; Zou, X.H.; Jiang, N.; Li, N.Q.; Zhu, H.N. Message Encoding/Decoding Using Unpredictability-Enhanced Chaotic VCSELs. *IEEE Photonics Technol. Lett.* **2012**, *24*, 1267–1269. [CrossRef]
27. Abdulameer, L.F.; Jignesh, J.D.; Sripati, U.; Kulkarni, M. BER performance enhancement for secure wireless optical communication systems based on chaotic MIMO techniques. *Nonlinear Dyn.* **2014**, *75*, 7–16. [CrossRef]
28. Jiang, N.; Zhao, A.; Wang, Y.; Liu, S.; Tang, J.; Qiu, K. Security-enhanced chaotic communications with optical temporal encryption based on phase modulation and phase-to-intensity conversion. *OSA Contin.* **2019**, *2*, 3422–3437. [CrossRef]
29. Kanakidis, D.; Bogris, A.; Argyris, A.; Syvridis, D. Numerical investigation of fiber transmission of a chaotic encrypted message using dispersion compensation schemes. *J. Light. Technol.* **2004**, *22*, 2256–2263. [CrossRef]
30. Canyelles-Pericas, P.; Burton, A.; Le-Minh, H.; Ghassemlooy, Z.; Busawon, K. Chaos synchronization on Visible Light Communication with application for secure data communications. In Proceedings of the 2013 Africon, Pointe aux Piments, Mauritius, 9–12 September 2013; pp. 1–5. [CrossRef]
31. Chen, B.; Zhang, L.; Lu, H. High Security Differential Chaos-Based Modulation with Channel Scrambling for WDM-Aided VLC System. *IEEE Photonics J.* **2016**, *8*, 7804513. [CrossRef]
32. Shu, X.; Wang, H.; Wang, J. Underwater chaos-based DS-CDMA system. In Proceedings of the 2015 IEEE International Conference on Signal Processing, Communications and Computing (ICSPCC), Ningbo, China, 19–22 September 2015; pp. 1–6. [CrossRef]
33. Bai, C.; Ren, H.P.; Grebogi, C.; Baptista, M.S. Chaos-Based Underwater Communication with Arbitrary Transducers and Bandwidth. *Appl. Sci.* **2018**, *8*, 162. [CrossRef]
34. Bai, C.; Ren, H.; Baptista, M.D.; Grebogi, C. Digital underwater communication with chaos. *Commun. Nonlinear Sci. Numer. Simul.* **2019**, *73*, 14–24. [CrossRef]

35. Boulkroune, A.; Bouzeriba, A.; Hamel, S.; Bouden, T. Adaptive fuzzy control-based projective synchronization of uncertain nonaffine chaotic systems. *Complexity* **2014**, *21*, 180–192. [CrossRef]
36. Jiang, X.; Shao, L.; Zhang, S.X.; Yi, X.; Wiersig, J.; Wang, L.; Gong, Q.; Lončar, M.; Yang, L.; Xiao, Y.F. Chaos-assisted broadband momentum transformation in optical microresonators. *Science* **2017**, *358*, 344–347. [CrossRef]
37. Zhang, J.Z.; Wang, A.B.; Wang, J.F.; Wang, Y.C. Wavelength division multiplexing of chaotic secure and fiber-optic communications. *Opt. Express* **2009**, *17*, 6357–6367. [CrossRef] [PubMed]
38. Zhao, Q.; Yin, H. Performance analysis of dense wavelength division multiplexing secure communications with multiple chaotic optical channels. *Opt. Commun.* **2012**, *285*, 693–698. [CrossRef]
39. LeCun, Y.; Bengio, Y.; Hinton, G. Deep Learning. *Nature* **2015**, *521*, 436–444. [CrossRef] [PubMed]
40. Krupitzer, C.; Müller, S.; Lesch, V.; Züfle, M.; Edinger, J.; Lemken, A.; Schäfer, D.; Kounev, S.; Becker, C. A Survey on Human Machine Interaction in Industry 4.0. *arXiv* **2020**, arXiv:2002.01025.
41. Károly, A.I.; Galambos, P.; Kuti, J.; Rudas, I.J. Deep Learning in Robotics: Survey on Model Structures and Training Strategies. *IEEE Trans. Syst. Man, Cybern. Syst.* **2021**, *51*, 266–279. [CrossRef]
42. Devlin, J.; Chang, M.W.; Lee, K.; Toutanova, K. BERT: Pre-training of Deep Bidirectional Transformers for Language Understanding. In Proceedings of the 2019 Conference of the North American Chapter of the Association for Computational Linguistics: Human Language Technologies, Minneapolis, MN, USA, 4 June 2019. [CrossRef]
43. Vaswani, A.; Shazeer, N.M.; Parmar, N.; Uszkoreit, J.; Jones, L.; Gomez, A.N.; Kaiser, L.; Polosukhin, I. Attention Is All You Need. In Proceedings of the 31st Conference on Neural Information Processing Systems, Long Beach, CA, USA, 4–9 December 2017.
44. Mobini, M.; Kaddoum, G.; Herceg, M. Design of a SIMO Deep Learning-Based Chaos Shift Keying (DLCSK) Communication System. *Sensors* **2022**, *22*, 333. [CrossRef]
45. Kanakidis, D. Secure Optical Communication Systems based on Chaotic Carriers. In Proceedings of the International Conference on Security and Cryptography (SECRYPT), Lisbon, Portugal, 8–10 July 2010.
46. Bülow, H. Experimental Demonstration of Optical Signal Detection Using Nonlinear Fourier Transform. *J. Light. Technol.* **2015**, *33*, 1433–1439. [CrossRef]
47. Kingma, D.P.; Ba, J. Adam: A Method for Stochastic Optimization. *arXiv* **2014**, arXiv:1412.6980.
48. Lawrance, A.; Ohama, G. Exact calculation of bit error rates in communication systems with chaotic modulation. *IEEE Trans. Circuits Syst. I Fundam. Theory Appl.* **2003**, *50*, 1391–1400. [CrossRef]
49. Fan, Q.; Zhou, G.; Gui, T.; Lu, C.; Lau, A.P.T. Advancing theoretical understanding and practical performance of signal processing for nonlinear optical communications through machine learning. *Nat. Commun.* **2020**, *11*, 3694. [CrossRef]
50. Wang, Z.; Yang, A.; Guo, P.; He, P. OSNR and nonlinear noise power estimation for optical fiber communication systems using LSTM based deep learning technique. *Opt. Express* **2018**, *26*, 21346–21357. [CrossRef] [PubMed]
51. Koike-Akino, T.; Wang, Y.; Millar, D.S.; Kojima, K.; Parsons, K. Neural Turbo Equalization: Deep Learning for Fiber-Optic Nonlinearity Compensation. *J. Light. Technol.* **2020**, *38*, 3059–3066. [CrossRef]
52. Wang, L.; Gao, M.; Zhang, Y.; Cao, F.; Huang, H. Optical Phase Conjugation with Complex-Valued Deep Neural Network for WDM 64-QAM Coherent Optical Systems. *IEEE Photonics J.* **2021**, *13*, 7200308. [CrossRef]
53. Li, C.; Wang, Y.; Wang, J.; Yao, H.; Liu, X.; Gao, R.; Yang, L.; Xu, H.; Zhang, Q.; Ma, P.; et al. Convolutional Neural Network-Aided DP-64 QAM Coherent Optical Communication Systems. *J. Light. Technol.* **2022**, *40*, 2880–2889. [CrossRef]
54. Xie, T.; Yu, J. Long-range photonics-aided 17.6 Gbit/s D-band PS-64QAM transmission using gate recurrent unit algorithm with a complex QAM input. *Opt. Express* **2023**, *31*, 40056–40069. [CrossRef]
55. Abbass, A.M.; Fyath, R.S. Performance investigation of geometric constellation shaping-based coherent WDM optical fiber communication system supported by deep-learning autoencoder. *Results Opt.* **2024**, *15*, 100629. [CrossRef]
56. He, L.; Lai, R.; Shao, S.; Li, Z. Design of a Chaotic Communication System Based on Deep Learning with Two-Dimensional Reshaping. *IEEE Trans. Veh. Technol.* **2024**, *73*, 10421–10434. [CrossRef]
57. Yang, Z.; Ke, J.; Zhuge, Q.; Hu, W.; Yi, L. Coherent chaotic optical communication of 30 Gb/s over 340-km fiber transmission via deep learning. *Opt. Lett.* **2022**, *47*, 2650–2653. [CrossRef]
58. Zhang, H.; Zhang, L.; Jiang, Y.; Wu, Z. Reliable and Secure Deep Learning-Based OFDM-DCSK Transceiver Design without Delivery of Reference Chaotic Sequences. *IEEE Trans. Veh. Technol.* **2022**, *71*, 8059–8074. [CrossRef]
59. Ke, J.; Yi, L.; Yang, Z.; Yang, Y.; Zhuge, Q.; Chen, Y.; Hu, W. 32 Gb/s chaotic optical communications by deep-learning-based chaos synchronization. *Opt. Lett.* **2019**, *44*, 5776–5779. [CrossRef]
60. Anh, D.V.; Tran Thi Thanh, T.; Huu, L.N.; Dung Truong, C.; Xuan, Q.N. Performance Analysis of High-Speed Wavelength Division Multiplexing Communication between Chaotic Secure and Optical Fiber Channels Using DP-16QAM Scheme. In Proceedings of the 2020 IEEE Eighth International Conference on Communications and Electronics (ICCE), Phu Quoc, Vietnam, 13–15 January 2021; pp. 33–38. [CrossRef]
61. Graves, A. *Long Short-Term Memory*; Springer: Berlin/Heidelberg, Germany, 2012; pp. 37–45. [CrossRef]
62. Zhou, H.; Zhang, S.; Peng, J.; Zhang, S.; Li, J.; Xiong, H.; Zhang, W. Informer: Beyond Efficient Transformer for Long Sequence Time-Series Forecasting. In Proceedings of the AAAI Conference on Artificial Intelligence, New York, NY, USA, 7–12 February 2020. [CrossRef]

63. Che, Z.; Purushotham, S.; Cho, K.; Sontag, D.; Liu, Y. Recurrent neural networks for multivariate time series with missing values. *Sci. Rep.* **2018**, *8*, 6085. [CrossRef] [PubMed]
64. Schulman, J.; Wolski, F.; Dhariwal, P.; Radford, A.; Klimov, O. Proximal Policy Optimization Algorithms. *arXiv* **2017**, arXiv:1707.06347.

photonics

MDPI

Article

Learning Gradient-Based Feed-Forward Equalizer for VCSELs

Muralikrishnan Srinivasan [1], Alireza Pourafzal [2], Stavros Giannakopoulos [3], Peter Andrekson [4], Christian Häger [2] and Henk Wymeersch [2,*]

1 Department of Electronics Engineering, Indian Institute of Technology (BHU), Varanasi 221005, India; muralikrishnan.ece@iitbhu.ac.in
2 Electrical Engineering, Chalmers University of Technology, 41296 Gothenburg, Sweden; alireza.pourafzal@chalmers.se (A.P.); christian.haeger@chalmers.se (C.H.)
3 Computer Science and Engineering, Chalmers University of Technology, 41296 Gothenburg, Sweden; stagia@chalmers.se
4 Photonics, Microtechnology and Nanoscience, Chalmers University of Technology, 41296 Gothenburg, Sweden; peter.andrekson@chalmers.se
* Correspondence: henkw@chalmers.se

Abstract: Vertical cavity surface-emitting laser (VCSEL)-based optical interconnects (OI) are crucial for high-speed data transmission in data centers, supercomputers, and vehicles, yet their performance is challenged by harsh and fluctuating thermal conditions. This paper addresses these challenges by integrating an ordinary differential equation (ODE) solver within the VCSEL communication chain, leveraging the adjoint method to enable effective gradient-based optimization of pre-equalizer weights. We propose a machine learning (ML) approach to optimize feed-forward equalizer (FFE) weights for VCSEL transceivers, which significantly enhances signal integrity by managing inter-symbol interference (ISI) and reducing the symbol error rate (SER).

Keywords: machine learning; optical communications; VCSEL-based optical interconnects; end-to-end learning

1. Introduction

Vertical cavity surface emitting laser (VCSEL)-based optical interconnects (OIs) serve as the primary connectivity solution in data centers, supercomputers, and vehicles, offering cost-effective and high-speed connections [1]. Given the harsh and dynamically changing environments in which these systems operate, VCSELs demand adaptive and resilient design strategies throughout the communication chain [2,3]. Among the many factors that influence the performance and reliability of VCSELs, temperature poses a particular challenge [4]. In short-range OIs, the optical links are positioned close to heat sources, which are typically the processing units, leading to rapid and substantial temperature variations. This scenario is common in data centers, where the compact and densely packed nature of systems often results in significant heat buildup [1]. Such temperature changes impact the operational characteristics of VCSELs in several ways, including increased threshold current, shifts in the emission wavelength of the VCSEL due to changes in the refractive index and the physical dimensions of the laser cavity, decreased output power due to decreased carrier density, and increased non-radiative recombination within the laser's active region [5].

The inherent nonlinear transfer characteristics of VCSELs, especially under significant temperature variations, necessitate sophisticated approaches to ensure optimal operation. For example, maintaining robust 100 Gbps links in such fluctuating environments requires implementing advanced equalization techniques [6]. Equalization can be implemented in two forms: post-equalization at the receiver, and pre-equalization at the transmitter. Pre-equalizers actively modify the signal before it encounters the distorting effects of the transmission medium and VCSEL nonlinearities [7,8]. This proactive approach allows for the correction of impairments before they occur, making it more efficient than attempting to reverse these effects at the receiver end. Moreover, pre-equalization helps to reduce the complexity and computational load on the receiver, which is particularly advantageous in high-speed applications where minimizing processing delays is crucial.

Equalizers have traditionally been designed based on mathematical models [9–11]. However, considerations of cost, energy efficiency, and temperature variations significantly impact communication capacity. Modeling individual components is already highly challenging; even if a model is available, it tends to be complex, as these models often involve the concatenation of numerous nonlinear, frequency-selective, and noisy submodels, which in turn precludes the possibility of designing an optimal transmitter and the corresponding optimal receiver.

machine learning (ML) provides an attractive alternative to traditional model-based approaches to overcome the three challenges of modeling, design, and adaptivity [12]. Classical models of components serve as a foundation for constructing neural network (NN) equivalents. In the context of optical communications, receiver-side algorithms encompassing equalization, synchronization, data detection, and decoding can be learned by mimicking conventional algorithms or utilizing deep neural networks (DNNs) from scratch [13–17]. Digital pre-equalization techniques have also gained popularity for enhancing performance the optical communication links [7,8,18,19]. The real-time deployment of NN-based digital equalizers hinges on computational complexity comparable to or lower than conventional digital signal processing (DSP) solutions. For instance, NN-based digital predistortion was designed using three convolutional layers in [20]. In the realm of pre-equalization methods with a view to reducing the complexity, the feed-forward equalizer (FFE) stands out as a prominent analog filter structure employed in transmitters. Operating as a finite impulse response (FIR) filter, the FFE optimally shapes the pulse response, aiming to eliminate inter-symbol interference (ISI) and reduce the link's symbol error rate (SER) performance.

Notably, no work has yet attempted to optimize FFE weights using ML in the context of VCSEL transceivers. Transmitter-side techniques such as pre-equalization introduce significant challenges due to the need for a corresponding adaptive receiver that must participate in the learning process [21]. One of the primary issues is learning the architecture on the transmit side, which often involves backpropagation through a mathematical model of the VCSEL. While differentiable models of VCSELs do exist, such as NN equivalents [6,20,22], the requirements for high-speed operation and precise control necessitate a more comprehensive modeling approach. Accurate modeling of VCSELs involves capturing both the small-signal and large-signal response, including thermal effects across varying temperatures and with limited training samples. However, these comprehensive models, which need to incorporate temperature dynamics explicitly, often lose their differentiability [23]. This complicates the application of standard ML approaches that require gradient calculations. This limitation presents an opportunity to explore novel representations of VCSELs that are both comprehensive and compatible with back-propagation.

Our work includes the integration of an ordinary differential equation (ODE) solver within the VCSEL-based OI chain framework [24,25]. This integration allows for simulating the dynamic behavior of VCSELs using the rate equations and ensures the availability of gradients at each step of the ODE. This gradient availability is essential for updating pre-equalizer weights in gradient-based learning methods.

The contributions of the paper are as follows:

1. **Integration of ODE-based ML for VCSEL Modeling:** We utilize the adjoint method [25,26] within an ML framework for backpropagation through the ODE solver. This approach directly integrates the VCSEL model and its dynamics, avoiding surrogate models and enabling optimization of transmitter components.
2. **Optimization of FFE Weights for VCSEL Transceivers:** Building on the ODE-based integration, we introduce an ML approach to optimize FFE weights for VCSEL transceivers. This method effectively manages ISI and SER, leading to improved overall performance.

The rest of this paper is organized as follows: Section 2 introduces the rate equations of VCSEL and discusses the intrinsic small-signal modulation response; Section 3 provides an overview of FFE; Section 4 describes the ML-based pipeline for training an FFE within an OI system; Section 5 discusses the numerical results; and Section 6 concludes the paper.

2. Rate Equations of VCSEL

The ability of VCSELs to effectively respond to current changes at the data rate is essential for ensuring dependable data transmission. To achieve this, a comprehensive understanding of the VCSEL's dynamic response is necessary. This dynamic response is governed by a set of rate equations that account for the intricate interactions between injected free carriers and photons within the cavity [27].

2.1. Parasitic Elements

Parasitic elements in VCSELs arise from their physical structure and manufacturing processes. These include imperfections at material interfaces such as the p–n junction and metallic contacts, which can lead to unwanted resistance. Parasitic capacitance formed at the interfaces between semiconductor layers and around the active region affects how quickly a VCSEL can respond to input signal changes, limiting the modulation speed. To account for these effects, a simple parallel RC circuit model with resistance (R_j) and capacitance (C_j) components is used in simulations, as shown in Figure 1. Here, I_{in} represents the VCSEL driving current, I is the injection current without any parasitic element, the transfer from I_{in} to I is the parasitic response, and the transfer from I to the optical output is determined by the rate equations of the VCSEL.

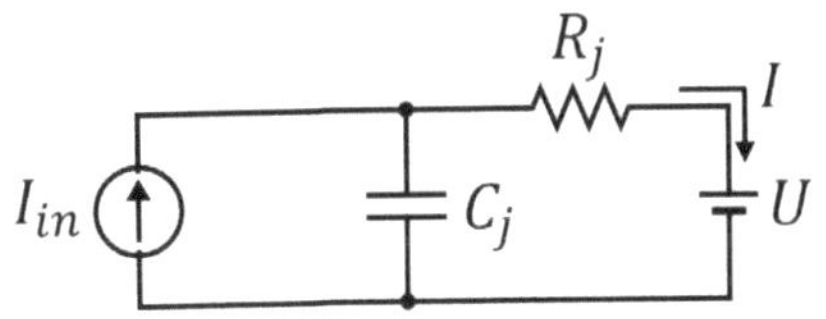

Figure 1. Schematic of the RC circuit model used to simulate the parasitic effects in VCSEL; I_{in} represents the VCSEL driving current, I is the injection current, and U is the device voltage.

2.2. Rate Equations

The laser's operation is modeled through the single-mode laser rate equations derived from a simplified VCSEL model [28], which provide a mathematical framework for understanding the interactions between carrier and photon dynamics within the laser cavity.

2.2.1. Carrier Dynamics Equation

The rate of change of carrier density N [m^{-3}] within the laser's active region is modeled by the following equation [27,28]:

$$\frac{dN}{dt} = \frac{I}{qV} - \frac{c}{n_{\text{geff}}} gS - \frac{N}{\tau_n}. \tag{1}$$

The rate of carrier injection $\frac{I}{qV}$ is driven by the injection current I [A], where q is the elementary charge and V [m^3] is the active volume. The term $\frac{c}{n_{\text{geff}}} gS$ represents the stimulated emission rate, where c denotes the speed of light in vacuum, n_{geff} is the effective modal refractive index, g is the optical gain per unit length, and S is the photon density; lastly, $\frac{N}{\tau_n}$ accounts for the carrier recombination losses, with τ_n [s] as the carrier lifetime, encompassing both radiative and non-radiative decay processes.

2.2.2. Photon Dynamics Equation

The photon density S [m^{-3}] that captures the dynamics of photon population within the laser cavity is provided by the following equation [27,28]:

$$\frac{dS}{dt} = \Gamma \frac{c}{n_{\text{geff}}} gS + \Gamma\beta \frac{N}{\tau_n} - \frac{S}{\tau_p}. \tag{2}$$

The stimulated emission rate is reduced by the internal quantum efficiency Γ. The second term $\Gamma\beta\frac{N}{\tau_n}$ introduces the contribution of spontaneous emission to the overall photon density, with β representing the spontaneous emission coupling factor. The photon losses are modeled by the final term $\frac{S}{\tau_p}$, where τ_p [s] is the the photon lifetime. It is important to note that we treat Γ, g, τ_n, and τ_p as temperature-dependent parameters.

2.2.3. Output Power Equation

The relationship between the output power P_o [W] and photon density S is expressed as [27,28]

$$P_o = S \cdot V \cdot h\nu \cdot \frac{\eta_{out}}{\tau_p \Gamma}. \tag{3}$$

The output power P_o of the VCSEL is directly proportional to the photon density S, and is calculated considering the active volume V and the energy per photon ($h\nu$). The efficiency of the laser output η_{out} expressed relative to the wavelength λ_{cav} [m] quantifies the energy conversion efficiency of the VCSEL, illustrating how the VCSEL converts electrical power into optical power at a specific wavelength.

2.3. Self-Heating

To analyze self-heating effects, an additional set of differential equations is used to monitor the internal temperature (T) of the VCSEL [23]:

$$\frac{\tau_{th}}{r_{th}} \frac{dT}{dt} = g_{gen} - g_{diss} \tag{4}$$

where g_{gen} [W] represents the rate of heat generation, calculated as

$$g_{gen} = U \cdot I_{in} - P_o, \tag{5}$$

where U [V] is the device voltage, I_{in} is the driving current (see Figure 1), and g_{diss} [W] denotes the rate of heat dissipation, provided by

$$g_{diss} = \frac{1}{r_{th}} (T - T_{amb}). \tag{6}$$

where τ_{th} denotes the thermal time constant, r_{th} [K/W] is the thermal impedance, and T_{amb} is the ambient temperature.

2.4. Dynamic Response of VCSEL

In this way, the rate equations establish a direct relationship between the excess carrier density in the active region and the photon density within the cavity when the current passes through the VCSEL. By perturbing these rate equations around a bias current I_b using first-order Taylor expansion and measuring the differential output power, we obtain the intrinsic small-signal modulation response, for which the two-pole transfer function is [27]

$$H_{\text{int}}(f) = \eta_d \frac{hc}{\lambda_0 q} \cdot \frac{f_r^2}{f_r^2 - f^2 + j\gamma \frac{f}{2\pi}}, \tag{7}$$

where η_d is the differential quantum efficiency, h is the Planck constant, c is the speed of light, λ_0 is the lasing wavelength in vacuum, q is the elementary charge, f_r is the resonance frequency, and γ is the damping factor. The small-signal modulation response is measured by S_{21}

$$S_{21} = 20\log_{10}\frac{|\ H_{\text{int}}(f)}{|\ H_{\text{int}}(0)}, \tag{8}$$

and is plotted in Figure 2 for increasing bias current $I_{b1} < I_{b2} < I_{b3}$ and two temperatures, 27 °C and 70 °C, showing the movement of f_r and that the VCSEL reaches a critically damped (flat) response at some current. The plot reveals shifts in resonance frequency and a nonlinear reduction in bandwidth, significantly influencing the frequency response and impacting data transmission capabilities. Pre-equalization at the transmitter, either analog or digital, is crucial to address impairments from the limited bandwidth of VCSELs.

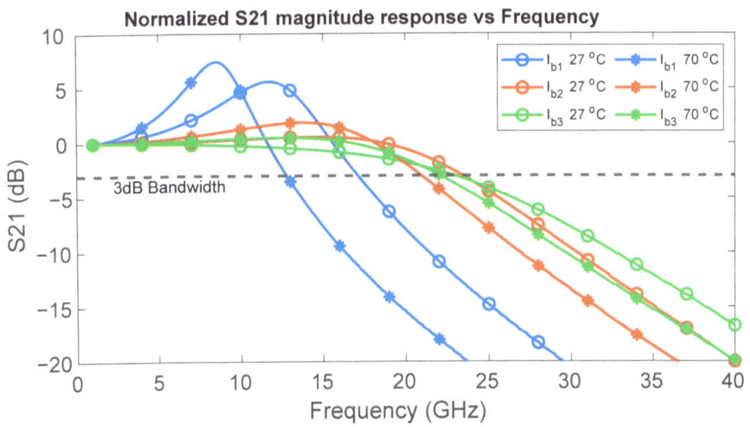

Figure 2. Simulated intrinsic VCSEL response (with parasitic effects neglected) for three representative bias currents ($I_{b1} < I_{b2} < I_{b3}$) and two temperatures (27 °C and 70 °C).

The rate equation is directly solved for the forward inference step using the ODE solver torchdiffeq in PyTorch, generating the output power for input current sequences and establishing the loss function [25]. This library not only facilitates the integration of rate equations but also supplies gradients at each ODE solver step for the back-propagation step. This capability is imperative for updating the transmitter FFE weights in the optimization process described in the next section.

3. FFEs Overview

FFEs use an FIR filter to shape the pulse response and ideally eliminate all ISI. The FFE consists of a series of weighted coefficients called taps. Each tap represents a particular weight applied to a delayed version of the input signal. The number of taps determines the complexity and effectiveness of the equalizer. The delay in an FFE refers to the time difference between the input signal and its delayed versions that are fed into the taps. This delay allows the FFE to capture and compensate for the effects of previous symbols on the current symbol. The output of the FFE at time instant t is provided by

$$I_p(t) = I_{in}(t) + \sum_{k=1}^{K} w_k I_{in}(t - t_k), \tag{9}$$

where $I_{in}(t)$ is the input current, w_k are the tap weights determining the contribution of each delayed input sample $I_{in}(t - t_k)$, and t_k are the corresponding delays.

A model block diagram indicating the position of the FFE in a VCSEL based OI is shown in Figure 3. The FFE is placed after the digital-to-analog converter (DAC) but before the VCSEL and its parasitic elements. The delay elements in FFEs can be implemented with synchronously clocked flip-flops, transmission lines, or analog delay elements. Coefficient summing and scaling can be achieved with scaled switched current sources either before or within the final driver stage.

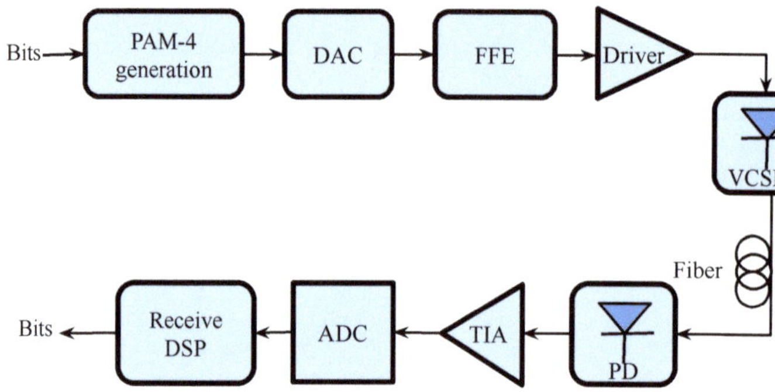

Figure 3. Model block diagram of a VCSEL-based OI system. DAC stands for digital-to-analog converter, ADC stands for analog-to-digital converter, PD stands for photodetector, and TIA stands for trans-impedance amplifier. The FFE weights are optimized in the paper.

However, FFEs have limitations, particularly in filtering out relaxation oscillations under varying biasing and data conditions. Relaxation oscillations or rapid fluctuations in laser output power, can degrade signal quality. Traditional FFE techniques may not fully compensate for these effects due to fixed or inadequately adaptive filter settings [29,30]. To overcome these limitations, we introduce an ML-based approach to dynamically optimize the FFE coefficients w_k. In this paper, we consider ideal driver/FFE electronics, as including transmitter and receiver non-idealities is beyond the scope of the current paper. The following section outlines the end-to-end pipeline for learning FFE weights within the OI system.

4. Pipeline for Learning FFE Weights

The end-to-end ML-based pipeline of the OI system and transmission chain, including the FFE, VCSEL, and the receiver, is shown in Figure 4. Detailed functionality from message encoding to output estimation is provided in the subsequent subsections.

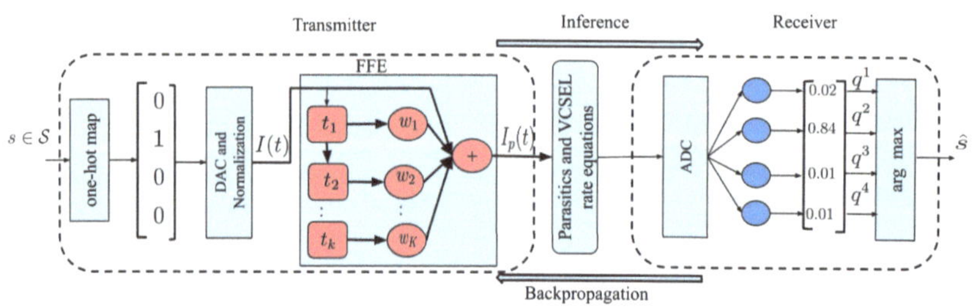

Figure 4. Model block diagram showing the placement of the FFE in the entire chain. Weights w_1 to w_K are learned in the paper.

4.1. Encoding and Input Transformation

The process starts by encoding a message $s \in \{1, \ldots, M\} \triangleq \mathcal{M}$, where $M = 4$. Each message s is encoded into a one-hot vector $\mathbf{x}$, where the s-th element is 1 and all other elements are 0. The output of the one-hot layer, ranging from $[0, 1]$, is scaled and shifted to the dynamic input current range of $[2, 12]$ mA. This ensures that the input remains above the VCSEL threshold current across all ambient temperatures, preventing the AE from arbitrarily increasing the bias current, which would lead to self-heating in a real system [28].

4.2. Signal Conversion and Transmission

After it is prepared, the input is sent to a DAC, converting the digital input into an analog signal for the FFE and VCSEL. The FFE adjusts the signal to compensate for potential distortions before it reaches the VCSEL. The fiber is modeled as a additive white Gaussian noise (AWGN) channel. The system can adapt to include additional features such as low-pass filtering and dispersion as well as intricate circuitry such as output driver circuits, which are beyond the current work's scope.

4.3. Output Processing and Estimation

At the receiving end, the photodiode output is processed through a fully-connected neural network layer with softmax activation. This converts the received signals into a probability vector $\mathbf{q} = [q_1, \ldots, q_M]^\top$, where the estimated message $\hat{s}$ is determined by selecting the highest probability from the softmax output, expressed as

$$\hat{s} = \arg\max_i q_i. \tag{10}$$

4.4. Optimization and Loss Minimization

The network optimization focuses on minimizing the categorical cross-entropy loss function, provided by

$$\mathcal{L} = \sum_{i=1}^{M} x_i \log(q_i), \tag{11}$$

where q_i for $i \in \{1, 2, \ldots, M\}$ is a predicted value and x_i is 1 for true classes and 0 otherwise. This measures the disparity between the predicted and true probability distributions, guiding the model towards accurate predictions by penalizing deviations from the true class probabilities. The loss $\mathcal{L}$ can be related to an achievable information rate using arguments from mismatched decoding [31].

4.5. Training FFE Weights with the Adjoint Method

The goal of the training is to find FFE weights w_k for $k = 1, \ldots, K$ that optimize performance in terms of the categorical loss function. A learning-based model is trained by adjusting its parameters to minimize the difference between its predictions and actual outcomes. Traditional backpropagation involves a backward pass through the network to update these parameters based on the gradient of the loss, leading to challenges with VCSEL components governed by differential equations. To address this, we propose using the adjoint method [26] for training the pipeline. Derived from the framework of neural ordinary differential equations (NODE) [25], this technique integrates ODEs as dynamically learnable components within the network. The adjoint method calculates the adjoint state during the backward pass, representing the gradient of the loss concerning the network state at any given time. By solving the reverse-time ODE for the adjoint state, it is possible to directly compute the gradients with respect to the differential equations governing the VCSEL.

To ensure robust learning, the training process utilized a dataset of 2.5×10^4 randomly selected message symbols processed in batches of 50 symbols over 7500 epochs. Training was conducted at an SNR of 18 dB and a temperature of 70 °C. Determining the number of taps is a critical factor, and is contingent on the desired equalization performance. A greater number of taps in the design enhances equalization performance, as it allows for fine-tuning the FFE frequency response; in turn, this fine-tuning enables more precise shaping of the system's limited bandwidth, leading to significantly improved overall bandwidth performance. The following section provides a detailed analysis of the numerical results and examines potential future extensions.

5. Numerical Results and Discussion

We performed different training for different FFE configurations ranging from two taps to five taps. The delay was chosen in multiples of T_s, where $T_s = 1/(15 * F_s)$ and $F_s = 56$ GBaud is the symbol rate, that is, $n_1 = 6T_s$, $n_2 = 9T_s$, $n_3 = 12T_s$, $n_4 = 15T_s$, and $n_5 = 18T_s$. During training, the FFE weights were optimized to minimize the error between the transmitted and received symbols. The optimized weights for various tap configurations are detailed in Table 1, reflecting the system's adaptation to diverse signal distortions encountered during the training phase.

The effectiveness of the optimized weights is demonstrated through eye diagrams in Figures 5 and 6a–d. Each diagram represents a different tap setting on the FFE, demonstrating how increasing the number of taps affects signal clarity. Figure 5 shows the eye diagram without the FFE. A clear trend is observed in Figure 6a–d, where the signal clarity improves as the number of taps increases. The average eye height increases from 0.6 mW for two taps to about 1.2 mW for five taps. Similarly, the average eye width is about 8.33 ps for two taps and 10.7 ps for five taps. The average jitter for two taps is 9.5 ps, while that for five taps is 7.14 ps. The added taps enhance the complexity of the weights, allowing for finer signal adjustments and reduced inter-symbol interference; however, as shown in Figure 6d, the improvement with five taps is minimal compared to the FFE with four taps.

Table 1. Optimized weights w_1 to w_5 for different tap settings.

Taps	w_1	w_2	w_3	w_4	w_5
2	0	−0.3438	0	0.0285	0
3	0	−0.4329	−0.3564	0.4961	0
4	0	−0.3357	−0.1708	0.0421	0.1947
5	−0.0281	−0.2644	−0.1067	−0.0223	0.1516

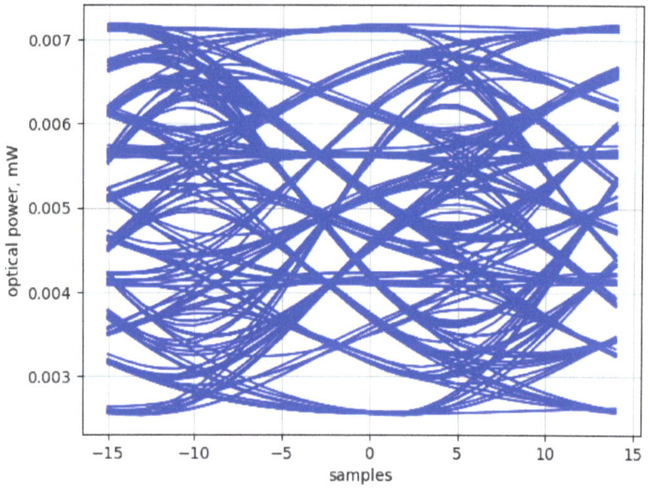

Figure 5. Eye diagram illustrating the signal quality in the absence of FFE.

The improvement in clarity observed in the eye diagram is correlated with a reduction in the SER. This relationship is illustrated in Figure 7, where higher SNRs lead to lower SERs, highlighting the advantages of more advanced pre-equalization techniques. For instance, using five learned taps provides a sensitivity gain of approximately 1 dB over configurations with only two taps. This gain is notable at a low SER level of 10^{-4}, indicating a substantial enhancement in the system's ability to accurately interpret the received symbols. Increasing the number of taps enables more precise adjustments of the equalizer's response, translating to improved performance metrics, such as lower SER at higher SNRs.

The proposed method trained at an SNR of 18 dB generalizes well across high SNR conditions, but may require additional training at lower SNRs to combat increased noise. To address SNR variability more robustly, a similar approach to the distance training method discussed in [32] could be adopted; in this approach, during training the SNRs are drawn from a Gaussian distribution with a mean of 18 dB and a certain standard deviation. Similarly, VCSELs, being temperature-sensitive, demand adaptive models for reliable performance across a wide operating range from −40 °C to +125 °C. Traditional methods often involve retraining FFEs for different temperatures or fine-tuning via transfer learning. Alternatively, a temperature-adaptive FFE that introduces temperature as an input to the neural network could be explored in future to enable dynamic adaptation without requiring retraining for each scenario. Similarly, to address nonlinearities such as VCSEL relaxation oscillations and temperature-induced variations, nonlinear equalizers based on other deep learning models could be trained to adapt to rapid shifts in operating conditions. This approach would enhance the system's robustness in extreme scenarios. In future work, the benefits of this approach can be explored for different symbol rates versus VCSEL bandwidths. Additionally, learning the delays along with the weights and including transmitter and receiver non-idealities could further enhance the adaptability of the system.

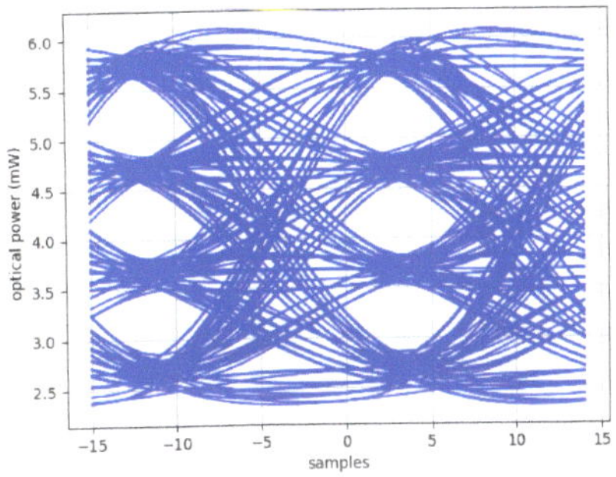

(a) With two taps: w_2 and w_4

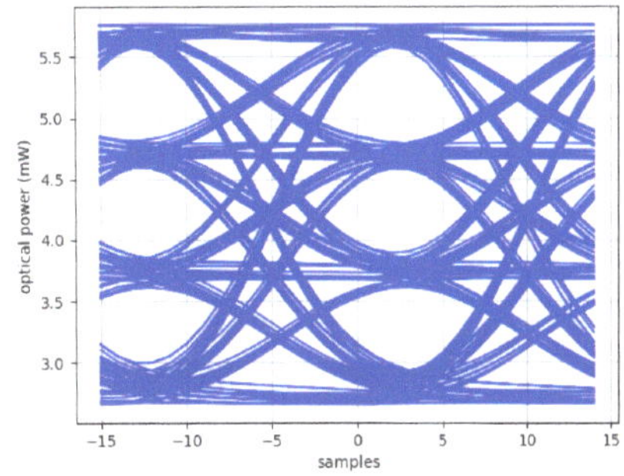

(b) With three taps: w_2, w_3, and w_4

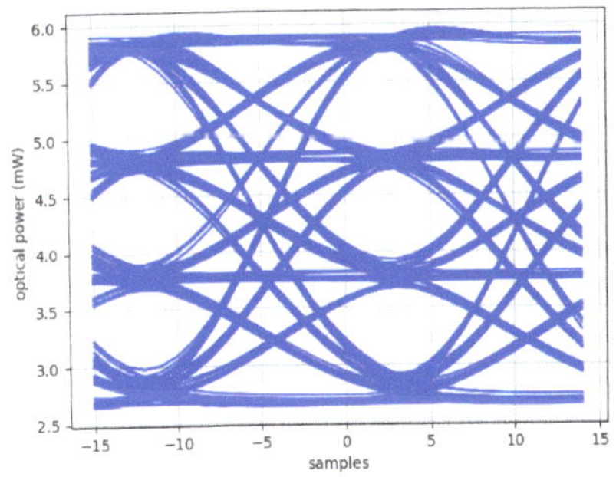

(c) With four taps: w_2, w_3, w_4, and w_5

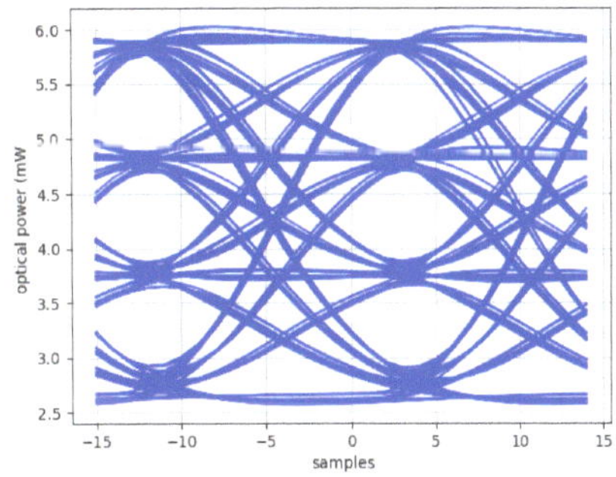

(d) With five taps: w_2, w_3, w_4, w_5, and w_6

Figure 6. Eye diagrams for different optimized FFE tap configurations.

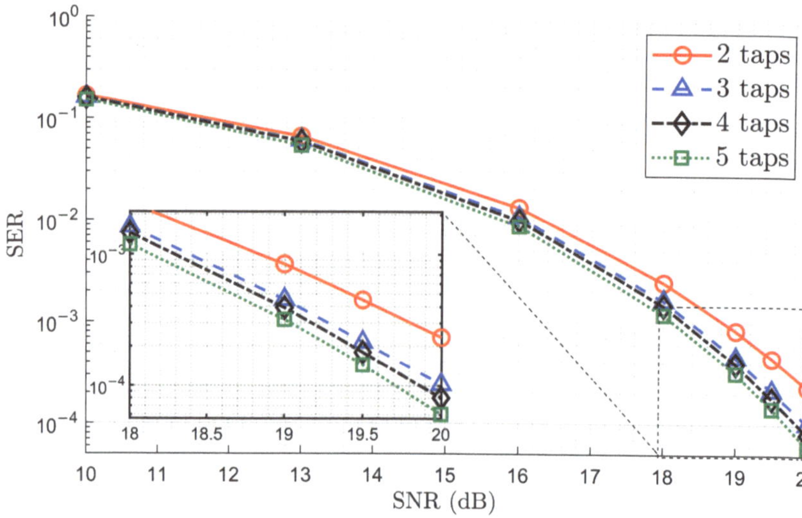

Figure 7. SER vs. SNR for different numbers of FFE taps.

6. Conclusions

In conclusion, we have developed an end-to-end pipeline for optimizing FFE weights within the OI system. By integrating the adjoint method with ODE solvers, we achieve gradient-based optimization of the FFE weights. The results demonstrate significant improvements in signal clarity and performance. Specifically, configurations with more taps enhance signal integrity, with the five-tap setup providing a 1 dB sensitivity gain over the two-tap setup and an SER of 10^{-4}. These findings validate the effectiveness of our approach and highlight the importance of the tap number in optimizing equalization strategies.

Author Contributions: Conceptualization, M.S., S.G. and H.W.; methodology, M.S.; software, M.S.; validation, M.S., A.P. and S.G.; formal analysis, M.S. and C.H.; investigation, M.S.; resources, S.G. and P.A.; data curation, A.P.; writing—original draft preparation, M.S. and A.P.; writing—review and editing, M.S., A.P., C.H., S.G. and H.W.; visualization, A.P.; supervision, H.W.; project administration, P.A., C.H. and H.W.; funding acquisition, P.A., C.H. and H.W. All authors have read and agreed to the published version of the manuscript.

Funding: This research was funded by the Swedish Foundation for Strategic Research (SSF, HOT-OPTICS Project).

Institutional Review Board Statement: Not applicable.

Informed Consent Statement: Not applicable.

Data Availability Statement: Data are contained within the article.

Conflicts of Interest: The authors declare no conflicts of interest; the funders had no role in the design of the study, in the collection, analysis, or interpretation of data, in the writing of the manuscript, or in the decision to publish the results.

References

1. Cheng, Q.; Bahadori, M.; Glick, M.; Rumley, S.; Bergman, K. Recent Advanced in Optical Technologies for Data Centers: A Review. *Optica* **2018**, *5*, 1354–1370. [CrossRef]
2. Rumley, S.; Bahadori, M.; Polster, R.; Hammond, S.D.; Calhoun, D.M.; Wen, K.; Rodrigues, A.; Bergman, K. Optical Interconnects for Extreme Scale Computing Systems. *Parallel Comput.* **2017**, *64*, 65–80. [CrossRef]
3. King, R. VCSEL Design for Automotive Datacom Experimental Results for 980 nm versus 850 nm. Technical Report, 2021. Available online: https://www.ieee802.org/3/cz/public/may_2021/king_3cz_01a_0521.pdf (accessed on 25 September 2024).

4. Aoki, T.; Kubota, R.; Hiiro, H.; Yoshimoto, S.; Yanagisawa, M. 50 Gb/s PAM-4 VCSELs operating up to 125 °C. In Proceedings of the Optoelectronics and Communications Conference (OECC), Hong Kong, China; 3–7 July 2021; Optica Publishing Group: Washington, DC, USA, 2021; p. T3D.4. [CrossRef]
5. Jasim, F.Z.; Omar, K.; Hassan, Z. Temperature effect on VCSEL output performance. *J. Optoelectron. Adv. Mater.* **2009**, *3*, 1136–1138.
6. Srinivasan, M.; Song, J.; Grabowski, A.; Szczerba, K.; Iversen, H.K.; Schmidt, M.N.; Zibar, D.; Schröder, J.; Larsson, A.; Häger, C.; et al. End-to-End Learning for VCSEL-based Optical Interconnects: State-of-the-Art, Challenges, and Opportunities. *J. Light. Technol.* **2023**, *41*, 3261–3277. [CrossRef]
7. Berenguer, P.W.; Nölle, M.; Molle, L.; Raman, T.; Napoli, A.; Schubert, C.; Fischer, J.K. Nonlinear Digital Pre-distortion of Transmitter Components. *IEEE J. Lightw. Technol.* **2015**, *34*, 1739–1745. [CrossRef]
8. Lavery, D.; Maher, R.; Liga, G.; Semrau, D.; Galdino, L.; Bayvel, P. On the Bandwidth Dependent Performance of Split Transmitter-receiver Optical Fiber Nonlinearity Compensation. *Opt. Express* **2017**, *25*, 4554–4563. [CrossRef] [PubMed]
9. Giannakopoulos, S.; He, Z.S.; Svensson, L.; Zirath, H. Scalable, Modular Feed-Forward Equalizer for Baseband Applications. In Proceedings of the 2022 17th European Microwave Integrated Circuits Conference (EuMIC), Milan, Italy, 26–27 September 2022; IEEE: Piscataway, NJ, USA, 2022; pp. 280–283.
10. Maeda, K.; Yamamoto, S.; Kohmu, N.; Nishimura, K.; Fukasaku, I. An Active-Copper-Cable with Continuous-Time-Linear-Equalizer IC for 30-AWG 7-meters Reach Interconnect of 400-Gbit/s QSFP-DD. In Proceedings of the 2019 IEEE Asia Pacific Conference on Circuits and Systems (APCCAS), Bangkok, Thailand, 11–14 November 2019; IEEE: Piscataway, NJ, USA, 2019; pp. 217–220.
11. Belfiore, G.; Khafaji, M.; Henker, R.; Ellinger, F. A 50 Gb/s 190 mW asymmetric 3-tap FFE VCSEL driver. *IEEE J.-Solid-State Circuits* **2017**, *52*, 2422–2429. [CrossRef]
12. Zhong, K.; Zhou, X.; Huo, J.; Yu, C.; Lu, C.; Lau, A.P.T. Digital Signal Processing for Short-reach Optical Communications: A Review of Current Technologies and Future Trends. *IEEE J. Lightw. Technol.* **2018**, *36*, 377–400. [CrossRef]
13. Shen, T.S.R.; Lau, A.P.T. Fiber Nonlinearity Compensation using Extreme Learning Machine for DSP-based Coherent Communication Systems. In Proceedings of the Optoelectronics and Communications Conference (OECC), Kaohsiung, Taiwan, 4–8 July 2011; pp. 816–817.
14. Gaiarin, S.; Pang, X.; Ozolins, O.; Jones, R.T.; Da Silva, E.P.; Schatz, R.; Westergren, U.; Popov, S.; Jacobsen, G.; Zibar, D. High Speed PAM-8 Optical Interconnects with Digital Equalization based on Neural Network. In Proceedings of the 2016 Asia Communications and Photonics Conference (ACP), Wuhan, China, 2–5 November 2016; pp. 1–3.
15. Ge, L.; Zhang, W.; Liang, C.; He, Z. Compressed Neural Network Equalization based on Iterative Pruning Algorithm for 112-Gbps VCSEL-enabled Optical Interconnects. *IEEE J. Lightw. Technol.* **2020**, *38*, 1323–1329. [CrossRef]
16. Deligiannidis, S.; Bogris, A.; Mesaritakis, C.; Kopsinis, Y. Compensation of fiber nonlinearities in digital coherent systems leveraging long short-term memory neural networks. *IEEE J. Lightw. Technol.* **2020**, *38*, 5991–5999. [CrossRef]
17. Argyris, A.; Bueno, J.; Fischer, I. Photonic Machine Learning Implementation for Signal Recovery in Optical Communications. *Sci. Rep.* **2018**, *8*, 8487. [CrossRef] [PubMed]
18. Paryanti, G.; Faig, H.; Rokach, L.; Sadot, D. A Direct Learning Approach for Neural Network based Pre-distortion for Coherent Nonlinear Optical Transmitter. *IEEE J. Lightw. Technol.* **2020**, *38*, 3883–3896. [CrossRef]
19. Wu, Y.; Gustavsson, U.; Amat, A.G.I.; Wymeersch, H. Low Complexity Joint Impairment Mitigation of I/Q Modulator and PA Using Neural Networks. *IEEE J. Sel. Areas Commun* **2022**, *40*, 54–64. [CrossRef]
20. Minelli, L.; Forghieri, F.; Shao, T.; Shahpari, A.; Gaudino, R. TDECQ-Based Optimization of Nonlinear Digital Pre-Distorters for VCSEL-MMF Optical Links Using End-to-end Learning. *J. Light. Technol.* **2023**, *42*, 621–635. [CrossRef]
21. Schaedler, M.; Kuschnerov, M.; Calabrò, S.; Pittalà, F.; Bluemm, C.; Pachnicke, S. AI-based Digital Predistortion for IQ Mach-Zehnder Modulators. In Proceedings of the 2019 Asia Communications and Photonics Conference (ACP), Chengdu, China, 2–5 November 2019; pp. 1–3.
22. Srinivasan, M.; Song, J.; Häger, C.; Szczerba, K.; Wymeersch, H.; Schröder, J. Learning optimal PAM levels for VCSEL-based optical interconnects. In Proceedings of the 2022 European Conference on Optical Communication (ECOC), Basel, Switzerland, 18–22 September 2022; IEEE: Piscataway, NJ, USA, 2022, pp. 1–4.
23. Grabowski, A.; Gustavsson, J.; He, Z.S.; Larsson, A. Large-Signal Equivalent Circuit for Datacom VCSELs. *IEEE J. Lightw. Technol.* **2021**, *39*, 3225–3236. [CrossRef]
24. Chen, R.T.Q. torchdiffeq, 2018. Available online: https://github.com/rtqichen/torchdiffeq (accessed on 25 September 2024).
25. Chen, R.T.Q.; Rubanova, Y.; Bettencourt, J.; Duvenaud, D. Neural Ordinary Differential Equations. In Proceedings of the NIPS'18: Proceedings of the 32nd International Conference on Neural Information Processing Systems, Montréal, QC, Canada, 3–8 December 2018; pp. 6572–6583.
26. Kidger, P.; Chen, R.T.; Lyons, T.J. "Hey, that's not an ODE": Faster ODE Adjoints via Seminorms. In Proceedings of the ICML, Virtual, 18–24 July 2021; pp. 5443–5452.
27. Coldren, L.A.; Corzine, S.W.; Mashanovitch, M.L. *Diode Lasers and Photonic Integrated Circuits*; John Wiley & Sons: Hoboken, NJ, USA, 2012; Volume 218.
28. Szczerba, K.; Kocot, C. Behavioral Modeling of VCSELs for High-speed Optical Interconnects. In Proceedings of the Vertical-Cavity Surface-Emitting Lasers XXII. International Society for Optics and Photonics, San Francisco, CA, USA, 31 January–1 February 2018; Volume 10552, p. 1055204.

29. Shivashankar, V.; Kottke, C.; Jungnickel, V.; Freund, R. Investigation of linear and nonlinear pre-equalization of VCSEL. In Proceedings of the Broadband Coverage in Germany; 11. ITG-Symposium, Berlin, Germany, 29–30 March 2017; VDE: Berlin, Germany, 2017; pp. 1–5.
30. Mowlavi, S.; Giannakopoulos, S.; Grabowski, A.; Svensson, L. A Review of IC Drivers for VCSELs in Datacom Applications. *IEEE Trans. Very Large Scale Integr. (VLSI) Syst.* **2023**, *32*, 42–54. [CrossRef]
31. Li, S.; Häger, C.; Garcia, N.; Wymeersch, H. Achievable information rates for nonlinear fiber communication via end-to-end autoencoder learning. In Proceedings of the 2018 European Conference on Optical Communication (ECOC), Rome, Italy, 23–27 September 2018; IEEE: Piscataway, NJ, USA, 2018; pp. 1–3.
32. Karanov, B.; Chagnon, M.; Thouin, F.; Eriksson, T.A.; Bülow, H.; Lavery, D.; Bayvel, P.; Schmalen, L. End-to-end deep learning of optical fiber communications. *J. Light. Technol.* **2018**, *36*, 4843–4855. [CrossRef]

hv photonics

MDPI

Article

Enhancing Tactile Internet Reliability: AI-Driven Resilience in NG-EPON Networks

Andrew Tanny Liem [1,*], I-Shyan Hwang [2,*], Razat Kharga [2] and Chin-Hung Teng [3]

1 Department of Computer Science, Universitas Klabat, Airmadidi 95371, Indonesia
2 Department of Computer Science and Engineering, Yuan Ze University, Taoyuan 32003, Taiwan; s1089107@mail.yzu.edu.tw
3 Department of Information Communication, Yuan Ze University, Taoyuan 32003, Taiwan; chteng@saturn.yzu.edu.tw
* Correspondence: andrew.heriyana@unklab.ac.id (A.T.L.); ishwang@saturn.yzu.edu.tw (I.-S.H.)

Abstract: To guarantee the reliability of Tactile Internet (TI) applications such as telesurgery, which demand extremely high reliability and are experiencing rapid expansion, we propose a novel smart resilience mechanism for Next-Generation Ethernet Passive Optical Networks (NG-EPONs). Our architecture integrates Artificial Intelligence (AI) and Software-Defined Networking (SDN)-Enabled Broadband Access (SEBA) platform to proactively enhance network reliability and performance. By harnessing the AI's capabilities, our system automatically detects and localizes fiber faults, establishing backup communication links using Radio Frequency over Glass (RFoG) to prevent service disruptions. This empowers NG-EPONs to maintain uninterrupted, high-quality network service even in the face of unexpected failures, meeting the stringent Quality-of-Service (QoS) requirements of critical TI applications. Our AI model, rigorously validated through 5-fold cross-validation, boasts an average accuracy of 81.49%, with a precision of 84.33%, recall of 78.18%, and F1-score of 81.00%, demonstrating its robust performance in fault detection and prediction. The AI model triggers immediate corrective actions through the SDN controller. Simulation results confirm the efficacy of our proposed mechanism in terms of delay, system throughputs and packet drop rate, and bandwidth waste, ultimately ensuring the delivery of high-quality network services.

Keywords: TI; smart resilience; NG-EPON; AI; SDN; SEBA; VOLTHA; RFoG; system performance

1. Introduction

The need for digital technology has grown significantly since the COVID-19 pandemic. On the other hand, the development of new technology such as the Internet of Things (IoT), fifth-generation (5G) cellular networks and the boom in Artificial Intelligence (AI) have created a new demand for network connectivity. Attention is now focused on the next-generation near future applications of the Tactile Internet (TI) that need to support the latency-sensitive human-to-machine/robot (H2M/R) applications such as Extended Reality (XR), tele-surgery, industry automation, and intelligent transport systems [1]. Statista has predicted that the number of connected devices will be more than 30.9 billion units by 2025 [2]. This means that there will be significant challenges to the network operators to provide robust and guaranteed services to the users.

One of the future applications that will need ultra-low latency and robustness is the TI. The TI has some similarities with and distinctions from the IoT or 5G. The 5G cellular networks are focused more on improving Human-to-Human (H2H) communications, whereas the IoT is dependent on Machine-to-Machine (M2M) communications to facilitate industrial automation systems or machine-centric activities [3]. However, the TI requires a human-centered design approach due to the inherent Human-in-the-Loop (HITL) nature of H2M/R interaction, such as tele-surgery types of applications [4].

In [5], they determined the QoS key performance indicator of the TI use cases. For example, the tele-operation scenario should have a latency below 1–10 ms and a reliability of 99.999%. In terms of M2M applications, such as self-driving cars and industrial automation, the required latency is 5–10 ms [6] and reliability is 99.999% (an average of less than 6 min downtime per year) [7]. These indicators show that the underlying network not only guarantees minimum latency but also ensures that the system is robust enough to have minimum reliability requirements.

Currently, wired and wireless communications networks are rapidly evolving in terms of their architecture and capabilities to challenge latency-sensitive H2M/R applications. In wired networks, especially optical fiber, Passive Optical Networks (PONs) have continually evolved over the years. PONs now offer bandwidth capacity and functionality, delivering low-latency, high-bandwidth, and cost-efficient services to large numbers of users. Moreover, most urban areas near residential and industrial premises have now deployed optical fiber [8].

Ethernet Passive Optical Network (EPON) technology is among the best PON technologies due to its lower cost, high bandwidth, and readiness to support efficient Quality-of-Services (QoS). The current standard of EPON is the IEEE 802.3ca, which was approved in 2020 as the next-generation EPON (NG-EPON), boosting the bandwidth of a single channel by a factor of 5 to 25 Gbps [9]. Moreover, the NG-EPON can have higher data rates by using channel bonding that can offer aggregated data rates of Nx25s Gbps. Consequently, a fully operating NG-EPON may deliver up to 50 Gbps for both upstream and downstream transmission [10]. Nevertheless, managing an NG-EPON that can handle the strict QoS from residential or industrial users is challenging. Industrial users usually have stringent QoS requirements, one of which is maintenance service [11]. This service includes ensuring that the network is fault-tolerant against any fiber fault. Any fiber cut or loss can significantly impact industrial systems, especially in terms of TI or H2M/R applications, which can involve life-and-death situations.

In general, different types of anomalies can affect the performance of NG-EPON. Some fiber failures can occur due to mechanical faults, optical faults, or electrical faults. Since a single fiber link can connect to the residential, industrial, or enterprise networks, carrying a mixture of data from personal to public or even 911 or TI data, any fiber failures can have enormous impacts and need to be responded to immediately [12]. Moreover, failure in optical network communication can be categorized as soft or minor failures and severe failures. Severe failures lead to immediate service loss due to fiber cuts, bends, and other problems. Minor failures can degrade the transmission quality due to signal overlap, laser deflection, filter switching, noise, and other problems [13]. Therefore, the network operators must ensure reliable data communication for high-speed Internet. Failures to handle this can lead to significant financial and data loss for both network operators and customers. At the same time, the network operators also need to reduce the operation and maintenance expenses (OPEX).

According to the Federal Communication Commission (FCC), more than one-third of fiber disruptions are caused by fiber-cable problems [14]. These issues can include failures of connectors or power supplies, fiber breaks, macro bends, or even problems with Optical Line Terminal (OLT) or Optical Network Unit (ONU) transceiver problems. Consequently, a remote and automatic monitoring or diagnosing mechanism for the fiber links would be very beneficial to reduce the mean time to repair (MTTR), increasing customer satisfaction.

The main contribution of this paper is as follows:

1. We propose a smart resilience mechanism architecture and operations in Next-Generation Ethernet Passive Optical Network (NG-EPON).
2. We introduce a novel Resilience Dynamic Bandwidth Allocation (RDBA) mechanism, ensuring the Quality-of-Services (QoS) of real-time and tactile internet applications.
3. We build a supervised AI model using Multi-Layer Perceptron (MLP) for detecting any anomalies or faults in the branches.

4. The extensive simulation results demonstrate that the resilience of AI-enhanced anomaly and fault detection effectively manages delay for real-time and tactile internet applications.

The remainder of this paper is organized as follows. Related work is presented in Section 2. The SDN-Enabled Broadband Access (SEBA) architecture is discussed in Section 3. Section 4 introduces the proposed smart resilience architecture. Section 5 presents the performance evaluation. Finally, Section 6 concludes our work.

2. Related Work

The objective of fiber monitoring is to detect any anomalies in the optical layer by analyzing the monitoring data. Several techniques are commonly used by engineers to identify fiber faults in Optical Distribution Networks (ODNs). For instance, one study [14] uses a Reference Reflector (RR) placed at the end of each fiber on the ODN and uses Optical Time-Domain Reflectometry (OTDR) to detect, locate, and estimate the reflectance of the connections and mechanical splices in the fiber links. Another approach uses binary-coded Fiber Bragg Granting (FBG) [11]. The FBG binary codes serve as indicators between one ONU to other ONUs by varying the wavelengths used by the FBGs to easily identify fault branches [11]. Some early studies have also proposed embedded OTDR called miniaturized OTDR integrated into the ONUs [15–18].

Furthermore, to consistently meet the Service Level Agreement (SLA), network operators need a mechanism to maintain service continuity even when there are fiber faults in the ODN. In EPON, network operators usually use protection mechanisms such as trunk protection or tree protection. Trunk protection primarily focuses on protecting the OLT and feeder fiber. In contrast, tree protection covers the entire area but is very costly. Dedicated protection might deliver more reliability for service continuity but cannot provide efficient resource utilization [19]. Several studies have shown the use of ring topology to minimize the cost of establishing redundant paths in traditional EPON while handling any fiber cut or failures within the network [20–22]. Apart from all the various approaches such as tree, trunk, star, ring, or bus protection mechanisms, some studies also used hybrid topologies that improve EPON network redundancy but increase the network complexity [23]. Moreover, some studies also use SDN capability and a bus protection line to enhance the resilience of the existing EPON system [24].

Recently, Artificial Intelligence (AI) entities have been able to perform operations analogous to human activities, such as learning and decision-making. AI-based techniques are already changing and improving industries, including telecommunications networks. These techniques range from performance monitoring and guaranteeing the transmission to optical network control and management in both transport and access networks [25]. Current studies related to fiber monitoring already use the Machine Learning (ML) approach to detect any anomaly in the optical networks [12,14,26,27]. These studies have shown that ML can detect and localize any fiber faults in the ODN. Although these studies have already proposed AI monitoring mechanisms, to the best of our knowledge, no studies have focused on integrated resilience that not only intelligently localizes any fiber faults in the ODN but also automatically recovers the network using AI mechanisms. Moreover, most studies only proposed the ML model without any simulation or experiment on the working PON systems. Table 1 presents a table of related work contributions.

Table 1. Related work contributions.

References	Contributions	Approach	Gap Addressed
[11]	Proposes Fiber Bragg Grating (FBG) binary codes for fault branch identification	FBG binary coding, wavelength variation	Lacks AI integration and automatic fault recovery

Table 1. *Cont.*

References	Contributions	Approach	Gap Addressed
[14]	Uses Reference Reflector (RR) and Optical-Time-Domain-Reflectometry (OTDR) for fault detection in ODN	OTDR detection, reflectance estimation	Does not cover AI-enabled resilience mechanisms or automatic fault recovery
[15–18]	Embedded OTDR in Optical Network Units (ONUs) for monitoring fiber faults	Miniaturized OTDR	Focuses on detection but lacks resilience and recovery mechanisms
[19]	Trunk and tree protection mechanisms in EPON networks	Protection for OLT, feeder fiber, and entire area	High cost and inefficient resource utilization
[20–22]	Ring topology for redundancy in EPON networks	Cost-effective redundant path establishment	Lacks integration with AI or dynamic adaptation
[23]	Hybrid topologies to improve EPON redundancy	Mixed network topologies	Increases network complexity
[24]	SDN-based protection lines to enhance EPON resilience	SDN	Does not incorporate AI for intelligent fault recovery
[25]	AI-based techniques for network control and performance monitoring	Learning and decision-making processes for networks	Does not include real-time resilience mechanisms for PON systems
[12,14,26,27]	Machine Learning (ML) techniques for fiber fault detection in ODN	ML-based anomaly detection	No focus on automatic network recovery
Proposed Approach	AI-enabled unified platform (SEBA) for smart resilience in NG-EPON	SEBA platform with SDN, automation, and AI-based fault detection and recovery	Fills the gap by providing automatic fault localization and recovery, ensuring service continuity

To realize this, our proposed architecture uses AI-enabled unified platforms to automate and adapt to changing circumstances and business needs. As Cisco's 2024 Global Networking Trends Report stated, in the next two years, network operators will use AI-enabled unified platforms to automate and adapt to changing circumstances and business needs [28]. SDN-Enabled Broadband Access (SEBA) is a unified cloud-native platform, providing scalable and flexible network management. SEBA is based on Software-Defined-Networking (SDN) principles, offering simpler, flexible, and easily customizable networks. Moreover, SEBA promotes interoperability between OLTs and ONUs from different manufacturers. SEBA is open-source, giving operators unprecedented flexibility in customizing SEBA for their access network, integrating it with the rest of their backend systems, implementing only the features they require, adding application programming interfaces (APIs) themselves, and not being bound by the timelines and prices of a traditional vendor [29].

Network Fault Detection and Localization

Commonly, to detect anomalies in the ODN, engineers are using OTDR, which is a technique based on the Rayleigh backscattering [12]. The concept is like a radar, so the OTDR will send a series of optical pulses into the ODN. Afterward, the backscattered signals will be recorded as a function of time that can be translated to the position of the optical fiber components such as the splitter, ONUs, and end connectors. This information is used for event analysis. Figure 1 illustrates the example of OTDR trace.

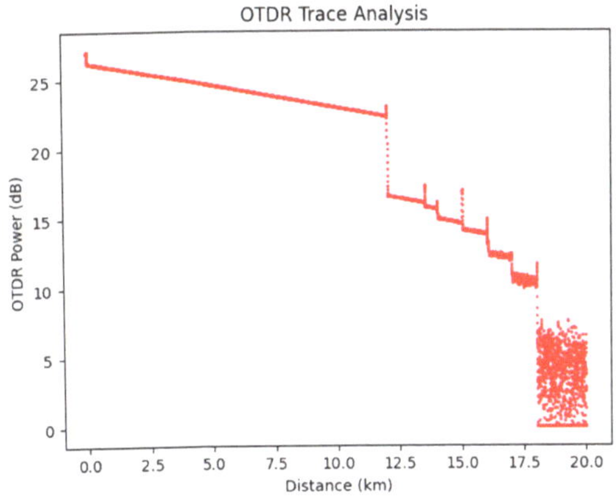

Figure 1. Example of OTDR traces.

As shown in Figure 1, we can see that the initial drop at the beginning of the figure represents the launch condition level of around 25 dB. Afterward, the downward-sloped line indicates the attenuation of the fiber (feeder fiber). At the end of the linear attenuation, a small peak signifies the splitter, connectors, ONUs, or other reflective events. The dense scattering at the end marks the termination of the fiber.

OTDR traces are usually difficult to interpret even for experienced engineers due to the noise that affects the signals. Analyzing these traces may be very challenging using conventional methods, especially to distinguish subscribers unambiguously [30]. It can be very time-consuming, since the engineer needs to remove the noise manually, which can increase the MTTR and reduce the detection and localization accuracy. One of the strategies to effectively manage and interpret the OTDR traces is for network operators to use baseline measurements, saving the measurements when the network is functioning normally. In this way, network operators create a reference point for future comparisons if faults occur in the ODN. Moreover, maintaining and organizing a database of reference points for all OTDR traces can help with quick retrieval and analysis during troubleshooting. Additionally, network operators must ensure that the network engineers are well-trained in interpreting OTDR traces and using the tools by conducting regular training sessions to stay updated. All these combined techniques still depend on the network engineers.

Furthermore, before any fault occurs in the ODN, some anomalies can also appear in the network condition. Network operators can use various visualization tools such as a Bit Error Rate (BER) analyzer, Optical Time Domain Visualizer, and Optical Spectrum Analyzer. These tools can show the performance of optical signal delivery. An eye diagram is used to measure the signal quality. Ideally, an eye diagram would consist of two parallel lines with instantaneous rise and fall times, making them virtually invisible. The eye diagram can show vital parameters such as timing jitter and inter-symbol interference [31]. Combining both OTDR trace analysis and the eye diagram can improve the early detection of faults in the ODN.

Consequently, in this paper, we propose automatic detection and localization using an ML algorithm by incorporating OTDR trace analysis data and eye diagram analyzer data. By incorporating ML algorithms, we can improve the accuracy and efficiency of detecting and localizing fiber faults. ML can process vast amounts of data, identifying patterns and detecting anomalies much faster with greater precision than network engineers. By leveraging ML, network operators can reduce their reliance on network engineers for fault detection and localization, leading to quicker resolutions and increased network reliability (as illustrated in the proposed Smart Resilience Architecture in "Figure 4").

3. SEBA Architecture

This section discusses the concept of SEBA architecture, shown in Figure 2, which consists of Virtual OLT Hardware Abstraction (VOLTHA), a Network Edge Mediator (NEM), and an SDN Controller.

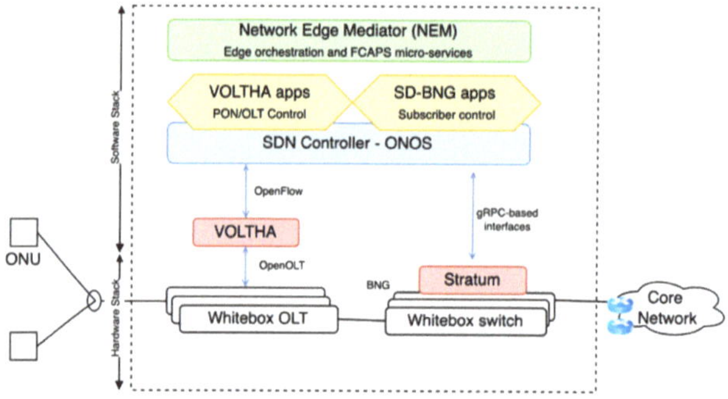

Figure 2. Generic SEBA architecture.

3.1. Virtual OLT Hardware Abstraction (VOLTHA)

In the central office, the white-box OLT will be used, incorporating Virtual OLT Hardware Abstraction (VOLTHA), allowing the Virtual OLT (vOLT) to be managed by the Software-Defined Networking (SDN) controller, i.e., Open Network Operating System (ONOS). The vOLT can have NetConf, OpenFlow Agent, OLT Application Programming Interface (API), and ONU Management and Control Interface (OMCI) stack-connected to the ONOS SDN controller. VOLTHA hides PON-level details from the SDN controller and abstracts each PON as a pseudo-Ethernet switch easily programmed by the SDN controller.

Figure 3 illustrates the operation architecture of VOLTHA. The process begins with VOLTHA activating the OLT, which has Network-to-Network Interface (NNI) ports on one side and PON ports on the other. Once activated, VOLTHA abstracts the OLT, including the connected NNI port, which is connected to a logical switch within VOLTHA. VOLTHA then informs ONOS of the existence of this logical switch. The ONUs are attached to the OLT through its PON ports, and the User Network Interface (UNI) port at the ONU is also added to the logical switch within VOLTHA. As an example, the Edgecore ASFvOLT16 White box OLT, which is used in industrial settings, supports [32]. Other vendors, such as CIG, Tellabs, and Iskratel offer similar OLT solutions.

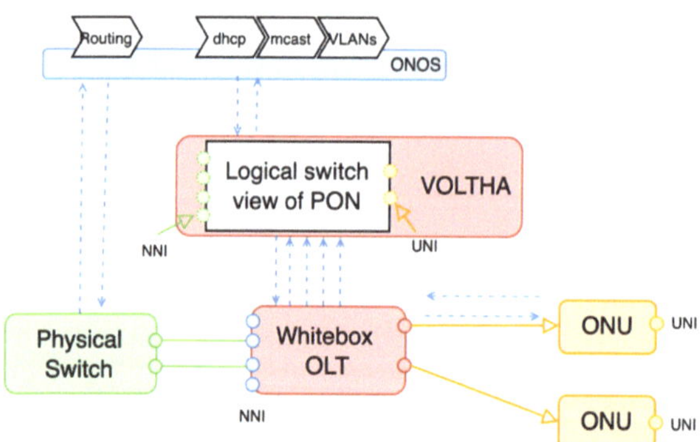

Figure 3. VOLTHA operation architecture.

3.2. Network Edge Mediator

The Network Edge Mediator (NEM) acts as the mediation layer between the access system and the service provider's back end, providing centralized management and orchestration capabilities. The NEM supports essential functionalities known as fault, configuration, accounting, performance, and security (FCAPS): (1) Fault management: detecting and resolving network faults; (2) Configuration: managing network element settings; (3) Accounting: collecting usage data for billing and auditing; (4) Performance: monitoring and analyzing network metrics to ensure service quality; and (5) Security: enforcing policies to protect network integrity. This ensures that the NEM facilitates seamless network control, configuration backup, and restoration [33].

4. Proposed Architecture

This section discusses the proposed smart resilience architecture that not only can detect and localize fiber faults but also automatically establish connections while waiting for the engineer to fix the fiber faults in the ODN. In this architecture, we use the SEBA for Residential Services Central Office Rearchitected as Datacenter (R-CORD) platform concept, which sits in the middle and provides management and abstraction solutions, enabling the use of white box hardware. White box hardware reduces both Capital expenditures (CAPEX) and Operation & Maintenance expenses (OPEX). In this way, we separate the software from the hardware, enhancing the agility that brings the best of the cloud Network Function Virtualization (NFV) and SDN together. The OLT and ONUs used in the proposed architecture are white box hardware, providing a highly flexible and cost-effective solution. The white box devices feature hardware platforms that can run third-party software, such as VOLTHA, which offers open programmability and interoperability.

Figure 4 shows the smart resilience architecture in NG-EPON. In the north part, the OLT is connected to VOLTHA, an SDN controller such as the ONOS and NEM. These components incorporate one another using APIs and Remote Procedure Calls (gRPC) to provide seamless communication between VOLTHA, SDN controllers, and the NEM. As already mentioned, VOLTHA will activate the OLT and add to its logical switch. Moreover, the ONUs will also be added to the logical switch by VOLTHA. The SDN controllers provide centralized control and management for dynamic traffic steering, automatic failover, and real-time network adjustment. The OTDR, located at the central office, detects and localizes fiber faults, while the BER analyzer at the business users' side captures eye diagrams to detect anomalies. Furthermore, in the south part, the users are categorized into two different groups: business users and residential users. Usually, business users have very strict SLAs and requirements. Therefore, as shown in the figure, business users such as ONU_1 and ONU_2 have the resilience area (indicated by the red dashed circle) which will be covered with the Radio Frequency over Glass (RFoG). The RFoG serves as a critical backup mechanism for business users in the event of a fiber fault. The RFoG allows RF signals to be transmitted over fiber optic cables, maintaining compatibility while providing the benefits of fiber optics, such as higher bandwidth and lower latency. In the proposed architecture, RFoG is activated as a secondary communication path when the primary fiber link experiences a fault or anomaly. The failover process is handled automatically by the ONU and the SDN controller, ensuring that the RFoG backup link is ready to carry traffic when needed. This mechanism will maintain continuous service, minimize downtime, and enhance overall network resilience.

In normal conditions, ONUs send/receive data using the primary optical path ($\lambda1$, $\lambda2$). The SDN controller monitors the network performance such as Bit Error Rate (BER) and OTDR trace analysis. Network operators oversee the network using a centralized platform, i.e., the NEM, which provides dashboards, alerts, and reports for network operators. In our proposed architecture, edge computing is realized in the NEM. This integration edge computing is to receive incoming data in real-time, identify any potential issues, and perform real-time analysis and alerts. Edge computing within the NEM can be implemented using high-performance servers equipped with GPUs for accelerated AI processing. Typically,

Kafka is used to stream the collected data (telemetry data) from the NEM to the edge computing device. One study [34] has shown that a Kafka-based framework is highly scalable and can support up to around 4000 messages per second with low CPU load and achieve an end-to-end latency of about 50 ms. The AI model deployed at the edge can detect anomalies in the network, predicting a variety of faults such as fiber cuts, partial fiber degradation, fiber bending, and faulty splitters. When anomalies are detected, the NEM communicates with the SDN controller to take corrective actions based on the AI predictions, such as activating backup conditions.

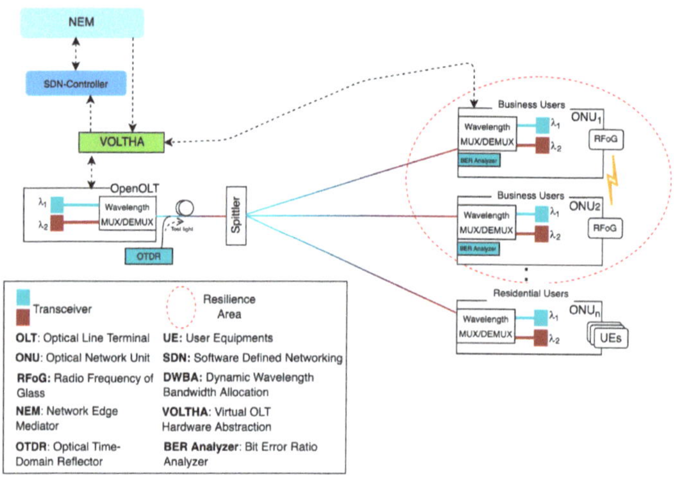

Figure 4. The smart resilience architecture in NG-EPON.

When faults or degradations occur in the ODN, including fiber cuts, fiber bending, and faulty splitters, the AI model identifies these anomalies and initiates a backup-mode plan. The OLT and ONU are notified via the NEM, and the OTDR is used to localize the fault within the network of the branches. When the ONU activates the backup mode, the RFoG becomes activated and ready to send the data to the nearest ONU (backup ONU) within its coverage. Simultaneously, the SDN controller updates the network configuration to handle the failover scenario. For instance, if partial fiber degradation is detected, the SDN controller may initially attempt to reroute traffic within the primary path. In the event of a complete fiber cut to ONU_1, the ONU_1 and SDN controller trigger the RFoG backup mechanism, routing data through ONU_2. This multi-layered approach ensures robustness against various types of failures.

Since there is no direct connection link between the affected ONU and OLT, a mechanism must be used so that the nearest ONU (backup ONU) can differentiate the incoming data from the OLT and send it to the affected ONU via RFoG. Similarly, the OLT needs to know that the data comes from the affected ONU. This can be achieved using data tagging such as a virtual local area network (VLAN).

In the proposed architecture, the VLAN tag table is established in the OLT and ONUs. This table can be changed over time and updated using the SDN controller, which dynamically updates the VLAN tag table and configurations based on network changes and faults. This ensures that the OLT and ONUs will map VLAN tags to their respective destinations. Table 2 shows an example of the VLAN tag table.

Table 2. VLAN Tag Example.

VLAN Tag	Source/Destination	Handling Instructions
100	Downstream to ONU_1	Forward to ONU_1 via RFoG
200	Downstream to ONU_2	Process locally (ONU_2 data)
101	Upstream from ONU_1	Forward to OLT via RFoG

4.1. Intelligence Fault Detection and Localization with Intelligence Diagnosis

As mentioned before, this paper focuses on fault detection and localization through OTDR trace analysis and the eye diagram evaluation. Figure 5 illustrates the comparison between normal and fault conditions from these perspectives. Figure 5a shows a clear eye opening, indicating minimal noise, jitter, and distortion. In contrast, Figure 5b depicts a situation with anomalies. When there are anomalies in noise, jitter, or distortion, the eye diagram shows that the eye opening is reduced vertically and horizontally, distorting the eye shape, which indicates a very high level of noise, higher jitter, and potential issues with the transmission channel. Figure 5c shows the power attenuation for different ONUs located at different distances in a normal trace event, while Figure 5d highlights the scenario where ONU_1 experiences a fiber fault. The OTDR trace results for ONU_1 show a loss, with no peak detected, indicating the presence of a fault. Typically, both the OTDR trace analysis and eye diagram are tested against a predefined mask. Any violation of these masks can indicate potential fiber faults within the ODN.

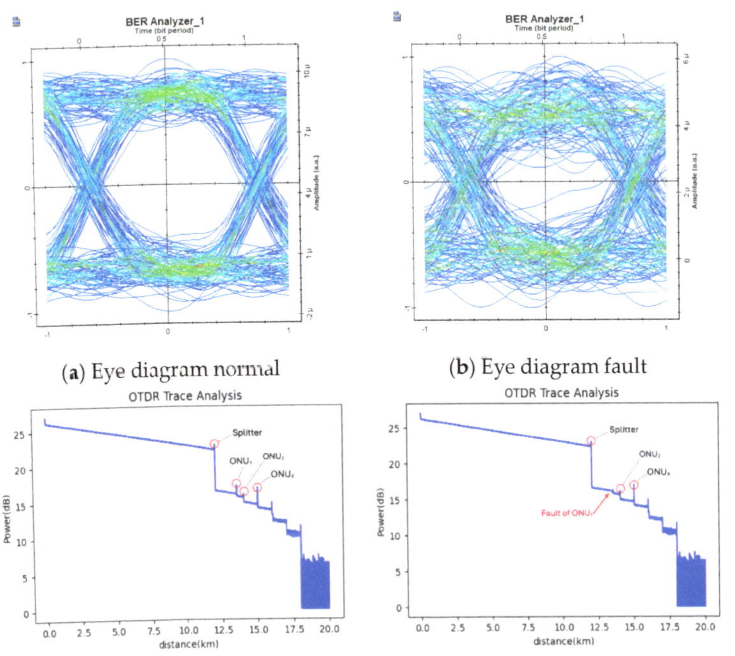

(**a**) Eye diagram normal (**b**) Eye diagram fault

(**c**) Normal trace (**d**) Fault trace

Figure 5. Comparison of normal vs. fault condition.

Consequently, in our proposed fault detection and localization approach, we use eye diagrams to complement the OTDR in identifying subtle degradation in signal quality, since OTDR alone only detects severe faults such as fiber cuts. The proposed ML model uses this combination of eye diagram and OTDR data to enhance the accuracy of prediction and localization. This leads to improved accuracy and efficiency, especially in identifying minor or soft faults that would not be captured by OTDR alone.

The proposed framework for fault detection and localization with intelligent diagnosis is shown in Figure 6, following the study in [12]. There are five main stages to realize the proposed framework, namely, (1) Data collection: The deployed ODN infrastructure is periodically monitored using OTDR and BER Analyzer. The generated OTDR traces and the eye diagram data are sent to the SDN controller; (2) Data processing: The collected data is pre-processed to normalize and standardize the features to a similar scale; (3) Anomaly detection: The processed data are compiled into a dataset, which is then used to train and evaluate a machine learning model designed to detect anomalies in the network; (4) Fiber fault diagnosis and localization ML model; (5) Mitigation and recovery from fiber failures plan: The plan will be formulated to address and fix the detected faults. Alerts are generated to notify engineers and customers of the issues. The SDN controller facilitates dynamic management and control of the network based on the machine learning model outputs.

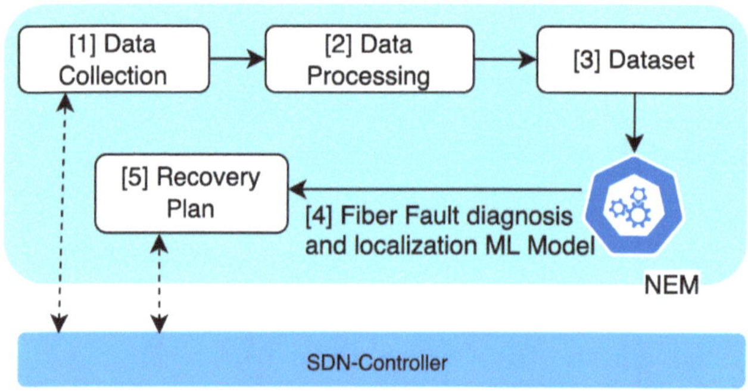

Figure 6. Proposed framework for fault detection and localization.

4.2. Simulation-Based Evaluation

To validate the proposed approach, the simulation-based evaluation setup was built using OptiSystem 21.0 software. OptiSystem is an innovative, rapidly evolving, and powerful software design tool that enables users to plan, test, and simulate almost every type of optical link in the transmission layer of a broad spectrum of optical networks, from LAN, SAN, and MAN to ultra-long-haul. It offers transmission layer optical communication system design and planning from component to system level and visually presents analysis and scenarios [35]. The setup comprises an OLT connected to the 8 ONUs with a passive splitter. The distance between the OLT and ONUs ranges from 15 to 20 km, with a feeder fiber length of 15 km and branch lengths varying from 2 km to 7 km. The optical transmitter frequency is set to 1550 nm with a power of 7 dBm, using NRZ modulation. The attenuation loss is 0.2 dB/km, and the splitter loss varies from 4 dB to 8 dB. Two scenarios were simulated: normal and faulty scenarios. For the faulty scenarios, different anomalies were introduced, including macro-bending, micro-bending, fiber cut, and bad splitter. The simulation generated 709.054 samples. The data set was constructed, normalized, and divided into a training (60%), a validation (20%), and a test set (20%) for fault and normal scenario eye diagrams using OTDR traces, obtained from [12]. It is worth mentioning that the eye diagrams were used for anomaly detection, while the OTDR was used to localize the fault. A BER analyzer was placed at the end of each branch to capture the eye diagrams. The dataset is balanced, with an approximately equal number of samples representing normal and faulty conditions. To mimic anomalies (such as fiber bending, bad splitter) and fiber faults, attenuators were placed at the 2 km, 3 km, 5 km, and 7 km branches, respectively. The termination at the end of the 7 km branch was removed to simulate a fiber fault. Figure 7 shows a simulation-based evaluation setup for generating faulty branch data using OptiSystem in the passive optical network. Furthermore, the normal samples are derived from the simulation-based evaluation setup conducted without any attenuator.

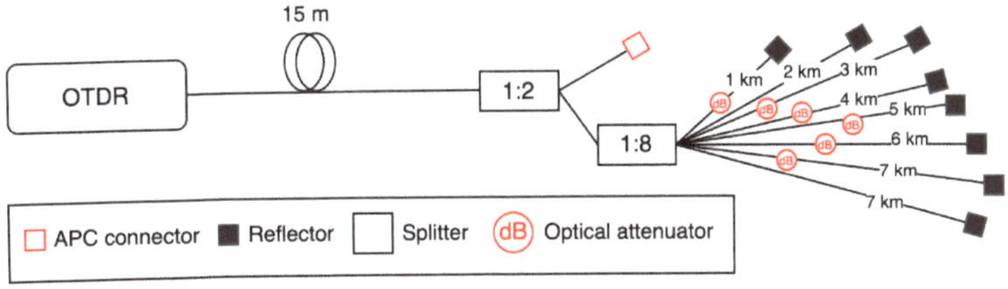

Figure 7. Simulation-based evaluation setup for generating faulty branch data using Optisystem.

4.3. Neural Network Architecture and Model Evaluation

We started by preprocessing the data, applying a standard scaler to normalize the features, and guaranteeing that all features are on a similar scale to enhance the model's performance. We then implemented a Multi-Layer Perceptron (MLP) neural network due to its simpler architecture, which requires less computational power compared with other machine learning algorithms, making it ideal for high-speed network environments.

As shown in Figure 8, our MLP model has an input layer, followed by three hidden layers. The input layer has two neurons (for time and amplitude/reflection) (indicated by blue), while the hidden layers have 8, 16, and 8 neurons, (indicated by green, red, and green), respectively. All layers use the ReLU activation function, except the output layer (indicated by blue), which has a single neuron and uses the sigmoid activation function for binary classification (e.g., fault or no-fault). In total, this model has 313 trainable parameters.

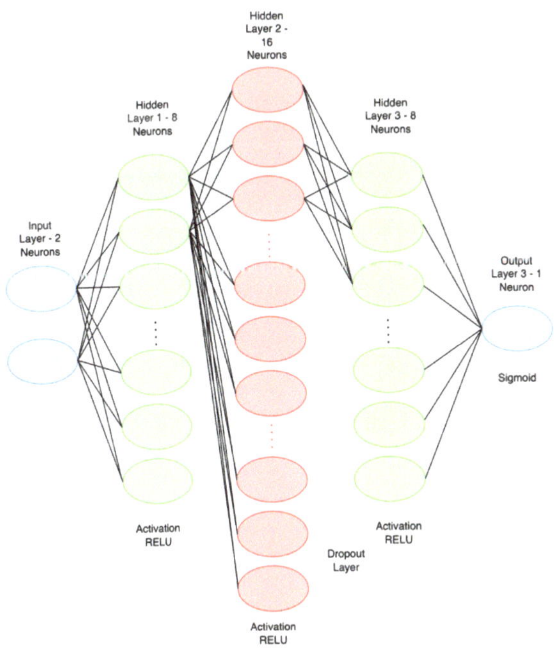

Figure 8. The proposed MLP model for fiber fault diagnosis and localization.

To assess the model's performance and robustness, we utilized stratified K-fold cross-validation, where each fold maintains the same class distribution as the original dataset. The training was conducted over 40 epochs with a batch size of 256, using 20% of the training data as a validation set to monitor for overfitting. Performance metrics such as accuracy, precision, recall, and F1-score were calculated for each fold. After completing all folds, we computed the average of these metrics to summarize the model's overall performance on unseen data. The model achieved an average accuracy of 0.8149, precision of 0.8433, average recall of 0.7818, and average F1-score of 0.8100. These results indicate that the model performs robustly in distinguishing between "Normal" and "Fault" classes. The high average precision suggests effective minimization of false positives, meaning the model reliably identifies true positives when making positive predictions. However, the slightly lower recall indicates that some fault instances may be missed, resulting in false negatives. The balanced average F1-score reflects a good trade-off between precision and recall, making the model suitable for applications where both types of errors are of concern.

4.4. Resilience Dynamic Bandwidth Allocation

The Resilience Dynamic Bandwidth Allocation (RDBA) uses an offline scheduler approach, where the OLT waits for report messages from all ONUs before performing dynamic bandwidth allocation (DBA). In this way, the OLT will have a holistic view of all ONU demands, ensuring fairness [36]. In the normal condition where no fault is detected, the OLT will assign the bandwidth allocation to ONUs based on the following Formula (1):

$$B_{available} = \frac{R_N(T_{max} - N{\cdot}G)}{N.512} \tag{1}$$

where R_N is the EPON line rate (in bits per second), T_{max} is maximum cycle time (in milliseconds), N is the total number of ONUs, G is the guard time, and 512 bits is the control message length. The minimum guaranteed bandwidth (B_{min}) of the ONU is calculated with the following Formula (2):

$$B_{min} = \frac{W_{max} \cdot W_{report}}{T_{max}}, \tag{2}$$

where W_{max} is the maximum timeslot of an ONU, W_{report} is the reserved window size of the report message (in bits). We limit each ONU timeslot to prevent upstream channel monopolization by heavily loaded ONUs. However, the W_{max} can also be set according to the SLA.

When the proposed ML identifies faults or anomalies in the ODN by analyzing data from OTDR traces and the BER analyzer, the NEM will inform the OLT using the SDN controller to switch from normal DBA to RDBA. Once the RDBA is activated, the OLT dynamically adjusts the bandwidth allocation to prioritize the backup ONU, ensuring it can handle the data from both its traffic and the affected ONU (i.e., the faulty ONU). The backup ONU receives additional bandwidth, scaled based on predefined factors, to maintain service continuity for both ONUs. This process ensures minimal service disruption even during fault conditions, as the RFoG link facilitates the rerouting of traffic from the affected ONU to the backup ONU.

Figure 9 shows the pseudocode of the proposed RDBA. In the normal condition, the OLT calculates the available bandwidth ($B_{available}$) and the guaranteed bandwidth (B_{min}) in each cycle. Under normal conditions in each cycle, the ONU gets the guaranteed bandwidth. If the guaranteed bandwidth (B_{min}) is greater than the reported bandwidth from the queue, the granted bandwidth (*GRANT_ONUi*) is set to the queue's requested bandwidth. Otherwise, the granted bandwidth is set to the remaining B_{min}. The remaining B_{min} is then updated by subtracting the granted bandwidth. In the restoration plan, when a fault occurs, the OLT will adjust for the backup ONU. If the current ONU is a backup ONU, the OLT sets the protection VLAN tag for the affected ONU. The B_{min} is then calculated for the backup ONU, but it will be multiplied by alpha (α). Here, α represents the additional bandwidth allocated to the backup ONU to ensure it can handle the increased traffic, as the affected ONU now routes all data through the backup ONU via RFoG. If the current ONU is not a backup ONU, the normal condition function is applied. Moreover, to verify that the total requested bandwidth from ONUs does not exceed $B_{available}$ due to the addition of variable α, the total_requested_bandwidth is calculated as follows (3):

$$\sum_{i \in activeONU} GRANT\_ONU_i + \sum_{i \in backupONU} GRANT\_ONU_i. \quad (3)$$

```
/* Pseudocode of RDBA
Calculate the B_available and B_min in each cycle */
Calculate total_requested_bandwidth

/* Normal Condition */
In each Cycle for every ONUi do {
  for each queue[j] do {
        GRANT_ONUi = min(REPORT_ONUi, B_min)
   B_min = B_min - GRANT_ONUi
  }
}

/* Restoration Plan */
In each Cycle for every ONUi do {
  if (ONUi == backup_ONUi) {
    Set protection VLAN tag of affected ONUi
    for each queue[j] do {
      GRANT_ONUi = min(REPORT_ONUi, α·B_min)
        B_min = B_min - GRANT_ONUi
    }
  }
  else {
                GRANT_ONUi = min(REPORT_ONUi, B_min)
        B_min = B_min - GRANT_ONUi
    }
 /* Proportional Adjustment */
  if (total_requested_bandwidth > B_available)
      Scaling_Factor = B_available / total requested bandwidth
      GRANT_ONUi = GRANT_ONUi · Scaling_Factor
  }
```

Figure 9. The pseudocode of RDBA.

5. Performance Evaluation

To validate the proposed model, we implemented the NG-EPON architecture using the OPNET simulator. All key components and protocols of NG-EPON, such as dynamic bandwidth allocation, cycle time, transmission capacity, guard time, etc., are fully modeled. The proposed system model consists of 32 ONUs and one OLT. The downstream and upstream channels between the OLT and ONU are configured to 1 Gbps. The distance from the OLT to the ONUs is uniformly distributed over 10 to 20 km. To generate Assured-Forwarding (AF), Best-Effort (BE), and Tactile-Internet (TI) traffic, we employ self-similarity and long-range dependence, generating highly bursty traffic with a Hurst parameter of 0.7 [17]. The packet size is uniformly distributed between 512 and 12,144 bits. The Expedited Forwarding (EF) traffic is modeled using a T1 circuit-emulated line with a constant frame rate (1 frame/125 μs) and a fixed packet size of 560 bits, which occupies approximately 14% of the total upstream bandwidth. The remaining traffic is distributed as 50% AF, 20% BE, and 30% TI for scenario I, and 40% AF, 20% BE, and 40% TI for scenario II. To evaluate the proposed mechanism, we construct different scenarios: (1) no-fault, (2) one fault, and (3) three faults.

The focus of the simulation is to evaluate the system's performance after faults are detected. Fault scenarios with one fault and three faults were introduced, and the system's performance was measured in terms of key metrics, such as mean packet delay, system throughput, packet drop rate, and bandwidth waste. These measurements help validate the resilience of the architecture in ensuring performance guarantees, particularly in terms of low-latency requirements for real-time traffic such as Tactile Internet (TI). While the optical network's physical characteristics (e.g., power levels, impairments) were not the focus of this simulation, the system response to fault scenarios was crucial in demonstrating the architecture's ability to maintain service continuity and minimize disruption. To further validate the system, we compared the performance of the proposed RDBA mechanism against a traditional DBA approach, which does not incorporate fault-tolerant features. In the baseline DBA approach, bandwidth is allocated without any resilience mechanisms to manage fault scenarios. The simulation parameters are summarized in Table 3.

Table 3. Simulation parameters.

Parameters	Value
Number of ONUs in the System	32
Upstream/downstream link capacity	1 Gbps
OLT-ONU distance (uniform)	10–20 km
Maximum transmission cycle time	1 ms
Guard time	1 μs
DBA computation time	10 μs
Control message length	0.512 μs
Number of Faults	1, 3 Faults
Traffic Proportion of Expedited Forwarding (EF)	14% of link capacity
Traffic Proportion of AF, BE, and TI Scenarios	(50%:20%:30%)/(40%:20%:40%)

5.1. Mean Packet Delay

Figure 10 shows the mean packet delay of Expedited Forwarding (EF), Assured Forwarding (AF), and Tactile Internet (TI) with different traffic proportions. Four scenarios are depicted: Normal: delay of no faults in the network (blue line); 1Fault_Average: delay with one fault in the network, which represents a single fault occurring in one branch of the ODN; 3Fault_Average: delay with three faults in the network, representing multiple faults distributed across different branches of the ODN; 1Fault_BackupNode: delay at a specific backup node handling the affected ONU with one fault; and 3Fault_BackupNode: delay at specific backup nodes handling the affected ONUs with three faults.

As seen in Figure 10a, the EF delay under normal conditions increases gradually with the traffic load, showing an expected behavior where higher traffic leads to higher delay. However, in the 1Fault_Average and 3Fault_Average scenarios, when the traffic loads are below 70%, the delay remains close to the normal operation but increases more significantly as the traffic load exceeds 70%. This highlights the compounded effect of multiple faults on the network performance. The green lines (1Fault_BackupNode and 3Fault_BackupNode) show that the EF delay at specific backup nodes handling the affected ONUs is slightly higher than the normal operation but much lower than the 1Fault_Average line, demonstrating the effectiveness of the backup node in mitigating the impact of the faults on the affected ONUs.

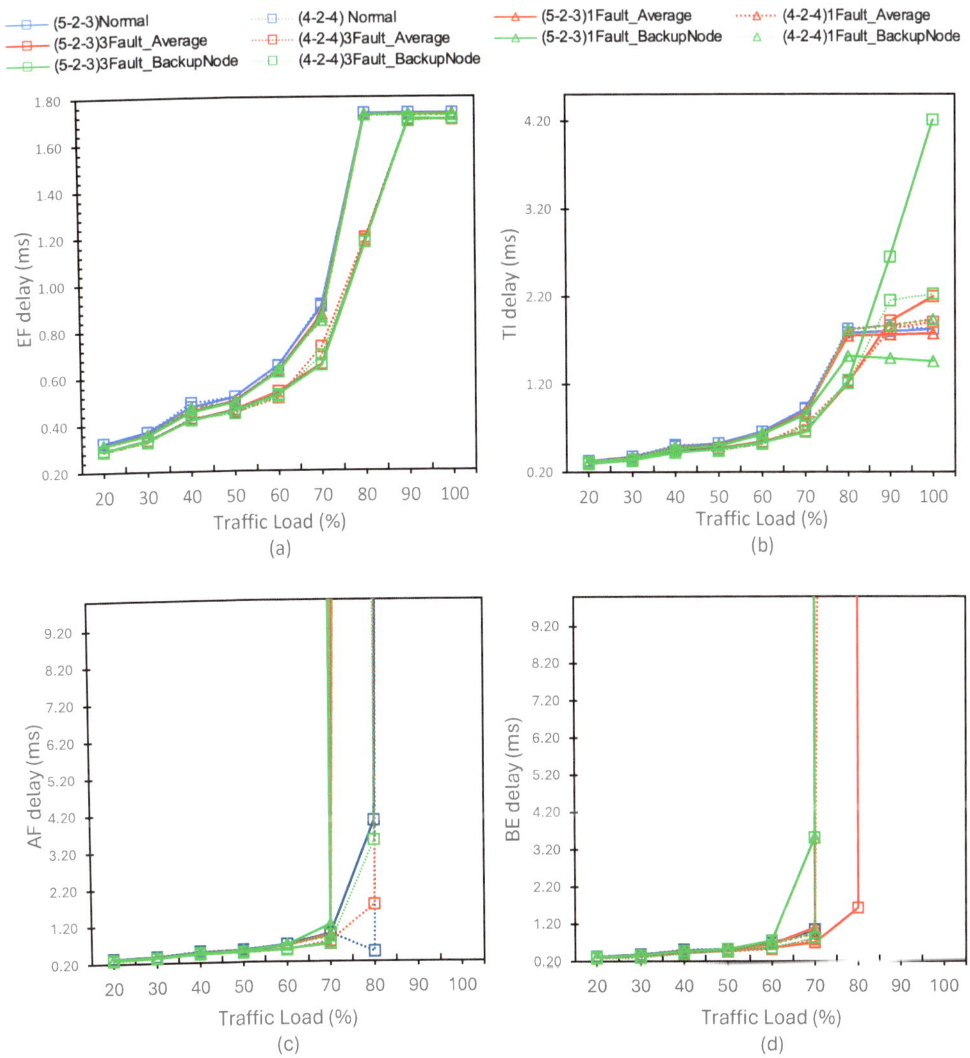

Figure 10. Mean packet delay of EF, TI, AF and TI traffic.

In terms of TI delay, shown in Figure 10b, when there is one fault in the network, the 1Fault_BackupNode and 3Fault_BackupNode manage to stay close to the normal operation levels, even at higher traffic loads. This again demonstrates the effectiveness of the backup node in mitigating the impact of the fault, ensuring that TI delay remains well below 2 ms up to 90% and slightly exceeds 2 ms at 100% load. In the 3Fault_BackupNode scenario, the delay remains relatively low at moderate traffic loads but spikes dramatically beyond 80% load, reaching up to 5 ms at 100% load. This indicates that while backup nodes help manage the delay better than without them, multiple faults still pose a significant challenge, especially under high-traffic conditions.

Figure 10c,d illustrate the AF and BE delay, respectively. AF delay, much like EF, shows a minimal increase with rising traffic loads in the normal scenario. BE traffic, typically given the lowest priority, shows non-congested conditions under normal conditions. However, as the traffic load increases, the limited available resources are allocated preferentially to higher-priority traffic; therefore, once the traffic load surpasses 70%, the resources available for AF and especially BE packets become increasingly constrained. When faults are present, resources are redistributed to maintain service levels for critical applications, exacerbating delays for AF and BE traffic.

Consequently, the proposed RDBA mechanism successfully ensures that delays for EF and TI packets remain below critical thresholds, i.e., below 2 ms [4,37], maintaining high QoS for real-time and tactile internet applications. The results show that under normal and fault conditions, the RDBA can keep the delays well managed. The RDBA prioritizes higher-priority traffic, which can lead to increased delays for AF and BE packets under fault conditions. The simulation results highlight the importance of having a robust DBA mechanism that incorporates resilient AI-enhanced fault detection and recovery to effectively manage delay, particularly for high-priority traffic such as EF and TI packets.

5.2. System Throughput

Figure 11 depicts the system throughput under normal and fault conditions. The system throughput of the network demonstrates a consistent increase as the traffic load rises, indicating the network's robust capacity to handle escalating demands. This pattern shows an efficient RDBA that successfully adapts to increasing traffic demands. Moreover, in fault conditions (1Fault and 3Fault Averages), there is an observed increase in throughput efficiency compared with normal conditions. This is because the overhead communication required for inactive or faulty ONUs decreases, allowing more bandwidth to be allocated to active connections, thus improving the overall efficiency of the NG-EPON systems.

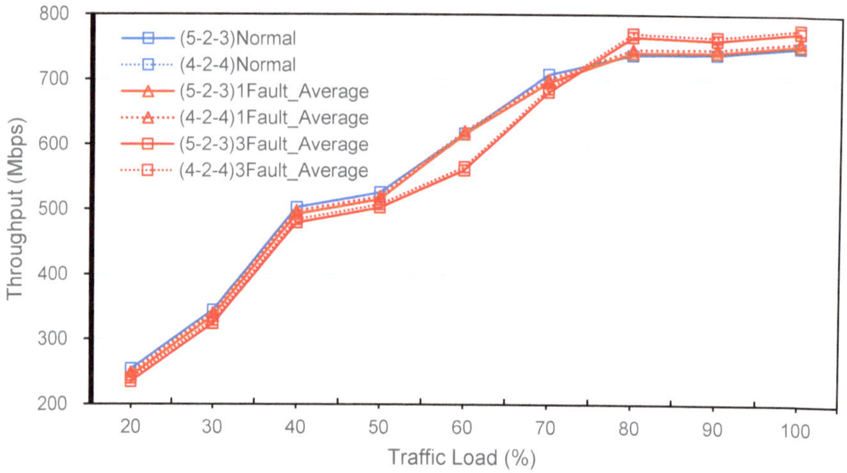

Figure 11. System throughput.

5.3. Packet Drop Rate

The packet drop rate shown in Figure 12 shows that the drop rates remain minimal at up to 70% traffic load across all scenarios, indicating healthy network functionality under moderate loads. However, as the load exceeds 80%, packet drop rates begin to rise, especially under conditions of three faults. The packet losses occur predominantly in AF and BE traffic categories, while EF and TI packets, given the highest priority in the network, experience no packet drop. This differentiation in packet treatment highlights the network's strategic prioritization, ensuring that critical real-time applications dependent on EF and TI traffic maintain uninterrupted service even as the system approaches or reaches full capacity.

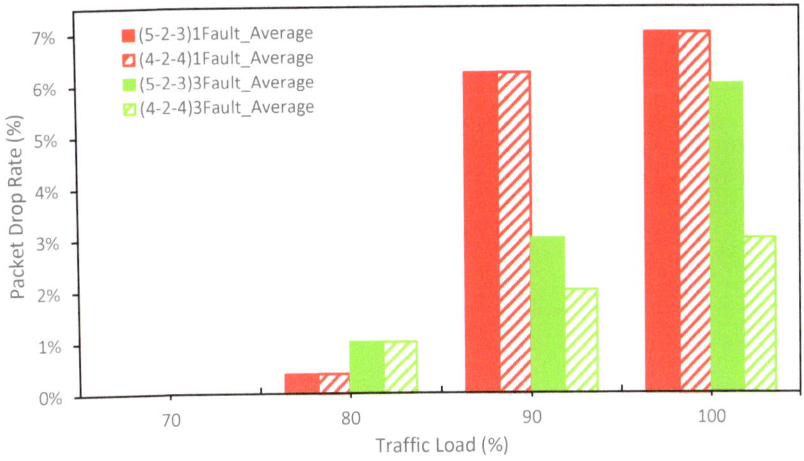

Figure 12. Packet drop rate.

5.4. *Bandwidth Waste*

Figure 13 showcases the trend of decreasing bandwidth waste as the traffic load increases with various scenarios including normal conditions and faults. At lower traffic loads, the bandwidth waste tends to be a surplus of allocated but unused bandwidth, leading to higher waste. As the traffic load increases, the demand for bandwidth rises, making the RDBA allocate nearly all available bandwidth to meet this demand, thereby minimizing waste. Thus, the RDBA demonstrates a robust capability to optimize resource management, particularly crucial when the network load reaches full capacity.

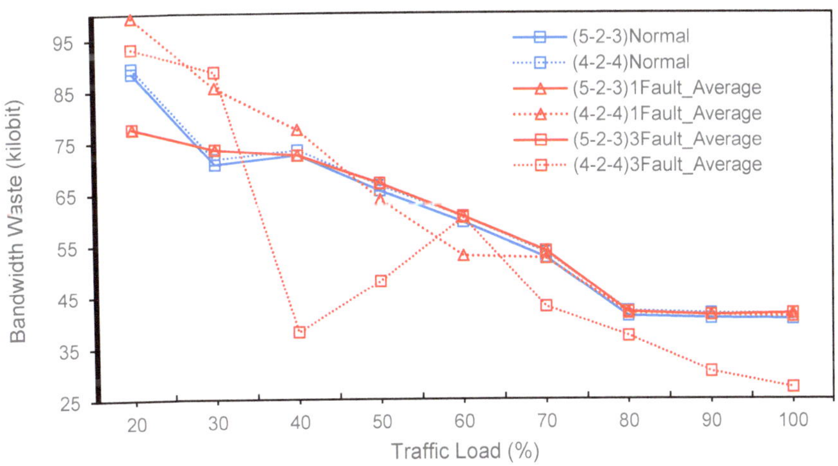

Figure 13. Bandwidth waste.

The results from the comparison show that the RDBA mechanism outperforms the baseline DBA, particularly under fault conditions. While the baseline DBA experiences significant delays in high-priority traffic (EF and TI) during fault scenarios, the RDBA mechanism mitigates these delays using backup nodes, ensuring that critical traffic maintains low-latency performance even when multiple faults are present. While the RDBA performs better in handling faults and maintaining service continuity, it introduces some complexity in terms of system management and leads to higher delays for low-priority traffic (AF and BE), particularly under high-load conditions. This trade-off highlights the need for balancing fault tolerance and resource management in heavily loaded networks.

6. Conclusions

In this paper, we propose a Smart Resilience in an NG-EPON AI-Enhanced fault tolerance system, trained on an NG-EPON topology to detect and localize faulty branch anomalies. The topology used for simulations consists of 32 ONUs connected to a central OLT, with fiber distances ranging from 10 to 20 km, representing a typical NG-EPON deployment. Faulty branch anomalies are detected using a combination of OTDR trace analysis at the central office and BER analysis at the ONUs, and the AI model identifies faults within the network. The proposed architecture and RDBA mechanisms perform effectively under different scenarios, including normal, one fault, and three faults. We validated the performance of the proposed method using simulation-based evaluation data derived from NG-EPON systems and the OPNET simulator. Our AI model is based on a neural network with three hidden layers, trained using datasets generated from OTDR traces and eye diagrams. Our simulations demonstrate that the proposed architecture and mechanism can maintain the system's performance even in the presence of faults. In future work, we aim to enhance our AI model's capability to operate in a more complex and autonomous environment, improving its ability to adapt dynamically to real-world, large-scale NG-EPON topologies. Furthermore, we will include a more detailed simulation-based evaluation scenario to fully quantify the benefits and limitations of our model in real-time fault scenarios.

Author Contributions: Conceptualization and methodology, A.T.L., R.K. and C.-H.T.; formal analysis A.T.L. and R.K.; validation and editing A.T.L., I.-S.H. and C.-H.T.; supervision and funding, I.-S.H. All authors have read and agreed to the published version of the manuscript.

Funding: This work was funded by the National Science Council under grants NSTC 113-2221-E-155-055.

Institutional Review Board Statement: Not applicable.

Informed Consent Statement: Not applicable.

Data Availability Statement: Data are contained within the article.

Acknowledgments: We would like to express our sincere gratitude to Khouloud Abdelli for her invaluable contribution to our research. Abdelli provided the critical data that formed the foundation of our study, and her support and insights were instrumental in the completion of this work. We are deeply appreciative of her assistance and dedication.

Conflicts of Interest: The authors declare no conflicts of interest.

Abbreviations

5G	Fifth-generation
AF	Assured-Forwarding
AI	Artificial Intelligence
API	Application Programming Interface
BE	Best-Effort
BER	Bit Error Rate
CAPEX	Capital expenditures
EF	Expedited Forwarding
EPON	Ethernet Passive Optical Network
FBG	Fiber Bragg Granting
FCAPS	fault, configuration, accounting, performance, and security
FCC	Federal Communication Commission
gRPC	Remote Procedure Calls
H2H	Human-to-Human
H2M/R	Human-to-Machine/Robot

HITL	Human-in-the-Loop
IoT	Internet of Things
M2M	Machine-to-Machine
ML	Machine Learning
MLP	Multi-Layer Perceptron
MTTR	Mean Time to Repair
NEM	Network Edge Mediator
NFV	Network Function Virtualization
NG-EPONs	Next-Generation Ethernet Passive Optical Networks
NNI	Network-to-Network Interface
ODN	Optical Distribution Networks
OLT	Optical Line Terminal
OMCI	ONU Management and Control Interface
ONOS	Open Network Operating System
ONU	Optical Network Unit
OPEX	Operation and Maintenance Expenses
OTDR	Optical Time-Domain Reflectometry
PON	Passive Optical Networks
QoS	Quality-of-Service
R-CORD	Residential Services Central Office Rearchitected as Datacenter
RDBA	Resilience Dynamic Bandwidth Allocation
RFoG	Radio Frequency over Glass
RR	Reference Reflector
SDN	Software-Defined Networking
SEBA	SDN-Enabled Broadband Access
SLA	Service Level Agreement
TI	Tactile Internet
UNI	User Network Interface
VLAN	Virtual Local Area Network
vOLT	Virtual OLT
VOLTHA	Virtual OLT Hardware Abstraction
XR	Extended Reality

References

1. Trisha, S.M.; Hossen, M. Auction-based Hybrid DBA Algorithm in EPON-WiMAX Converge Network for 5G and Beyond (5 GB) Communications. *Opt. Fiber Technol.* **2023**, *81*, 103542. [CrossRef]
2. Statista. Global IoT and Non-IoT Connections 2010–2025. Available online: https://www.statista.com/statistics/1101442/iot-number-of-connected-devices-worldwide/ (accessed on 25 July 2024).
3. Saad, W.; Bennis, M.; Chen, M. A Vision of 6G Wireless Systems: Applications, Trends, Technologies, and Open Research Problems. *IEEE Netw.* **2020**, *34*, 134–142. [CrossRef]
4. Promwongsa, N.; Ebrahimzadeh, A.; Naboulsi, D.; Kianpisheh, S.; Belqasmi, F.; Glitho, R.; Crespi, N.; Alfandi, O. A Comprehensive Survey of the Tactile Internet: State-of-the-Art and Research Directions. *IEEE Commun. Surv. Tutor.* **2021**, *23*, 472–523. [CrossRef]
5. Holland, O.; Steinbach, E.; Prasad, R.V.; Liu, Q.; Dawy, Z.; Aijaz, A.; Pappas, N.; Chandra, K.; Rao, V.S.; Oteafy, S.; et al. The IEEE 1918.1 'Tactile Internet' Standards Working Group and its Standards. *Proc. IEEE* **2019**, *107*, 256–279. [CrossRef]
6. Wang, X.; Ji, Y.; Zhang, J.; Bai, L.; Zhang, M. Low-Latency Oriented Network Planning for MEC-Enabled WDM-PON Based Fiber-Wireless Access Networks. *IEEE Access* **2019**, *7*, 183383–183395. [CrossRef]
7. Amjad, Z.; Sikora, A.; Hilt, B.; Lauffenburger, J.-P. Low Latency V2X Applications and Network Requirements: Performance Evaluation. In Proceedings of the 2018 IEEE Intelligent Vehicles Symposium (IV), Changshu, China, 26–30 June 2018; pp. 220–225. [CrossRef]
8. Dias, I.; Ruan, L.; Ranaweera, C.; Wong, E. From 5G to Beyond: Passive Optical Network and Multi-Access Edge Computing Integration for Latency-Sensitive Applications. *Opt. Fiber Technol.* **2023**, *75*, 103191. [CrossRef]
9. *IEEE 802.3ca-2020*; IEEE Standard for Ethernet Amendment 9: Physical Layer Specifications and Management Parameters for 25 Gb/s and 50 Gb/s Passive Optical Networks. IEEE: Piscataway, NJ, USA, 2020; pp. 1–267. [CrossRef]
10. Zhang, C.; Yang, M.; Zheng, W.; Zheng, Y.; Wu, Y.; Zhang, Y. Analysis of Wavelength Deployment Schemes in Terms of Optical Network Unit Cost and Upstream Transmission Performance in NG-EPONs. *J. Opt. Commun. Netw.* **2021**, *13*, 214. [CrossRef]
11. Ab-Rahman, M.S.; Ridzuan, A.M.; Kaharudin, I.H.; Hwang, I.S. Real Time FTTH Network Monitoring using Binary Coded Fiber Bragg Grating. *Optik* **2022**, *251*, 168408. [CrossRef]

12. Abdelli, K.; Cho, J.Y.; Azendorf, F.; Griesser, H.; Tropschug, C.; Pachnicke, S. Machine Learning-based Anomaly Detection in Optical Fiber Monitoring. *J. Opt. Commun. Netw.* **2022**, *14*, 365. [CrossRef]
13. Villa, G.; Tipantuña, C.; Guamán, D.S.; Arévalo, G.V.; Arguero, B. Machine Learning Techniques in Optical Networks: A Systematic Mapping Study. *IEEE Access* **2023**, *11*, 98714–98750. [CrossRef]
14. Abdelli, K.; Grießer, H.; Ehrle, P.; Tropschug, C.; Pachnicke, S. Reflective Fiber Fault Detection and Characterization using Long Short-Term Memory. *J. Opt. Commun. Netw.* **2021**, *13*, E32. [CrossRef]
15. Honda, N.; Iida, D.; Izumita, H.; Azuma, Y. In-Service Line Monitoring System in PONs Using 1650-nm Brillouin OTDR and Fibers With Individually Assigned BFSs. *J. Light. Technol.* **2009**, *27*, 4575–4582. [CrossRef]
16. Fouli, K.; Chen, L.R.; Maier, M. Time-, Wavelength-, and Code-Domain Optical Reflection Monitoring for Next-Generation Access-Metro Networks. *Comput. Commun.* **2011**, *34*, 1011–1021. [CrossRef]
17. Liem, A.T.; Hwang, I.S.; Nikoukar, A. An Autonomous Recovery Mechanism Against Optical Distribution Network Failures in EPON. *Opt. Fiber Technol.* **2014**, *20*, 552–560. [CrossRef]
18. Liem, A.T.; Hwang, I.S.; Nikoukar, A.; Ab-Rahman, M.S. Distribution Drop Fiber in-Service Fault Management in Enhanced EPON System. *Opt. Switch. Netw.* **2015**, *17*, 52–63. [CrossRef]
19. Ashraf, M.; Idrus, S.; Iqbal, F.; Butt, R.; Faheem, M. Disaster-Resilient Optical Network Survivability: A Comprehensive Survey. *Photonics* **2018**, *5*, 35. [CrossRef]
20. Yeh, C.H.; Chow, C.W.; Liu, Y.L. Self-Protected Ring-Star-Architecture TDM Passive Optical Network with Triple-Play Management. *Opt. Commun.* **2011**, *284*, 3248–3250. [CrossRef]
21. Kanungoe, A.; Mukhopadhyay, A.; Das, G.; Banerjee, R.; Das, R. A New Protection Scheme for A Combined Ring-Star based Hybrid WDM/TDM PON Architecture. *Opt. Switch. Netw.* **2015**, *18*, 153–168. [CrossRef]
22. Hsu, C.-H.; Jiang, S.-Y.; Hsieh, S.-E.; Yeh, C.-H.; Lai, Y.-T.; Chen, L.-Y.; Liaw, S.-K.; Chow, C.-W. Hybrid Self-Protected Fiber-FSO WDM-PON System with Fiber Breakage Prevention. *Photonics* **2022**, *9*, 822. [CrossRef]
23. Zentani, A.; Zulkifli, N.; Ramli, A. Network Resiliency and Fiber Usage of Tree, Star, Ring and Wheel based Wavelength Division Multiplexed Passive Optical Network Topologies: A Comparative Review. *Opt. Fiber Technol.* **2022**, *73*, 103038. [CrossRef]
24. Pakpahan, A.F.; Hwang, I.S.; Yu, Y.M.; Hsu, W.H.; Liem, A.T.; Nikoukar, A. Novel Elastic Protection Against DDF Failures in An Enhanced Software-Defined SIEPON. *Opt. Fiber Technol.* **2017**, *36*, 51–62. [CrossRef]
25. Mata, J.; de Miguel, I.; Durán, R.J.; Merayo, N.; Singh, S.K.; Jukan, A.; Chamania, M. Artificial Intelligence (AI) Methods in Optical Networks: A Comprehensive Survey. *Opt. Switch. Netw.* **2018**, *28*, 43–57. [CrossRef]
26. Nyarko-Boateng, O.; Adekoya, A.F.; Weyori, B.A. Predicting the Actual Location of Faults in Underground Optical Networks using Linear Regression. *Eng. Rep.* **2021**, *3*, eng212304. [CrossRef]
27. Straub, M.; Reber, J.; Saier, T.; Borkowski, R.; Li, S.; Khomchenko, D.; Richter, A.; Färber, M.; Käfer, T.; Bonk, R. ML Approaches for OTDR Diagnoses in Passive Optical Networks—Event Detection and Classification: Ways for ODN Branch Assignment. *J. Opt. Commun. Netw.* **2024**, *16*, C43. [CrossRef]
28. Cisco. 2024 Global Networking Trends Report. Available online: https://www.cisco.com/c/dam/en/us/solutions/enterprise-networks/2024-global-networking-trends.pdf (accessed on 25 May 2024).
29. Das, S. From CORD to SDN Enabled Broadband Access (SEBA) [Invited Tutorial]. *J. Opt. Commun. Netw.* **2021**, *13*, A88. [CrossRef]
30. Bentz, C.; Baudzus, L.; Krummrich, P. Signal to Noise Ratio (SNR) Enhancement Comparison of Impulse-, Coding- and Novel Linear-Frequency-Chirp-Based Optical Time Domain Reflectometry (OTDR) for Passive Optical Network (PON) Monitoring Based on Unique Combinations of Wavelength Selective Mirrors. *Photonics* **2014**, *1*, 33–46. [CrossRef]
31. Almarhabi, A.; Alhazmi, H.; Samarkandi, A.; Yao, Y.D. Modulation Classification Based on Eye Diagrams and Deep Learning. In Proceedings of the 2022 31st Wireless and Optical Communications Conference (WOCC), Shenzhen, China, 11–12 August 2022; pp. 35–40. [CrossRef]
32. Edge Core Networks Corporation. Edgecore Networks. Available online: https://www.edge-core.com/solution-inquiry.php?cls=5&id=51 (accessed on 27 May 2024).
33. Open Networking Foundation. Reference Design SDN Enabled Broadband Access (SEBA). *March 2021*. Available online: https://opennetworking.org/reference-designs/seba/ (accessed on 28 May 2024).
34. Sgambelluri, A.; Pacini, A.; Paolucci, F.; Castoldi, P.; Valcarenghi, L. Reliable and Scalable Kafka-based Framework for Optical Network Telemetry. *J. Opt. Commun. Netw.* **2021**, *13*, E42. [CrossRef]
35. Optiwave Sytstem. OptiSystem Overview. *Optiwave*. Available online: https://optiwave.com/optisystem-overview/ (accessed on 23 June 2024).
36. Liem, A.T.; Hwang, I.S.; Ganesan, E.; Taju, S.W.; Sandag, G.A. A Novel Temporal Dynamic Wavelength Bandwidth Allocation Based on Long-Short-Term-Memory in NG-EPON. *IEEE Access* **2023**, *11*, 82095–82107. [CrossRef]
37. Simsek, M.; Aijaz, A.; Dohler, M.; Sachs, J.; Fettweis, G. 5G-Enabled Tactile Internet. *IEEE J. Select. Areas Commun.* **2016**, *34*, 460–473. [CrossRef]

photonics

MDPI

Article

The Stability Optimization of Indoor Visible 3D Positioning Algorithms Based on Single-Light Imaging Using Attention Mechanism Convolutional Neural Networks

Wenjie Ji [1], Lianxin Hu [1], Xun Zhang [1,2,*], Jiongnan Lou [1], Hongda Chen [1] and Zefeng Wang [1]

1 School of Information Engineering, Huzhou University, Huzhou 313000, China; 2022388109@stu.zjhu.edu.cn (W.J.); 03080@zjhu.edu.cn (L.H.); 03261@zjhu.edu.cn (H.C.); zefeng.wang@zjhu.edu.cn (Z.W.)

2 Institut Supérieur d'Électronique de Paris, LISIT-ECoS, 75020 Paris, France

* Correspondence: xun.zhang@isep.fr

Abstract: In recent years, visible light positioning (VLP) techniques have been gaining popularity in research. Among them, the scheme of using a camera as a receiver provides a low-cost, high-precision positioning capability and easy integration with existing multimedia devices and robots. However, the pose changes of the receiver can lead to image distortion and light displacement, significantly increasing positioning errors. Addressing these errors is crucial for enhancing the accuracy of VLP. Most current solutions rely on gyroscopes or Inertial Measurement Units (IMUs) for error optimization, but these approaches often add complexity and cost to the system. To overcome these limitations, we propose a 3D positioning algorithm based on an attention mechanism convolutional neural network (CNN) aimed at reducing the errors caused by angles. We designed experiments and comparisons within a rotation angle range of ±15 degrees. The results demonstrate that the average error for 2D positioning is within 5.74 cm and the height error is within 3.92 cm, and the average error for 3D positioning is within 6.8 cm. Among the four groups of experiments for 3D positioning, compared to the traditional algorithm, the improvements were 7.931 cm, 15.569 cm, 6.004 cm, and 16.506 cm. The experiments indicate that it can be applied to high-precision visible light positioning for single-light imaging.

Keywords: CNN; VLC; VLP; OCC; indoor positioning; attention mechanism

1. Introduction

Over the years, positioning technology has played a crucial role in various fields such as transportation, production, navigation, and daily life. In outdoor settings, wireless signals can be transmitted in open spaces, a challenge often addressed using technologies like the Global Positioning System (GPS). However, the performance of a GPS for indoors is significantly degraded due to transmission hindrance. With indoor positioning widely applied in spaces such as shopping malls, museums, warehouses, and parking lots, the demand for low-cost, high-precision positioning technology has rapidly increased [1]. Currently, indoor positioning methods include Wireless Fidelity (Wi-Fi), Zigbee, Bluetooth, Ultra-Wideband (UWB), Infrared (IR), ultrasound, and Radio-Frequency Identification (RFID) [2]. With the widespread adoption of LED lighting in recent years and considering factors such as infrastructure coverage, cost, positioning accuracy, information security, and abundant spectrum resources, VLP has emerged as a research hotspot [3].

Visible light positioning systems require either multiple lights or a single light to be realized. Work [4–9] proposed positioning algorithms based on multiple lights, inferring position information by leveraging the geometric relationships or deformations between lights to achieve indoor positioning. However, the multiple light positioning system is demanding in terms of layout and lacks robustness and flexibility. Some of the works also discuss single-light positioning, but it does not guarantee high precision or requires high-cost receiving devices. In [10,11], a positioning method based on a single LED with a beacon was proposed, utilizing the fundamental principle of geometric relationships among multiple lights for auxiliary positioning. Work [12,13] proposed a method using binocular cameras and a single light to obtain the position, but it incurs high application costs and loses the good adaptability of monocular camera applications. We proposed a trilateral positioning method using a single rectangular light in previous work [14].

During positioning, the unavoidable random angles of the receiver can cause errors in the results. In work [2,3,15–17], algorithms relying on a single light and gyroscopes or IMU sensors for positioning were proposed. Work [16] proposed a single-light positioning method using PD, camera, and gyroscope. It used a hybrid RSS/AOA-based algorithm to achieve 3D positioning. In work [18], the gyroscope angle information is used to reconstruct the image and reduce the positioning error caused by the change of camera pose. However, it only tested for an angle of five. In practical application scenarios, the receiver often encounters significant angles. In summary, most of the single-light positioning methods rely heavily on sensors to handle the smaller angular changes at the receiver.

So, based on the above analysis, we propose a high-precision 3D positioning algorithm based on single-light imaging. The main contributions of this study are as follows:

1. We use the multi-head attention mechanism (MHA) and residual convolutional neural network (resnet50) to form a new model MHA-Resnet50, which effectively avoids model overfitting and makes training more efficient.
2. The dependencies between image features and pose are automatically learned by the model, and the predicted coordinates are regressed.
3. The reduced use of IMU sensors simplifies the algorithm and enhances the robustness of the positioning system.

Accurate positioning at a ± 15 angle is achieved with a low-resolution image of 1280×720. The results are better compared to the proposed methods of works [18,19].

The rest of this paper is organized as follows. Section 2 briefly describes the system and analyzes the issues that need to be addressed. Section 3 introduces the proposed MHA-Resnet50. Section 4 describes the experimental environment, and how the model is trained. The positioning results of the model are also analyzed in comparison with the original algorithm. Finally, Section 5 concludes the paper.

2. Visible Light Positioning System and Issue Analysis

2.1. System Overview

Figure 1 illustrates an indoor VLP system utilizing a single LED. The transmitter consists of a rectangular LED and a signal modulator mounted on the ceiling parallel to the floor. Each LED is assigned a unique ID associated with its actual spatial position, and these correspondences are stored in a database. After modulation processing, the LED repetitively transmits its ID. The receiver side is a camera connected to a computer for capturing video frames. Through image frame processing, the LED ID can be decoded and the corresponding LED world coordinates will be identified. At the same time, the 3D relative coordinate position of the camera with respect to the LED can be calculated by MHA-Resnet50. The specific modeling method and prediction process of MHA-Resnet50 are illustrated in Section 3. Next, the spatial coordinate position of the camera can be calculated by combining the world coordinates of the LED.

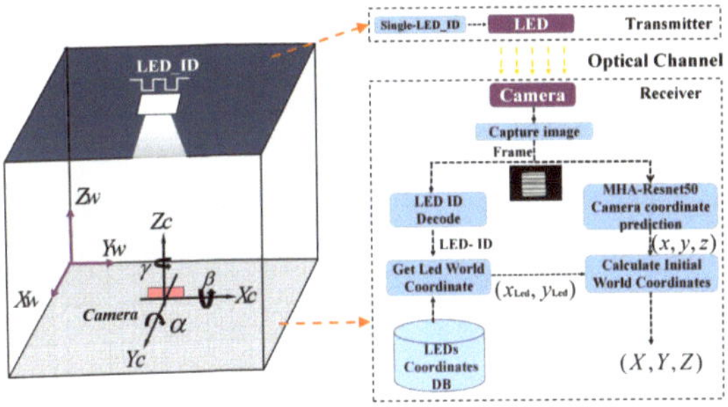

Figure 1. Visible light positioning system.

2.2. Foundation

In order to conduct the analysis easily, in this paper, the complex nonlinear model of the camera lens system is simplified to a simple pinhole camera model. As shown in Figure 2, the relationship between the world coordinate system (X_w, Y_w, Z_w) and the pixel coordinate system (u, v) can be expressed as (1):

$$Z_C \begin{bmatrix} u \\ v \\ 1 \end{bmatrix} = \begin{bmatrix} \frac{f}{dx} & 0 & u_0 & 0 \\ 0 & \frac{f}{dy} & v_0 & 0 \\ 0 & 0 & 1 & 0 \end{bmatrix} \begin{bmatrix} R & T \\ 0 & 1 \end{bmatrix} \begin{bmatrix} X_W \\ Y_W \\ Z_W \\ 1 \end{bmatrix} \tag{1}$$

where Zc represents the Z coordinate of the camera in the world coordinate system, (X_w, Y_w, Z_w) is its actual position in space, and u and v are the coordinates of the point on the image. Focal length f and physical dimensions d_x, d_y of the pixels constitute the intrinsic matrix that connects the pixel coordinate system to the camera coordinate system [14,20]. Similarly, translation vector T and rotation matrix R constitute the extrinsic matrix that connects the camera to the world coordinate system. R is given by (2).

$$\begin{aligned} R &= R_X(\alpha) R_Y(\beta) R_Z(\gamma) \\ &= \begin{bmatrix} 1 & 0 & 0 \\ 0 & \cos\alpha & \sin\alpha \\ 0 & -\sin\alpha & \cos\alpha \end{bmatrix} \begin{bmatrix} \cos\beta & 0 & -\sin\beta \\ 0 & 1 & 0 \\ \sin\beta & 0 & \cos\beta \end{bmatrix} \begin{bmatrix} \cos\gamma & \sin\gamma & 0 \\ -\sin\gamma & \cos\gamma & 0 \\ 0 & 0 & 1 \end{bmatrix} \end{aligned} \tag{2}$$

α, β, γ denote the pitch, roll, and azimuth.

The original positioning algorithm is implemented based on the imaging principles shown in Figure 2. Its 3D positioning is explained below.

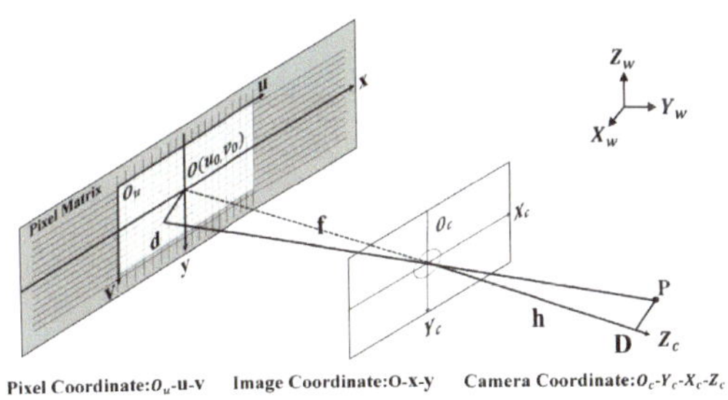

Figure 2. Imaging principle.

Figure 3a is an image of visible light communication which contains much positional information. The pixel coordinates (u_{cen},v_{cen}) of the center of the light mass can be obtained directly in the image. Its image coordinates (x_{cen},y_{cen}) can be obtained by Equation (3).

$$\begin{bmatrix} x \\ y \\ 1 \end{bmatrix} = \begin{bmatrix} \frac{1}{dx} & 0 & u_0 \\ 0 & \frac{1}{dy} & v_0 \\ 0 & 0 & 1 \end{bmatrix}^{-1} \begin{bmatrix} u \\ v \\ 1 \end{bmatrix} \tag{3}$$

Then, the length of d_{cen} can be obtained by the calculation of Equation (4).

$$d_{cen} = \sqrt{x_{cen}^2 + y_{cen}^2} \tag{4}$$

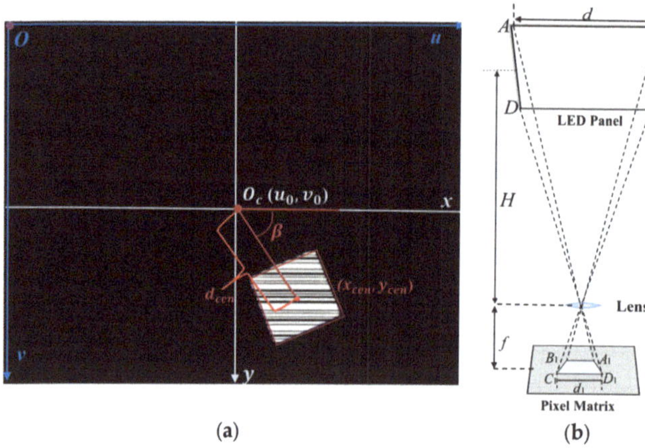

Figure 3. Original positioning principle; (**a**) is the principle of 2D positioning, (**b**) is the principle of height calculation.

According to the image coordinates of the light, the pinch angle β in Figure 3a can be calculated. After the length d_{cen} and the angle β are known, the 2D coordinate of the light relative to the camera is obtained utilizing the imaging relation in Figure 2.

Figure 3b illustrates a conventional method of height calculation. The four outer vertices $A(x_1, y_1, z_1)$, $B(x_2, y_2, z_2)$, $C(x_3, y_3, z_3)$, and $D(x_4, y_4, z_4)$ of a rectangular light lie in a plane with equal Z coordinates. The corresponding projected points of the four vertices in the image are $A_1(x_{11}, y_{11})$, $B_1(x_{12}, y_{12})$, $C_1(x_{13}, y_{13})$, and $D1(x_{14}, y_{14})$. According to the monocular camera imaging geometry, the vertical distance H between the rectangular light and the camera lens can be calculated using Equation (5):

$$H = \frac{fd}{\max(d_i)}, (i = 1, 2, 3, 4) \tag{5}$$

where d_i is the edge length of the rectangular light in the image. The camera deflection angle causes the geometric projection of the rectangular light to deform, so the maximum d_i is chosen to minimize the error.

From the above description, it is clear that 3D positioning functionality can be achieved using a camera and a rectangular LED. In Section 4 of the experiment, we conducted a comparative analysis between this original algorithm and the proposed algorithm, providing a comprehensive evaluation.

2.3. Issue Analysis

The above positioning method is implemented under the ideal condition where the receiver remains level with the light, as shown in Part 2 of Figure 4. The imaging position at this point correctly reflects the relative positions of camera and light. However, when an angular deflection of the receiver occurs, the imaging of the light in the image undergoes a positional shift and deformation. This results in positioning errors, as shown in Figure 4, Part 1 and Part 3. The receiver angle deviation can have an impact on the 2D position and altitude calculation, which in turn leads to errors in 3D positioning. This is detailed and analyzed below.

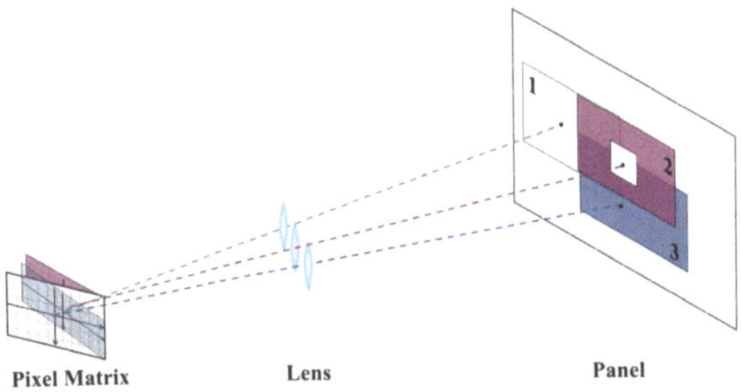

Figure 4. Light position information in the pixel matrix; 1 is left deflection, 2 is normal, 3 is down deflection.

2.3.1. D Error

In order to analyze the specific effects, three scenarios are constructed in Figure 5. The position of the LED in the camera coordinate system reflects the relative position with respect to the camera. Therefore, analyzing the position error of the LED in the camera coordinate system reflects the position error of the camera.

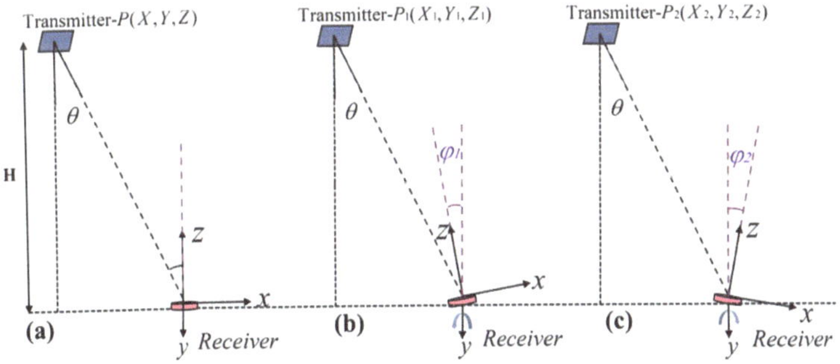

Figure 5. Coordinate change due to angle change; (**a**) is initial scene, (**b**,**c**) are after rotation.

Figure 5a shows that the initial coordinates of the LED in the camera coordinate system are $P(X, Y, Z)$ without any rotation around the Y-axis. The LED coordinates could be various because of the camera's rotation angle. Figure 5b,c shows two rotation scenarios in different directions. From Figure 5b to Figure 5a, the coordinates of the camera coordinate system of the LED are changed from $P_1(X_1, Y_1, Z_1)$ to $P(X, Y, Z)$. The mathematical equation is expressed as follows.

$$\begin{bmatrix} X \\ Y \\ Z \end{bmatrix} = R_Y(\varphi_1) \begin{bmatrix} X_1 \\ Y_1 \\ Z_1 \end{bmatrix} = \begin{bmatrix} \cos\varphi_1 & 0 & \sin\varphi_1 \\ 0 & 1 & 0 \\ -\sin\varphi_1 & 0 & \cos\varphi_1 \end{bmatrix}^{-1} \begin{bmatrix} X_1 \\ Y_1 \\ Z_1 \end{bmatrix} \tag{6}$$

After the camera is rotated, the change in height between it and the LED is so slight that it can be ignored. Therefore, the 2D coordinate error obtained before and after camera rotation can be expressed as follows.

$$Eerror = \sqrt{(X_1 - X)^2 + (Y_1 - Y)^2} \tag{7}$$

2.3.2. Height Error

When the angle of the receiver is changed, both perspective transformations and affine transformations result in the distortion of the light image. When projecting a rectangle in three-dimensional space onto a two-dimensional image plane, the lengths, angles, and coordinate positions of the edges change due to rotation and scaling.

Combining the 2D error with the height error, we can mathematically represent the 3D positioning error as (8).

$$Eerror = \sqrt{(X_1 - X)^2 + (Y_1 - Y)^2 + (Z_1 - Z)} \tag{8}$$

2.4. Solution Concept

Above, we describe the basic optical imaging positioning algorithm and the effect of the angle on its results. The implementation of these original algorithms is based on imaging principles, which only use the modulated LEDs as beacons for assisted positioning, and do not really incorporate the characteristics of optical camera communication.

For this reason, we started with the signal frames and found properties that allow for positioning. As shown in Figure 6b, we increase the brightness and exposure of the original frame. It was found that the light signal stripes were still present outside the area of the LED, just hard to distinguish with the naked eye. The light intensity of the stripes in the picture is diffusely attenuated, with different attenuation characteristics at different positions. The light intensity weakening in the picture is reflected in the gray value of the reduction, based on the characteristics of the change in grayscale for positioning. As shown in Figure 6c, we change the camera pose and the imaging areas in the signal frame, which undergo deformation. Through the principles of perspective transformation and affine transformation, it is known that different deformations represent different camera poses and are regular.

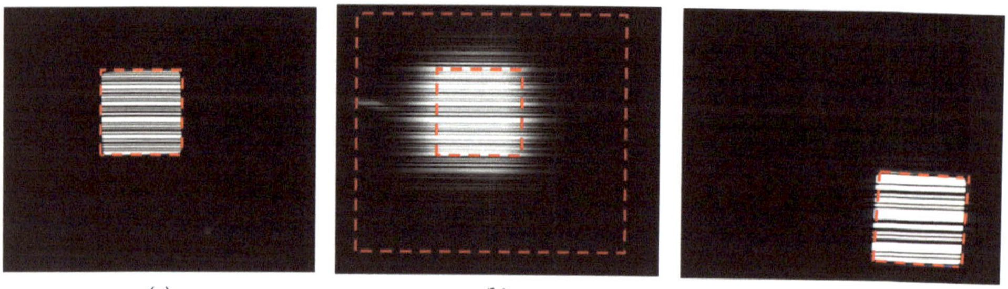

(a) (b) (c)

Figure 6. Images of different parameters; (**a**) is the original image, (**b**) brightness up by 20, exposure up by 5, (**c**) is rotated at an angle of 10.

Combining these findings, we consider fusing two features, light intensity and imaging distortion, to achieve positioning that can cope with angular variations. These features are difficult to extract using conventional image-processing algorithms and the workload is enormous. For this reason, we consider extracting these features using a convolutional neural network and propose the MHA-Resnet50 model.

3. Advanced MHA-Resnet50 Model

As shown in Figure 7, the backbone network of the model is Resnet50, which incorporates a multi-head attention mechanism in the middle. The input to MHA-Resnet50 is the signal frames from the camera at multiple angles and coordinates. After extracting features by multilayer convolution, a regressor is used to predict the 3D coordinates of the camera. Its implementation is described in detail below.

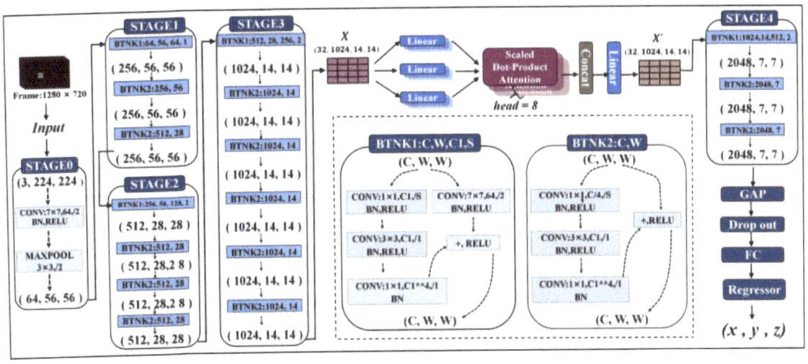

Figure 7. MHA-Resnet50 model structures.

Initially, signal frames with a resolution of 1280 $\times$ 720 are normalized, activated, and subjected to max-pooling operations.

In STAGE 1, the model primarily extracts low-level features, with feature map sizes large enough to capture rich spatial detail information.

In STAGE 2, the model reduces the spatial size of the feature maps while increasing the number of channels. This enables the identification of more complex shape features in the light signal regions of the frames.

In STAGE 3, the model further increases the number of channels and compresses the spatial size of the feature maps.

After STAGE 3, the network has formed a rich set of abstract features. Introducing the multiple attention mechanism in this stage can effectively highlight the light intensity change features by assigning weights. It helps the model to better understand the key information in the image.

In STAGE4, the feature maps are then subjected to a convolution operation. A global average pooling process is then performed to reduce the number of model parameters. To further enhance the generalization ability of the model and reduce the risk of overfitting, we add the dropout module. It can randomly drop some features to reduce the model's dependence on specific features [21].

Next, the feature vectors are linearly processed by the FC layer and passed to the regressor xgboost, which finally predicts the 3D coordinates.

3.1. Resnet 50 Model

ResNet50 is a classic convolutional neural network structure in the field of deep learning belonging to the category of residual networks. ResNet50 introduces the residual connection mechanism, which establishes direct connections between different layers. This allows features learned in shallower layers to be passed directly to deeper layers. Consequently, the gradient can still propagate efficiently even as the network becomes deeper. Thus, the problem of vanishing gradient is avoided and the stable training of deep neural networks is ensured.

3.2. Multi-Head Attention Mechanism

This part provides the motivation for using the multi-attention mechanism in the RenNet50 backbone network through test results and details how the mechanism reduces the risk of overfitting and improves the training efficiency.

3.2.1. Motivation

As shown in Figure 8a, when ResNet50 is used to process signal frames to extract features, its convolution operation uses a convolution kernel to slide over the image and compute a weighted sum of the partial area to extract features. As the convolutional layers are stacked layer by layer, high-level features are gradually extracted from the low-level features. In Section 4.2.4, we used the ResNet50 model for training and extracted features from the convolutional layer to draw feature maps and heat maps. The results show that the model only extracts the deformation features of the imaging region and fails to extract the change features of the light intensity. In order to extract both features simultaneously, we use the attention mechanism.

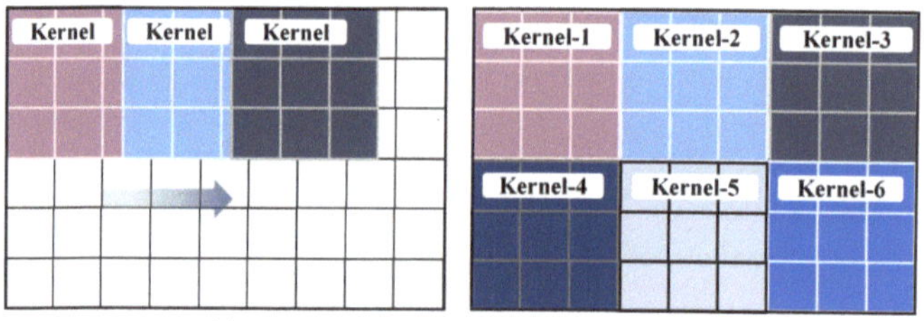

Figure 8. Convolution; (**a**) is ResNet50, (**b**) is MHA-ResNet50.

Attentional mechanisms simulate the perceptions of cognitive functions that are integral to humans. An important characteristic of perception is that humans do not process all information immediately. Instead, we selectively focus on a portion of the information when and where it is needed. Meanwhile, other perceptible information is ignored [22]. This mechanism can help the neural network to process the input data more efficiently. It can distribute different attentional weights between different positions or different features. Consequently, it improves the model's ability to perceive and understand the input data and its representation [23].

As shown in Figure 6, a large number of light intensity features are not obvious without manually adjusting the image parameters. Therefore, we need to assign higher weights to light intensity features during feature extraction. Meanwhile, as shown in Figure 6b, the feature area of the signal black and white stripes is large. If we want to fuse the deformation features and the light intensity features of the stripes to jointly characterize the position information, we need to extract these features at the same time and enhance the model's understanding of the dependency between the long-range features. For this purpose, we use the multi-head attention mechanism. As shown in Figure 8b, the multi-head attention mechanism uses multiple convolution kernels simultaneously to acquire features from different positions of the image while performing convolution. It will adaptively assign attentional weights based on the input features, and it can assign higher weights for features with insignificant light intensity variations. In Section 4.2.4, we used the MHA-ResNet50 model for training and extracting features from the convolutional layer to draw feature maps and heat maps. The results showed that a large number of light intensity features in the signal frames were extracted.

3.2.2. Multi-Head Attention Principle

In this research, we use the 8-head attention mechanism, the principle of which is shown in Figure 9.

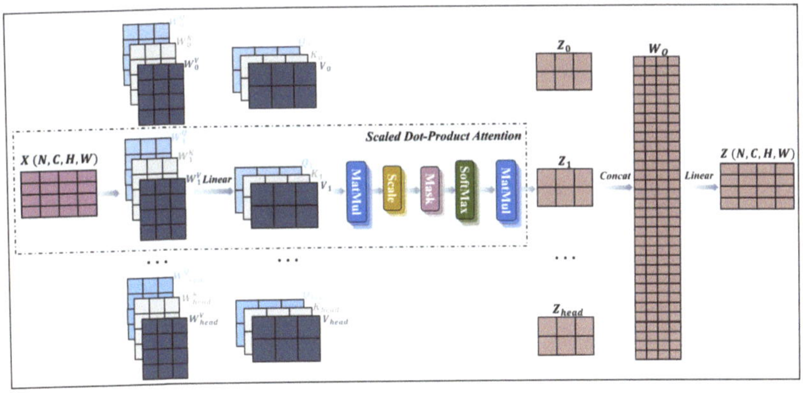

Figure 9. Multi-head attention principle.

The essence of a multi-head attention mechanism is indeed the combination of multiple self-attention mechanisms. Each attention head independently computes self-attention, capturing different features and contextual information within the input sequence. Through concatenation and linear transformation, these information streams are integrated, thereby enhancing the model's expressive power and performance. In each self-attention mechanism, there exists a query matrix Q, a key matrix K, and a value matrix V. The specific implementation of the self-attention mechanism is the Scaled Dot Product Attention (SDA). Its input is a four-dimensional tensor $X(N, C, H, W)$. X is linearly transformed to Q, K, and V, respectively, by Equation (8):

$$Q = XW^Q, K = XW^K, V = XW^V \tag{9}$$

where W^Q, W^K, and W^V are the weight matrices of queries, keys, and values with dimensions $C \times dk$, $C \times d_k$, and $C \times d_v$. Subsequently, the dot product between the query and the key is computed and divided by a scaling factor $\sqrt{d_k}$ to prevent the value from being too large.

$$A = \frac{QK^T}{\sqrt{d_k}} \tag{10}$$

The dot product results are normalized using softmax to obtain the attention weight matrix.

$$S = softmax(A) \tag{11}$$

The summation is weighted to obtain the expression for attention.

$$Attention = SV \tag{12}$$

Then, the SDA is collapsed and expressed as follows.

$$Attention(Q, K, V) = softmax(\frac{QK^T}{\sqrt{d_k}})V \tag{13}$$

The concept of the multi-head attention mechanism is to employ the different parameters W^Q, W^K, and W^V to successively perform linear transformations on the matrices Q, K, and V. The results of these linear transformations are then input into the SDA. The computation result is denoted as $head_i$, and its expression is given by the following.

$$head_i = Attention(Q_i, K_i, V_i), i = 1, \ldots, h \tag{14}$$

The computed $head_i$ are concatenated into a matrix. It is transformed linearly with the matrix Wo to convert the output of the multi-head attention into a four-dimensional tensor Z. The mathematical representation is as in (17), where h is the number of heads.

$$Z = Mutilhead(Q, K, V) = Concat[head_1; head_2; \ldots; head_h]Wo \tag{15}$$

4. Experiment and Discussion

4.1. Laboratory Testbed

Our algorithm has been tested in an experimental environment. As shown in Figure 10a, all experiments were conducted in an indoor enclosed area of 2.6 m × 2.6 m × 2.2 m. The ground test area was 1 m × 1 m, divided equally into a 10 cm × 10 cm grid with a total of 121 test points. As shown in Figure 10b, to improve the efficiency of the test, we developed the operator interface for the VLP test using Python. The interface allows for the real-time intuitive monitoring of parameter changes during LED recognition and positioning.

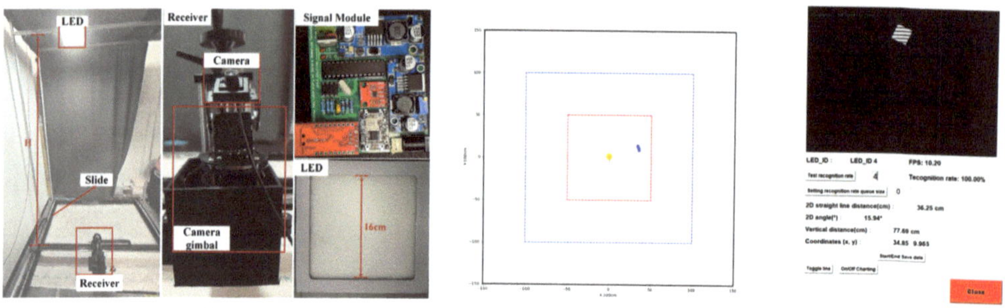

(**a**) (**b**)

Figure 10. Testbed; (**a**) is testbed, (**b**) is operator interface.

The transmitter device of the testbed consists of a rectangular LED and a signal modulation module, and is mounted parallel to the ground above the center of the test area. The signal modulation module consists of several off-the-shelf modules, specifically expressed as follows: the microcontroller (MCU) compiles the binary modulation signal into the digital-to-analog conversion module (DAC) after the DAC processing output analog signal. This signal is then fed into the inverting input of the operational amplifier. The MCU inputs a DC bias voltage to the DAC and then inputs the DC bias to the positive phase input of the operational amplifier. The modulated signal and the DC bias are coupled and amplified by the operational amplifier, ensuring that the signal voltage amplitude reaches the operating voltage of the LED. An on-off keying (OOK) modulation scheme is employed to generate control signals for modulating the LED. The detailed parameters of the LED and the modules are shown in Table 1.

Table 1. Parameters of transmitter circuit.

Controller	DAC Module	Power/W	Voltage/V	Dimension/MM
ATmega328P-PU	MCP4725	10	12	160 × 160
Boost Module	**Buck Module**	**NMOSFET**	**Operational Amplifier**	
LM2587	LM2596SDC	TRFB4110	OPA551	

The camera at the receiver was connected to the computer. Before conducting the test, we calibrated the camera so that it could correctly identify the correct position of the LED in the image. Its detailed parameters are shown in Table 2.

Table 2. Parameters of camera.

Resolution	Pixel Size	Image Sensor	Format	F
1280 × 720	2.9 µm × 3.0 µm	IMX335	JPG	2.2 mm
Shutter	**Brightness**	**Contrast**	**Gamma value**	**Image gain**
1/2.8CMOS	26	61	500	128

4.2. Model Training Results and Comparative Analysis

4.2.1. Model Parameter Setting

In this study, we built the basic software environment using Torch 2.0.1, Python 3.9, Torch-vision 0.13, and Torch-audio 0.12. The model training was performed on an NVIDIA RTX 4090 graphics processing unit. To ensure consistency and fairness in comparison, the model training parameters were uniformly set. Specifically, we set the batch size to 32 and employed the Adam optimizer for all model training processes. The initial learning rate was set to 0.002. All experimental models underwent training for 200 epochs, during which the probability p of the dropout module was set to 0.3.

In the data preprocessing stage, two image enhancement operations were used to improve the performance and robustness of the model. First, a normalization operation was used to normalize the image data to a range with a mean of [0.485, 0.456, 0.406] and a standard deviation of [0.229, 0.224, 0.225]. This ensures that the data distribution is close to zero mean and unit variance, thus accelerating the convergence of model training. Next, we employed a color dithering operation to increase the diversity of the data. This was achieved by randomly adjusting the brightness, contrast, saturation, and hue of the image. These adjustments improve the model's ability to adapt to different lighting conditions, shooting environments, and color distributions. The combination of the two operations can effectively enhance the feature representation of image data.

To evaluate the performance of the MHA-Resnet50 model and assess the magnitude of positioning errors, we utilized the root mean square error (RMSE) as the loss function. It provides a comprehensive measure of prediction error, and a reduction in RMSE can improve the model's prediction accuracy. In this study, its mathematical expression is given as follows:

$$E_{RMSE} = \sqrt{\frac{1}{N}\sum_{i=1}^{N}\left[\left(x_{true}^{i} - x_{pred}^{i}\right)^2 + \left(y_{true}^{i} - y_{pred}^{i}\right)^2 + \left(z_{true}^{i} - z_{pred}^{i}\right)^2\right]} \tag{16}$$

where N is the number of samples per point, $(x_{true}^i, y_{true}^i, z_{true}^i)$ is the true coordinates of the ith sample, and $(x_{pred}^i, y_{pred}^i, z_{pred}^i)$ is the predicted value of the model for the ith sample.

4.2.2. Data Acquisition

We fixed the LED at a height of 160 cm and 180 cm, respectively. The pitch angle α of the camera was fixed to 0 and the roll angle β was set to -15, -10, -5, 0, 5, 10, 15 at each height, respectively. Finally, a total of 14 sets of data were taken, each set of data had 121 collection points and each point captured 50 frames of images with a resolution of 1280 × 720. A database was created by recording the coordinate position of the images and the information of the rotation angle. The label of the ith point is noted as $C\alpha$:

$$C\alpha = (x_{\alpha,i}, y_{\alpha,i}, z_{\alpha,i}) \tag{17}$$

where $i = 1, \ldots, 121$. In the model training process, the first 40 images of each point are taken as the training set, totaling 67,760 images, and the last 10 images are taken as the validation set, totaling 16,940 images.

4.2.3. Training Results and Comparison

In this part, we analyze the loss curves of the MHA-Resnet50 and set up two sets of tests to compare with multiple models.

In the first group of tests, we compared the MHA-Resnet50 model with widely used convolutional neural network architectures including DenseNet121, MobileNetv2, ResNet50, and ResNet101. Based on the RMSE curves in Figure 11a and the MSE and MAE parameters in Table 3, we found that the proposed MHA-Resnet50 has a large advantage over the other four models. The RMSE is reduced by about 14.478 cm, 14.318 cm, 13.559 cm, and 14.855 cm, respectively. To explore the optimization effect of the multi-head attention mechanism, we incorporated the MHA attention mechanism into the four network structures and conducted the second group of comparative tests.

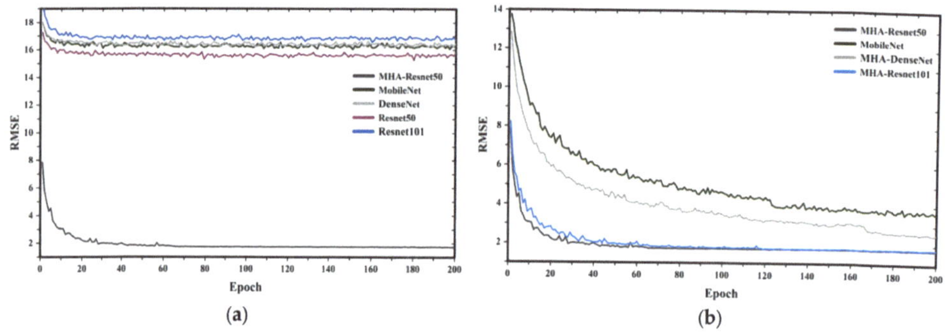

Figure 11. RMSE comparison; (**a**) is the first group, (**b**) is the second group.

Table 3. Model training results of first group.

Model	MSE	RMSE	MAE
DenseNet121	262.41043	16.199087	18.95478
MobileNetv2	257.25336	16.03912	18.806831
Resnet50	233.47543	15.279902	17.386572
Resnet101	274.74146	16.575327	19.563368
MHA-Resnet50	2.960944	1.7207394	0.24662831

The results of the second group of tests are in Figure 10b and Table 4. It can be found that, after incorporating the attentional mechanism, the RMSE of the four models was reduced by 13.696 cm, 12.49 cm, 13.559 cm, and 14.843 cm, respectively. This demonstrates that the addition of the attention mechanism can improve the performance of different models. Moreover, our proposed MHA-Resnet50 exhibits the best coordinate prediction performance.

Table 4. Model training results of second group.

Model	MSE	RMSE	MAE	FLOPs	Params
MHA-DenseNet121	6.2677374	2.503545	1.1392384	2.90GFLOPs	6.96 M
MHA-MobileNetv2	12.592225	3.5485525	2.3558052	326.46MFLOPs	2.23 M
MHA-Resnet50	2.960944	1.7207394	0.24662831	4.13GFLOPs	23.52 M
MHA-Resnet101	3.032443	1.7413912	0.29298633	7.87GFLOPs	42.51 M

4.2.4. Comparison and Analysis of Feature

In this part, we further compare the advantages of using the multi-attention mechanism and verify that MHA-Resnet50 is better than other well-established models.

As shown in Figure 12, we used the eight models trained above to process a frame and plotted the feature map and heat map before and after using the multi-head attention mechanism, respectively. The feature map consists of the superposition of the features of multiple channels in the convolutional layer of the model, which represents the various features captured in the image. The heat map is a further interpretation of the feature map, which visualizes the degree of attention paid by the model to the different regions of the input image. In the feature map and heat map, the color changes from blue to red to indicate that its feature value is gradually getting bigger.

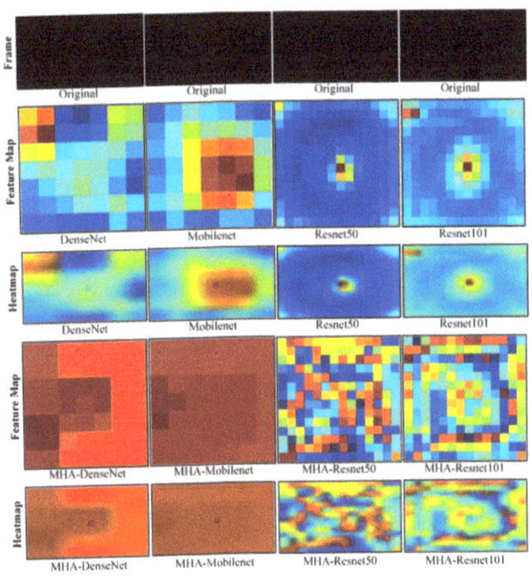

Figure 12. Comparison of feature distributions, color indicates the feature value.

At first, the results of the models were compared before and after the addition of the multi-head attention mechanism. The features of the four well-established models were mostly focused on the imaging region of the LEDs before the addition of the multi-head attention mechanism, and the features of light intensity changes were not extracted. This indicates that the model only focuses on partial features. After the addition, the feature distribution is more balanced. The model extracts a large number of light intensity change features and the imaging region features are also still obvious. This verifies that the multi-attention mechanism pays attention to all regions of the whole image during feature extraction and assigns different weights.

MHA-DenseNet, MHA-MobileNet, and MHA-ResNet were compared after the addition of the multi-attention mechanism. The feature map and heat map colors of MHA-DenseNet and MHA-MobileNet are basically red. This indicates that, although the two models are able to extract features, they have poor importance differentiation ability and can only extract extensive features at a shallower level. The color distribution of MHA-ResNet is more balanced and diverse, which indicates that it can capture both partial details and global features, and the weights are assigned according to importance through the attention mechanism.

Finally, the feature map and heat map of the MHA-ResNet50 and MHA-ResNet101 models were compared. We found that the differences are not significant, and the performance of the two models is close according to the training results in Table 4. For this, we further compared the Floating Point Operations Per Second (FLOPs) and the Params, as shown in Table 4. MHA-ResNet50 clearly has lower requirements on computational resources and memory.

4.3. Comparative Analysis of Positioning Tests and Results

In this part, we design a test to evaluate the positioning performance of the model in terms of 2D, height, and 3D.

4.3.1. 2D Positioning Test

In order to test the 2D positioning capability of the proposed algorithm, we compared it with the original algorithm. The camera pitch angle α was fixed to 0 and the LED height was set to 160 cm. For the roll angle β, in addition to the seven angles used for model training, we set six angles of 2.5, −2.5, 7.5, −7.5, 12.5, and −12.5 to test the generalization ability of the model. Similarly, for each angle, 121 test points were taken uniformly in a 1 m × 1 m area. At each point, coordinates were calculated using the original and proposed algorithms, respectively. To minimize the impact of random errors, the results were averaged over 10 consecutive frames for each point.

As shown in Figure 13, we depict the variation curves of the average positioning error of the original and proposed algorithms in 13 sets of tests. When using the original algorithm, the positioning error increases with the angle and the average error is 13.69 cm. In contrast, when using the model for positioning, the error remains stable within the range of 0.82823 cm to 5.73876 cm across all 13 sets. It was not affected by the increase in angle and the average error was 2.185 cm. In addition, for the seven angles used for model training, the average error was 1.114 cm. For the six angles not used for model training, the average error was 4.497 cm. Figure 14 compares the cumulative distribution function (CDF) of the positioning errors of the two algorithms. Figure 15 shows the error distribution, which is relatively uniform. The results show a significant improvement in the error of the modeling algorithm compared to the original algorithm. Overall, the proposed algorithm exhibits a certain degree of generalization capability and resistance to angle variations in 2D positioning.

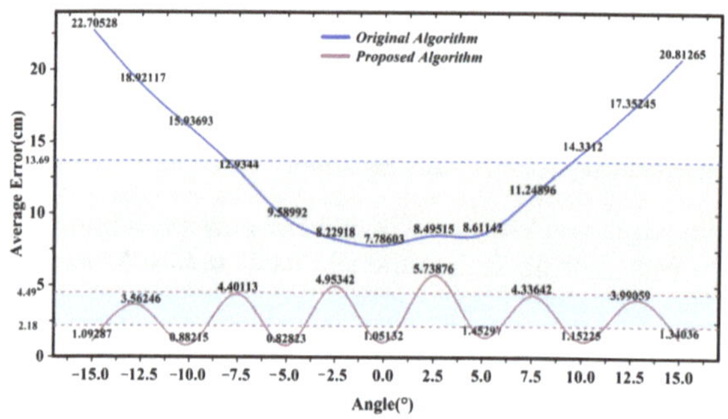

Figure 13. Height 160 cm, average positioning error at different angles.

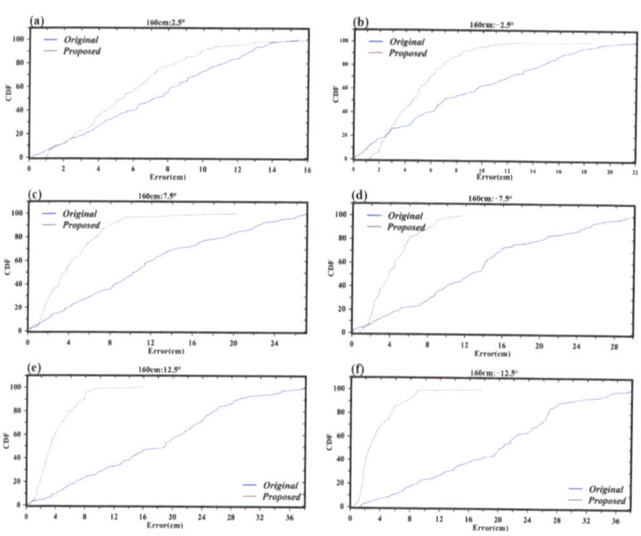

Figure 14. The CDF of 2D positioning errors for different angles; (**a**) 2.5°, (**b**) −2.5°, (**c**) 7.5°, (**d**) −7.5°, (**e**) 12.5°, (**f**) −12.5°.

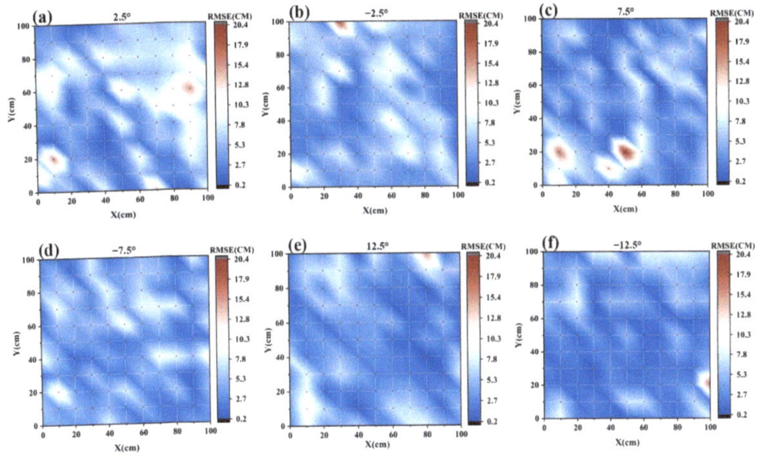

Figure 15. 2D positioning error distribution for different angles; (**a**) 2.5°, (**b**) −2.5°, (**c**) 7.5°, (**d**) −7.5°, (**e**) 12.5°, (**f**) −12.5°.

4.3.2. Height Test

In order to test the capability of the model to predict heights, we similarly conducted a comparison between the original and the proposed algorithm. The camera pitch angle α was fixed at 0°, and the LED heights were set to 160 cm and 170 cm, respectively. Notably, images with a height of 170 cm were not used for model training. For each height, the roll angle β was set to 0 and 12.5, resulting in a total of four sets of tests. Similarly, for each angle, 121 test points were uniformly distributed within a 1 m × 1 m area. At each point, the coordinates were calculated using both the original and the proposed algorithm. To minimize the impact of random errors, the results were averaged over 10 consecutive frames for each point.

Figure 16 shows the error comparison results of the four sets of tests. At a height of 160 cm, as the angle increases from 0 to 12.5, the average error of the original algorithm increases from 4.6 cm to 7.767 cm, while the proposed model algorithm maintains average errors of 1.101 cm and 0.844 cm, respectively. At a height of 170 cm, as the angle increases from 0 to 12.5, and the average error of the original algorithm increases from 9.385 cm to 13.439 cm, whereas the proposed model algorithm achieves average errors of 3.912 cm and 3.736 cm, respectively. According to the results, it can be observed that the error of the original height computation algorithm increases with both the angle and the height. However, with the proposed model algorithm, the error does not increase with the angle. Since the data for the 170 cm height was not used for model training, the error shows a slight increase. In summary, the proposed algorithm exhibits robustness against angle variations and demonstrates a certain degree of model generalization capability in height computation.

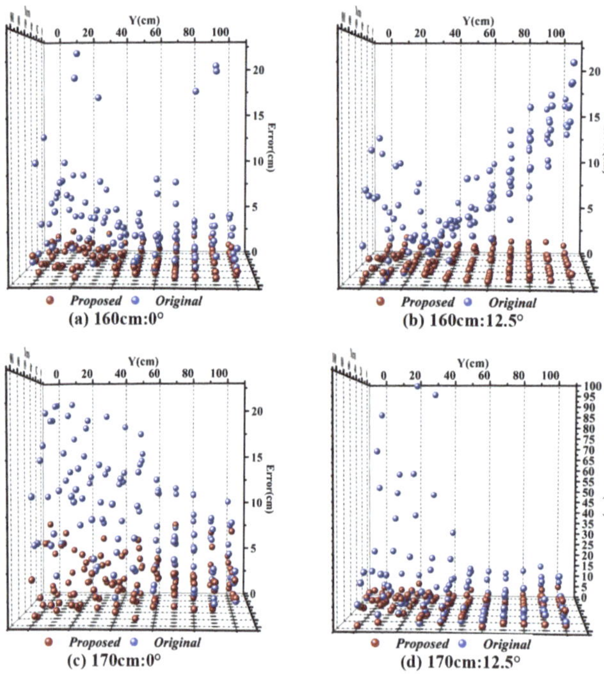

Figure 16. Error comparison at different heights and angles; (**a**) 160 cm:0°, (**b**) 160 cm:12.5°, (**c**) 170 cm:0°, (**d**) 170 cm:12.5°.

4.3.3. 3D Positioning Test

In this part, we tested and compared the 3D positioning capability of the model with the same experimental environment settings as during the height test. The CDF of 3D errors for the four sets of tests is shown in Figure 17.

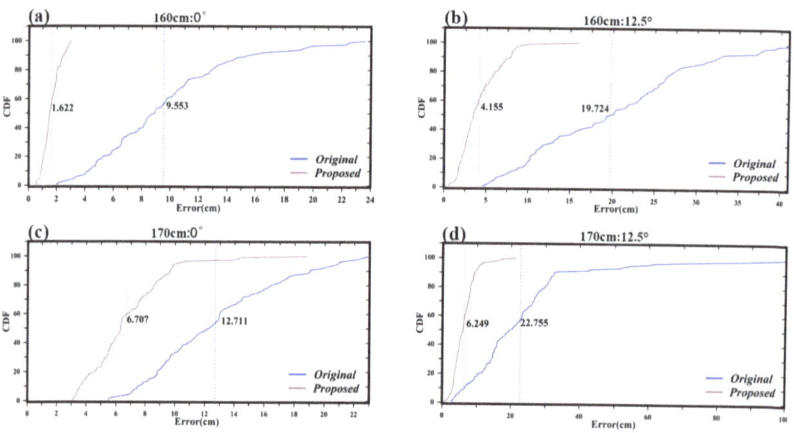

Figure 17. The CDF of 3D positioning errors at different angles and heights; (**a**) 160 cm:0°, (**b**) 160 cm:12.5°, (**c**) 170 cm:0°, (**d**)170 cm:12.5°.

Firstly, considering the test results at a height of 160 cm, as the angle changes from 0 degrees to 12.5 degrees. The 3D average error using the original algorithm increases from 9.553 to 19.724 cm, while the 3D average error of the proposed modeling algorithm only increases from 1.622 cm to 4.155 cm. Since the image data with an angle of 12.5 are not involved in the model training, they are combined with the experimental results of 2D positioning above. We can see that the positioning error of 4.155 cm is within a reasonable range.

Looking again at the two sets of tests for the 170 cm height, we should note that we did not use image data with a height of 170 cm and an angle of 12.5 for model training. When the angle is changed from 0 degrees to 12.5, the 3D average error using the original algorithm increases from 12.711 cm to 22.755 cm. While the proposed model algorithm's average error increases only slightly from 6.707 cm to 6.249 cm, the errors are reduced by about 6.004 cm and 16.506 cm compared to the original algorithm, respectively. Since the image data in these two sets of tests were not involved in the model training, the results can correctly reflect the 3D positioning ability of the model. In summary, our proposed modeling algorithm is resistant to angular changes and has some model generalization ability when performing 3D positioning.

4.4. Discussion

As shown in Table 5, our proposed positioning method eliminates the reliance on IMU when addressing positioning errors caused by camera pose variations. Additionally, we used a medium to low-resolution camera and were able to maintain 3D positioning errors within 7 cm. Compared with [2,3,18], although [2] has a high positioning accuracy, it is poorly convincing with a test deflection angle of only five. In addition, our method only uses the camera and has a lower system cost.

Table 5. Comparison with traditional algorithms.

Require LEDs	Angle (°)	Resolution	Receiver Type	RMSE (cm)	Method	System Cost
1	(−5,5)	1280 × 960	Camera + IMU	2.67	[18]	★★★
1	(−40,40)	2048 × 1536	Camera + IMU	10	[2]	★★★
1	0	Unspecified	Camera + IMU	11.2	[3]	★★★
2	(−40,40)	4032 × 3024	Camera	7.9	[8]	★★
1	(−15,15)	1280 × 720	Camera	6.2	Proposed	★★

The number of stars in the table represents the level of power consumption, with more stars indicating higher power consumption. Compared to [8], our advantage is in the low resolution of the image and the use of only one light. Importantly, the proposed algorithm relies on light intensity variations and the imaging deformation feature and still has a greater potential to ensure highly accurate positioning when coping with greater angular deviations.

5. Conclusions

In visible light positioning systems based on a camera and a single light, changes in camera orientation significantly impact positioning accuracy. Currently, most solutions employ IMU sensors for accuracy compensation. To address this issue, this paper proposes an optimization algorithm based on a convolutional neural network with a multi-head attention mechanism capable of predicting position coordinates when the camera undergoes orientation changes. We design multiple sets of experiments to evaluate the ability of the model to perform 2D positioning, height calculation, and 3D positioning in real time. The experimental results demonstrate that, within an angular range of ±15 degrees, using images with a resolution of 1280 × 720, the model achieves a 2D positioning error within 5.738 cm, height error within 3.912 cm, and 3D positioning error within 6.707 cm. Compared to the original algorithm, the positioning accuracy has been significantly improved. Importantly, this reduces the complexity and cost of the system. Although the maximum angle in the experiments was set to 15 degrees, the algorithm predicts positions based on the relationship between angles and image features. Therefore, it has the potential to handle larger angular deviations. In conclusion, the proposed method provides a reference for the development of indoor positioning technology based on visible light.

In our future work, we will add more training data with light intensity interference caused by varying lighting conditions and changes in LED characteristics to enhance the model's robustness. Additionally, we will explore using lower-resolution signal frames to reduce the algorithm's computational complexity and cost. Finally, we will test larger camera pose variations to evaluate the method's limits.

Author Contributions: Conceptualization, W.J. and X.Z.; methodology, W.J. and X.Z.; software, W.J.; validation, W.J.; investigation, W.J. and X.Z.; resources, Z.W., L.H. and X.Z.; data curation, H.C. and W.J; writing—original draft preparation, W.J. and J.L.; writing—review and editing, W.J., X.Z. and L.H.; visualization, W.J. and X.Z.; supervision, X.Z.; project administration, W.J. and L.H.; funding acquisition, Z.W., L.H. and W.J. All authors have read and agreed to the published version of the manuscript.

Funding: Science and Technology Innovation Activity Program for College Students in Zhejiang Province, China, 2024: [Grant No. 2024R430C027]. Huzhou Key Laboratory of Waters Robotics Technology, Huzhou Science and Technology Bureau: [Grant No. 2022-3]. Zhejiang-French Digital Monitoring Lab for Aquatic Resources and Environment, Department of Science and Technology of Zhejiang Province: [Grant No. 2023-70].

Institutional Review Board Statement: Not applicable.

Informed Consent Statement: Not applicable.

Data Availability Statement: The data that support the findings of this study are available from the corresponding author upon reasonable request.

Conflicts of Interest: The authors declare no conflicts of interest.

References

1. Chen, J.; Zeng, D.; Yang, C.; Guan, W. High accuracy, 6-DoF simultaneous localization and calibration using visible light positioning. *J. Light Technol.* **2022**, *40*, 7039–7047. [CrossRef]
2. Cheng, H.; Xiao, C.; Ji, Y.; Ni, J.; Wang, T. A single LED visible light positioning system based on geometric features and CMOS camera. *IEEE Photonics Technol.* **2020**, *32*, 1097–1100. [CrossRef]
3. Hao, J.; Chen, J.; Wang, R. Visible light positioning using a single LED luminaire. *IEEE Photonics J.* **2019**, *11*, 1–13. [CrossRef]
4. Lin, B.; Ghassemlooy, Z.; Lin, C.; Tang, X.; Li, Y.; Zhang, S. An indoor visible light positioning system based on optical camera communications. *IEEE Photonics Technol.* **2017**, *29*, 579–582. [CrossRef]
5. Li, Y.; Ghassemlooy, Z.; Tang, X.; Lin, B.; Zhang, Y. A VLC smartphone camera based indoor positioning system. *IEEE Photonics Technol.* **2018**, *30*, 1171–1174. [CrossRef]
6. Kuo, Y.-S.; Pannuto, P.; Hsiao, K.-J.; Dutta, P. Luxapose: Indoor positioning with mobile phones and visible light. In Proceedings of the 20th Annual International Conference on Mobile Computing and Networking, Maui, HI, USA, 7–11 September 2014; pp. 447–458.
7. Rahman, M.S.; Haque, M.M.; Kim, K.-D. Indoor positioning by LED visible light communication and image sensors. *Int. J. Electr. Comput. Eng.* **2011**, *1*, 161. [CrossRef]
8. Yoshino, M.; Haruyama, S.; Nakagawa, M. High-accuracy positioning system using visible LED lights and image sensor. In Proceedings of the 2008 IEEE Radio and Wireless Symposium, Orlando, FL, USA, 22–24 January 2008; pp. 439–442.
9. Xu, J.; Gong, C.; Xu, Z. Experimental indoor visible light positioning systems with centimeter accuracy based on a commercial smartphone camera. *IEEE Photonics J.* **2018**, *10*, 1–17. [CrossRef]
10. Zhang, R.; Zhong, W.-D.; Kemao, Q.; Zhang, S. A single LED positioning system based on circle projection. *IEEE Photonics J.* **2017**, *9*, 1–9. [CrossRef]
11. Li, H.; Huang, H.; Xu, Y.; Wei, Z.; Yuan, S.; Lin, P.; Wu, H.; Lei, W.; Fang, J.; Chen, Z. A fast and high-accuracy real-time visible light positioning system based on single LED lamp with a beacon. *IEEE Photonics J.* **2020**, *12*, 1–12. [CrossRef]
12. Moon, M.-G.; Choi, S.-I.; Park, J.; Kim, J.Y. Indoor positioning system using LED lights and a dual image sensor. *IEEE Photonics J.* **2015**, *19*, 586–591. [CrossRef]
13. Guan, W.; Zhang, X.; Wu, Y.; Xie, Z.; Li, J.; Zheng, J. High precision indoor visible light positioning algorithm based on double LEDs using CMOS image sensor. *Appl. Sci.* **2019**, *9*, 1238. [CrossRef]
14. Zhao, H.; Zhang, X.; Bader, F.; Zhang, Y. Non-Point Visible Light Transmitter Localization based on Monocular Camera. In Proceedings of the 2021 IEEE International Symposium on Broadband Multimedia Systems and Broadcasting (BMSB), Chengdu, China, 4–6 August 2021; pp. 1–5.

15. Qin, C.; Zhan, X. VLIP: Tightly coupled visible-light/inertial positioning system to cope with intermittent outage. *IEEE Photonics Technol.* **2018**, *31*, 129–132. [CrossRef]
16. Bera, K.; Parthiban, R.; Karmakar, N. A Truly 3D Visible Light Positioning System using Low Resolution High Speed Camera, LIDAR, and IMU Sensors. *IEEE Access* **2023**, *11*, 98585. [CrossRef]
17. Hou, Y.; Xiao, S.; Bi, M.; Xue, Y.; Pan, W.; Hu, W. Single LED beacon-based 3-D indoor positioning using off-the-shelf devices. *IEEE Photonics J.* **2016**, *8*, 1–11. [CrossRef]
18. Huang, H.; Feng, L.; Ni, G.; Yang, A. Indoor imaging visible light positioning with sampled sparse light source and mobile device. *Chin. Optics* **2016**, *14*, 090602. [CrossRef]
19. Kim, J.Y.; Yang, S.H.; Son, Y.H.; Han, S.K. High-resolution indoor positioning using light emitting diode visible light and camera image sensor. *IEEE Photonics J.* **2016**, *10*, 184–192. [CrossRef]
20. Naseer, T.; Burgard, W. Deep regression for monocular camera-based 6-dof global localization in outdoor environments. In Proceedings of the 2017 IEEE RSJ International Conference on Intelligent Robots and Systems (IROS), Vancouver, BC, Canada, 24–28 September 2017; pp. 1525–1530.
21. Wang, S.; Manning, C. Fast dropout training. In Proceedings of the International Conference on Machine Learning, Atlanta, GA, USA, 16–21 June 2013; pp. 118–126.
22. Philipp, G.; Song, D.; Carbonell, J.G. Gradients Explode-Deep Networks are Shallow-ResNet Explained. In Proceedings of the 6th International Conference on Learning Representations ICLR Workshop Track, Vanvouver, BC, Canada, 30 April–3 May 2018.
23. Guo, M.-H.; Xu, T.-X.; Liu, J.-J.; Liu, Z.-N.; Jiang, P.-T.; Mu, T.-J.; Zhang, S.-H.; Martin, R.R.; Cheng, M.-M.; Hu, S.-M. Attention mechanisms in computer vision: A survey. *Comput. Vis. Media* **2022**, *8*, 331–368. [CrossRef]

photonics

MDPI

Article

Neural Network Equalisation for High-Speed Eye-Safe Optical Wireless Communication with 850 nm SM-VCSELs

Isaac N. O. Osahon [1,2], Ioannis Kostakis [3], Denise Powell [4], Wyn Meredith [4], Mohamed Missous [3,5], Harald Haas [2], Jianming Tang [1] and Sujan Rajbhandari [6,*]

1 DSP Centre of Excellence, School of Computer Science and Electronic Engineering, Bangor University, Bangor LL57 1UT, UK; ino20@cam.ac.uk (I.N.O.O.); j.tang@bangor.ac.uk (J.T.)
2 LiFi Research and Development Centre, Electrical Engineering Division, University of Cambridge, 9 J J Thomson Avenue, Cambridge CB3 0FA, UK; huh21@cam.ac.uk
3 Integrated Compound Semiconductors Ltd., Manchester M17 1RW, UK; i.kostakis@icsld.com (I.K.); m.missous@icsld.com (M.M.)
4 Compound Semiconductor Centre Ltd., St. Mellons, Cardiff CF3 0LW, UK; dpowell@compoundsemicentre.com (D.P.); wmeredith@compoundsemicentre.com (W.M.)
5 Department of Electrical and Electronic Engineering, University of Manchester, Oxford Road, Manchester M13 9PL, UK
6 Institute of Photonics, University of Strathclyde, Glasgow G1 1RD, UK
* Correspondence: sujan.rajbhandari@strath.ac.uk

Abstract: In this paper, we experimentally illustrate the effectiveness of neural networks (NNs) as non-linear equalisers for multilevel pulse amplitude modulation (PAM-*M*) transmission over an optical wireless communication (OWC) link. In our study, we compare the bit-error-rate (BER) performances of two decision feedback equalisers (DFEs)—a multilayer-perceptron-based DFE (MLPDFE), which is the NN equaliser, and a transversal DFE (TRDFE)—under two degrees of non-linear distortion using an eye-safe 850 nm single-mode vertical-cavity surface-emitting laser (SM-VCSEL). Our results consistently show that the MLPDFE delivers superior performance in comparison to the TRDFE, particularly in scenarios involving high non-linear distortion and PAM constellations with eight or more levels. At a forward error correction (FEC) threshold BER of 0.0038, we achieve bit rates of ~28 Gbps, ~29 Gbps, ~22.5 Gbps, and ~5 Gbps using PAM schemes with 2, 4, 8, and 16 levels, respectively, with the MLPDFE. Comparably, the TRDFE yields bit rates of ~28 Gbps and ~29 Gbps with PAM-2 and PAM-4, respectively. Higher PAM levels with the TRDFE result in BERs greater than 0.0038 for bit rates above 2 Gbps. These results highlight the effectiveness of the MLPDFE in optimising the performance of SM-VCSEL-based OWC systems across different modulation schemes and non-linear distortion levels.

Keywords: optical wireless communications; vertical-cavity surface-emitting lasers; multilevel pulse amplitude modulation; digital equalisation; neural network; multilayer perceptron

1. Introduction

Optical wireless communication (OWC) stands as a pivotal technology that offers substantial opportunities to meet the demanding requirements of 6G and beyond and can serve as both access and cross-haul links; its versatile applications extend to device-to-device (D2D) communications and the Internet-of-Things (IoT) [1]. Two prominent OWC variants, visible light communication (VLC) and infrared OWC (IOWC), have showcased remarkable bit rates exceeding 20 Gbps for one optical wavelength [2], with VLC achieving up to 35 Gbps through wavelength division multiplexing (WDM) [3]. While VLC excels in scenarios requiring illumination, IOWC presents enhanced versatility and is particularly suitable for applications like LiFi uplinks and data centres. Integrating IOWCs with high-speed fibre communication systems has become increasingly significant, offering terabits-per-second (Tbps) indoor wireless access and interfacing seamlessly with fibre-to-the-home access networks [4]. Despite these advantages, the mobility of high-speed IOWC systems is constrained due to the use of narrow-divergence transmitters and receivers with narrow fields-of-view (FOVs). To address this limitation, implementing robust user tracking, localisation mechanisms, and optical beam steering is essential for the practical deployment of IOWC, thereby ensuring its effectiveness and widespread application [4,5].

Vertical-cavity surface-emitting lasers (VCSELs) stand out as important light sources for intensity-modulated/direct-detection (IM/DD)-based OWC systems, primarily owing to their distinct advantages over other optical sources. Notably, VCSELs boast cost-effectiveness in terms of fabrication and require lower electrical power consumption compared to edge-emitting lasers [6]. Their superiority extends to offering higher modulation bandwidths and superior emission coherence in comparison to light-emitting diodes (LEDs). Leveraging their vertical emission and streamlined fabrication, VCSELs can be seamlessly integrated into two-dimensional (2D) arrays, facilitating spatial diversity and enabling the deployment of multiple-input–multiple-output (MIMO) systems [7]. The significance of VCSELs is underscored by experimental endeavours in gigabit OWC, where individual VCSEL sources operating at wavelengths of $\sim$650 nm [8,9], $\sim$850 nm [10], and 1310 nm [2] have achieved bit rates of up to 25 Gbps using IM/DD. The pioneering fabrication of VCSELs within the 850–980 nm band marks a significant milestone in optical communications, and their continued dominance in the market is attributed to the alignment of these wavelengths with the high responsivity of cost-effective silicon photodetectors (PDs) [6]. However, the paramount consideration lies in ensuring the safety of the 850 nm system for the human eye, especially for protecting the retina. To meet the stringent class-1 laser safety requirements, compliance with the accessible emission limit (AEL) becomes imperative, and this is ~-1.10 dBm, as specified by the International Electrotechnical Commission (IEC 60825-1) for a point source operating at 850 nm [11].

In the pursuit of optimising the information capacity of VCSEL-based OWC systems, extensive research has been dedicated to exploring spectral-efficient modulation schemes. Among these, multilevel pulse amplitude modulation (PAM-M) has emerged as a prominent contender. PAM involves transmitting symbols through pulses in a signal waveform, with the amplitudes of the pulses representing different symbols. A PAM transmission with M distinct amplitudes or "levels" represents $\log_2 M$ bits in each level, providing a robust mechanism for conveying information. PAM is the preferred choice for many commercial OWC systems due to its simplicity in handling real-valued symbols and its spectrum efficiencies that are comparable to other advanced modulation schemes such as orthogonal frequency division multiplexing (OFDM) [12]. However, PAM is not without challenges, mainly when deployed in high-speed communication scenarios, as it is susceptible to intersymbol interference (ISI) and system non-linearities [13]. Consequently, in numerous optical communication systems, PAM is often accompanied by an equaliser, such as a decision feedback equaliser (DFE) [13,14]. This strategic pairing ensures the mitigation of potential issues, enhancing the reliability and performance of VCSEL-based OWC systems employing PAM.

While DFEs effectively address ISI, non-linearity renders them inefficient, particularly for higher modulation levels. Therefore, non-linear DFEs like the Volterra-series equaliser and neural networks (NNs) are utilised to overcome ISI and non-linear effects. For instance, ref. [15] demonstrated that recurrent NNs outperform conventional DFEs in non-linear channels. Feedforward NNs, such as multilayer perceptrons (MLPs), radial basis functions (RBFs), and long short-term memory (LSTM), have been tested on optical channels using PAM-M [8,16,17]. Specifically, ref. [16] showed that MLP outperforms both conventional DFEs and Volterra-series equalisers.

To the best of the authors' knowledge, few experimental studies have reported using 850 nm VCSELs with PAM-M for OWC transmissions, especially with launched optical powers below the AEL. The work outlined in [18] that uses 850 nm VCSELs reported a bit rate of 25 Gbps, but this was with a 7×7 array with an optical power of $\sim$22 dBm.

The study in [8] employed a deep LSTM NN for effective ISI compensation in VCSEL-based OWC, achieving a data rate of 13.5 Gbps. However, the experimental work was limited to PAM-2, and the NN used at least 100 neurons per layer.

This paper presents the first experimental evaluations of directly modulated eye-safe single-mode VCSELs (SM-VCSELs) operating at 850 nm for high-speed OWC links. The evaluation incorporates PAM-M and employs an NN equaliser conducted over a 2.5 m OWC link. The SM-VCSEL is launched with a transmitted optical power of −1.47 dBm, which is securely below the AEL of −1.10 dBm, ensuring its adherence to the class-1 laser safety standards and thus complying with eye safety. Compared to multimode VCSELs (MM-VCSELs), SM-VCSELs offer advantages such as narrower spectral widths and lower-intensity noise [6]. We obtain a bit rate of about 29 Gbps at a link distance of 2.5 m with PAM-4 modulation and an NN equaliser. The main contributions of this work are summarised as follows:

1. We show the first use of NN equalisers in an eye-safe 850 nm SM-VCSEL-based OWC that employs PAM-M schemes of up to 16 levels. The equalisers demonstrate superior performance to the conventional DFEs, particularly for high-level PAM {8,16}, which requires less bandwidth but suffers from higher device non-linearity. This is especially beneficial in OWC links using photodiodes with larger detection areas that offer higher power margins but lower bandwidths. While some studies have employed NN equalisers for OWC with PAM up to eight levels [8,17,19,20], this paper is, to the best of the authors' knowledge, the first experimental work to achieve a multigigabit OWC link with PAM-16 and NN equalisers by leveraging the SM-VCSELs' high signal-to-noise ratio (SNR).
2. Previous work on NN equalisation for VCSEL-based OWC employed multiple hidden layers with over 100 neurons per layer. The NN equaliser for this study require less computational complexity, utilising only one hidden layer with six neurons.
3. We highlight the non-linearity compensation capability of the NN equaliser by examining two modulation conditions for the VCSELs, with the first condition exhibiting lower non-linearity than the second. Notably, NN equalisers perform better in the second scenario. Additionally, for low-level PAM {2,4}, NN equalisers provide superior bit rate performance compared to DFEs with MM-VCSELs, despite MM-VCSELs having higher non-linearity than SM-VCSELs.

This study parallels the experimental report in [21], where we achieved a bit rate of 38 Gbps using optimised OFDM with bit and power loading. However, the goal of this work is to demonstrate the rates achievable using simple PAM modulation schemes and to experimentally show the effectiveness of NNs in compensating for non-linearity and ISI. While [21] aims to maximise the achievable bit rate through an optimised OFDM scheme, this work focuses on the non-linear behaviour of VCSELs and explores possible mitigation using an NN-based non-linear equaliser with multilevel PAM. Moreover, we employ lower modulation amplitudes in this study, as PAM has a lower peak-to-average-power ratio than OFDM.

2. Experiment Setup

Figure 1a depicts a photo of the VCSEL-based OWC system, showcasing its functionality over a transmission distance of 2.5 m. The essential instruments and devices governing the OWC system are highlighted in Table 1. Figure 1a is complemented by Figure 1b, which illustrates a schematic diagram presenting the PAM generation and decoding process employed within the IM/DD system. It begins with the offline generation of a random bit sequence using MATLAB, which is subsequently mapped to PAM symbols through gray coding. These PAM symbols undergo upsampling and processing via a non-return-to-zero (NRZ) rectangular pulse-shaping filter. The resulting signal is then loaded to an arbitrary waveform generator (AWG) for the creation of an analogue waveform, which, in turn, drives the VCSEL directly through a bias tee (Tektronix PSPL5542, 50 GHz).

Table 1. OWC system parameters.

Device/System	Module Description	Setup/Parameters
AWG	Keysight M8195A	BW = 23 GHz, SR = 60 GSa/s, 8-bit Res. DAC
Photodetector	Newport 818-BB-45A	BW = 9 GHz, ∅ 60 μm, ∼550 V/W Gain at 850 nm
Optics Lens	Thorlabs ACL2520U	∅ 25 mm, Focal length = 20 mm
Oscilloscope	Tektronix DPO71254C	BW = 12.5 GHz, SR = 50 GSa/s, 8-bit Res. ADC

BW—Bandwidth, SR—Sampling Rate, and Res.—Resolution.

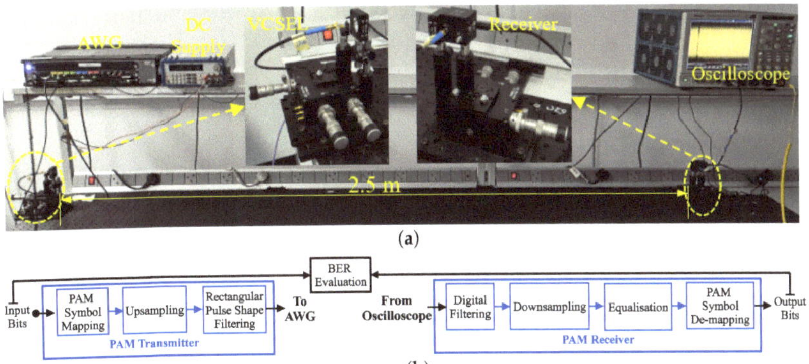

Figure 1. Illustration of experiment setup: (**a**) Photo of VCSEL-based OWC link. (**b**) Schematic for multilevel PAM.

The VCSEL output undergoes collimation using an aspheric lens (Thorlabs ACL2520U). At the receiver, the optical signal is collected through an aspheric lens and is focused onto a PIN PD module (818-BB-45A). To facilitate optimal OWC link alignment and receiver optical power adjustment, both the VCSEL and PD are mounted on 3D-axis stages (Thorlabs NanoMax 300). The received signal is then captured by an oscilloscope. Then, standard digital signal processing (DSP) techniques are employed, including synchronisation, digital filtering, and equalisation. Finally, the transformed PAM symbols are de-mapped into the received bitstream, facilitating the evaluation of the bit-error-rate (BER) performance.

The model that would best describe the output signal for the IM/DD OWC link assuming that most of the system non-linearity comes from the VCSEL is defined as [22,23]:

$$y(t) = \mathcal{G}_{pd}\mathcal{G}_{owc}f_{L-I-V}\left((V_{bias} + s(t)) \otimes h_{vcsel}(t)\right) \otimes h_{owc}(t) \otimes h_{pd}(t) + n(t), \quad (1)$$

where $\mathcal{G}_{pd}$ and $\mathcal{G}_{owc}$ denote the gains from the photodetector and OWC link, respectively; $f_{L-I-V}(.)$ denotes the non-linear transfer function that converts the input voltage signal to the optical signal; V_{bias} denotes the bias voltage to drive the VCSEL, and $s(t)$ denotes the bipolar input voltage signal; $h_{vcsel}(t)$, $h_{owc}(t)$ and $h_{pd}(t)$ denote the impulse responses for the VCSEL, OWC link and photodetector, respectively; $n(t)$ denotes the noise. This memory-based model is particularly useful for high-speed OWC links with significant system non-linearities.

2.1. Single-Mode and Multimode VCSELs Comparison

The SM-VCSEL is custom-designed and fabricated by Integrated Compound Semiconductor (ICS) from a GaAs/AlGaAs-based epitaxial structure supplied by the Compound Semiconductor Centre (CSC), whereas the MM-VCSEL is an off-the-shelf device (OPTEK OPV310). Measurements characterising both devices, including light–current–voltage (L-I-V) curves, magnitude responses and noise power spectral densities (PSDs), can be found in [21]. The L-I-V curves show that the SM-VCSEL exhibits higher dynamic resistance and emits lower optical power due to its smaller oxide aperture, which ensures single-mode emission. To maintain eye-safe conditions, we drive the SM-VCSEL at approximately 2.2 mA, resulting in an optical power of around 0.7 mW (−1.47 dBm). Similarly, the MM-VCSEL is driven at about 5 mA to emit approximately 3 mW (4.7 dBm) of optical power. This ensures comparable bandwidths between the two VCSELs for meaningful performance comparisons. The measured −3 dB bandwidths of SM-VCSEL and MM-VCSEL are approximately 6.2 GHz and 6.4 GHz, respectively.

We compare both VCSELs for the OWC link based on their measured noise power spectral densities (PSDs) at a received optical power (ROP) of approximately −1.75 dBm. The noise PSDs for the SM-VCSEL and MM-VCSEL average around −137 dB/Hz and −129 dB/Hz, respectively. In comparison, the OWC system with a ROP of −1.75 dBm has thermal and shot noises estimated at −146 dB/Hz and −151 dB/Hz, respectively. Therefore, the primary noise source for the OWC link is the relative intensity noise (RIN) from the VCSELs, as the noise PSDs from both VCSELs exceeds that of shot or thermal noise.

We focus more on the eye-safe SM-VCSEL for PAM-M since it has a lesser RIN than the MM-VCSEL and offers a higher data rate for similar operating conditions [6,21]. To illustrate the performance of the equalisers for the SM-VCSEL, we consider two setups with the SM-VCSEL-based OWC link as follows:

- Setup-I: amplitude of 0.5 Vpp corresponding to ∼70% modulation index at 100 MHz, maximum ROP of −1.75 dBm;
- Setup-II: amplitude of 0.7 Vpp corresponding to ∼95% modulation index at 100 MHz, maximum ROP of −3.08 dBm.

While Setup-II offers a higher SNR than Setup-I, it does so at the expense of higher non-linearity. This is demonstrated in Figure 2, which illustrates the non-linearity of the SM-VCSEL (with both setups) and the MM-VCSEL (with Setup-I) through their total harmonic distortion (THD). The THD indicates the level of additional harmonics per frequency, with higher THD signifying greater non-linear distortion from these harmonics [24]. The MM-VCSEL exhibits approximately 3.5 dB more THD than the SM-VCSEL with both VCSELs at ∼70% modulation index. Moreover, the SM-VCSEL has about 2.2 dB more THD when Setup-II is used instead of Setup-I.

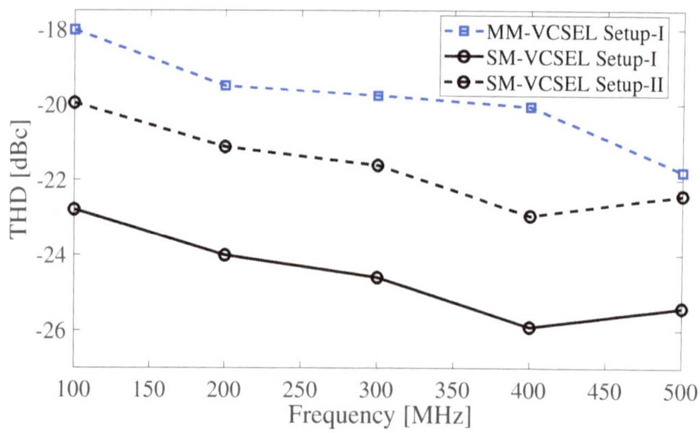

Figure 2. THD vs. frequency for the VCSELs.

2.2. NN Equalisation

In channel equalisation tasks, various NN architectures can be employed [25]. However, our work uses multilayer perceptrons (MLPs), as they are among the oldest and the most common types of feedforward NNs, making them a good representation of other NNs [26]. MLP is a collection of perceptrons organised into layers, as illustrated in Figure 3. Each perceptron computes a linear combination of its inputs and incorporates an externally applied bias. The result of this combination is applied to an activation function [26]. The expression for the output of a perceptron is:

$$y = f_{avn}\left(w_{bias} + \sum_{i=0}^{m-1} w_i x_i\right), \tag{2}$$

where m denotes the number of inputs to the neuron, and $f_{avn}(.)$ is the activation function of the neuron; x_i denotes the neuron inputs, w_i denotes the weight for each input x_i, and w_{bias} denotes the neuron bias.

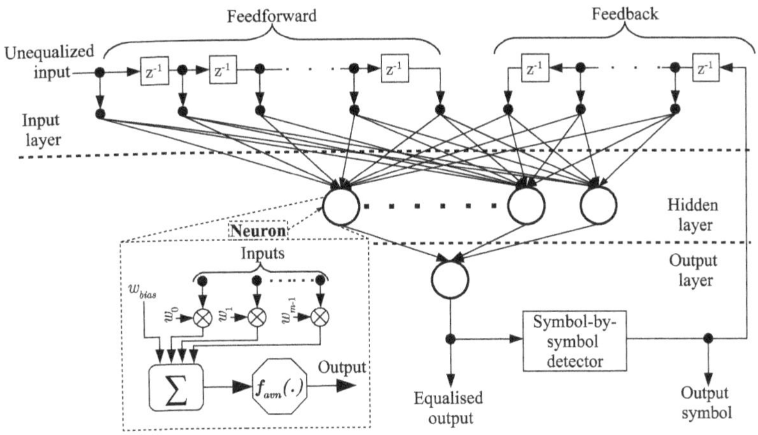

Figure 3. Illustration of the NN-based decision feedback equaliser.

A three-layer architecture suffices for the MLP for equalisation due to the universal approximation theorem [25,26]. This theorem asserts that for non-linear input–output mapping, if the activation function of neurons in the hidden layer is monotonically continuous, bounded and non-constant, a finite number of neurons in the layer can approximate the mapping effectively. The sigmoid function, which satisfies these conditions, is commonly employed as the activation function for neurons in the hidden layer. It is defined in (3) as:

$$f_{sgm}(x) = \frac{2a_1}{1+\exp(-a_2 x)} - a_1 = a_1 \tanh(a_2 x), \tag{3}$$

where a_1 and a_2 are suitably chosen constants. In line with the study in [16], tan-sigmoid ($a_1 = a_2 = 1$) is employed in this study as the activation function. The activation function for the neuron in the output layer is the linear function ($f_{lin}(x) = x$). The feedback inputs can be sent to the MLP equaliser like with a conventional DFE, as illustrated in Figure 3. This MLP-based DFE (MLPDFE) can be expressed as [26]:

$$\tilde{z}_n = \sum_{c=1}^{N_{hn}}\left[w_c \tanh\left(w_{c,-1} + \sum_{a=0}^{N_{ft}-1} w_{c,a} y_{n-a} + \sum_{b=1}^{N_{bt}} w_{c,b}\hat{z}_{n-b}\right)\right] + w_{-1}, \tag{4}$$

where $\tilde{z}_n$ denotes the output of the MLPDFE; $\{y_n, \ldots, y_{n+1-N_{ft}}\}$ is the unequalised input sequence, and $\{\hat{z}_{n-1}, \ldots, \hat{z}_{n-N_{bt}}\}$ is the set of previously detected symbols; w_a and w_b are the co-efficients of the feedforward and feedback tap weights, respectively; N_{ft} and N_{bt} denote the number of feedforward and feedback taps, respectively, for the DFE; N_{hn} is the number of hidden-layer neurons; $w_{c,a}$ and $w_{c,b}$ are the synaptic weights for processing the feedforward and feedback inputs, respectively, for a hidden-layer neuron (with index c); w_c denotes the weight used to process the neuron at the output layer; $w_{c,-1}$ and w_{-1} denote the biases of a hidden-layer neuron and the output layer neuron, respectively.

Compared to the conventional DFE, the MLPDFE requires more computational complexity and memory due to the additional hidden-layer neurons: each has a non-linear function. Based on the analysis done in [27], the MLPDFE needs about N_{hn} times more real-valued multiplications than the conventional DFE. The performance of MLP is significantly impacted by the number of neurons in the hidden layer. Too few hidden neurons mean less computational complexity but could result in poor error performance. Conversely, an excessive number of neurons might lead to irregular error performance due to overfitting [25]. The back-propagation (BP) algorithm is popular for training the MLP equaliser. While there are multiple variants of the BP algorithm, the Levenberg–Marquardt BP (LMBP) algorithm is chosen in this study due to its superior convergence and better mean-square-error (MSE) performance [26]. However, LMBP uses batch training, which requires more computing memory than other BP algorithms that employ online training. The parameters for the MLP equaliser are highlighted in Table 2. Increasing the equaliser parameters beyond those in Table 2 offers negligible improvement at the cost of increased complexity.

Table 2. Equaliser parameters.

Parameter	Symbol	Value
Number of Forward Taps	N_{ft}	32
Number of Feedback Taps	N_{bt}	8
Number of Hidden-Layer Neurons	N_{hn}	6
Number of Training Symbols	N_{tr}	4000
Number of Bits for BER Testing		10^6

3. Communication Performance and Analysis

To assess the effectiveness of the NN equaliser in handling PAM-M, this section presents the BER results from the SM-VCSEL-based OWC experiment setups (Setup-I and Setup-II) discussed in Section 2. The bit rates range from 2 Gbps to 40 Gbps in order to explore multigigabit communication over the SM-VCSEL-based OWC link. For comparison purposes, the link is evaluated (1) without equalisation (unequalised), (2) with the transversal DFE (TRDFE), and (3) with the MLPDFE. The parameters for both the TRDFE and the MLPDFE are summarised in Table 2.

The BER results are presented across various bit rates for the equalisers employing PAM-2 and PAM-4 schemes for the SM-VCSEL and the MM-VCSEL in Figure 4. A BER of $\sim 3.8 \times 10^{-3}$, which is the second-generation super forward error correction (S-FEC) limit and has a 7% overhead and $\sim$6 dB coding gain [28], is used to estimate the system's bit rate performance. For PAM-2, the MLPDFE output layer uses the sigmoid activation function due to its "binary" nature and superior performance compared to the linear activation function [16]. However, the linear activation function is better for PAM at four or more levels. Without equalisation at this FEC limit, the PAM-2 scheme with the SM-VCSEL achieves bit rates of $\sim$8 Gbps and $\sim$9 Gbps for Setup-I and Setup-II, respectively. Conversely, the bit rates achieved with PAM-4 are $\sim$5 Gbps and $\sim$4.5 Gbps for Setup-I and Setup-II, respectively. A notable observation is that Setup-II demonstrates superior performance with PAM-2, while Setup-II showcases better results with PAM-4. This discrepancy arises due to PAM-2's lower susceptibility to non-linearity than PAM-4, leveraging the high SNR specifically achievable in Setup-II.

With PAM-2, the MLPDFE demonstrates a marginal BER performance improvement compared to the TRDFE across all bit rates. Both equalisers significantly enhance the OWC link, enabling bit rates of $\sim$23 Gbps and $\sim$28 Gbps at the 7% S-FEC limit for Setup-I and Setup-II, respectively. With PAM-4, the MLPDFE and TRDFE exhibit similar BER performances for bit rates exceeding 25 Gbps, achieving $\sim$27 Gbps and $\sim$29 Gbps at the 7% S-FEC limit for Setup-I and Setup-II, respectively. However, the MLPDFE notably outperforms the TRDFE for bit rates below 25 Gbps, i.e., at low BER values. For instance, at a BER of 10^{-6} with Setup-I, the MLPDFE achieves bit rates of $\sim$11 Gbps, while the TRDFE offers $\sim$6 Gbps. This is because non-linearity is the dominant performance-limiting factor at lower bit rates, while SNR becomes the limiting factor at higher bit rates. Hence, this result clearly demonstrates the effectiveness of the NN in compensating for non-linearity, which is further evident at higher modulation levels. It is worth mentioning that both equalisers exhibit better performance with PAM-4 on Setup-I than in Setup-II at bit rates below 25 Gbps. Specifically, the MLPDFE achieves ∼7 Gbps and ∼10 Gbps at a BER of 10^{-6} for Setup-I and Setup-II, respectively. Additionally, both equalisers showcase enhanced performance with PAM-2 on Setup-II relative to Setup-I at lower bit rates, as the MLPDFE attains ∼10 Gbps and ∼15 Gbps for Setup-I and Setup-II, respectively, at a BER of 10^{-6}.

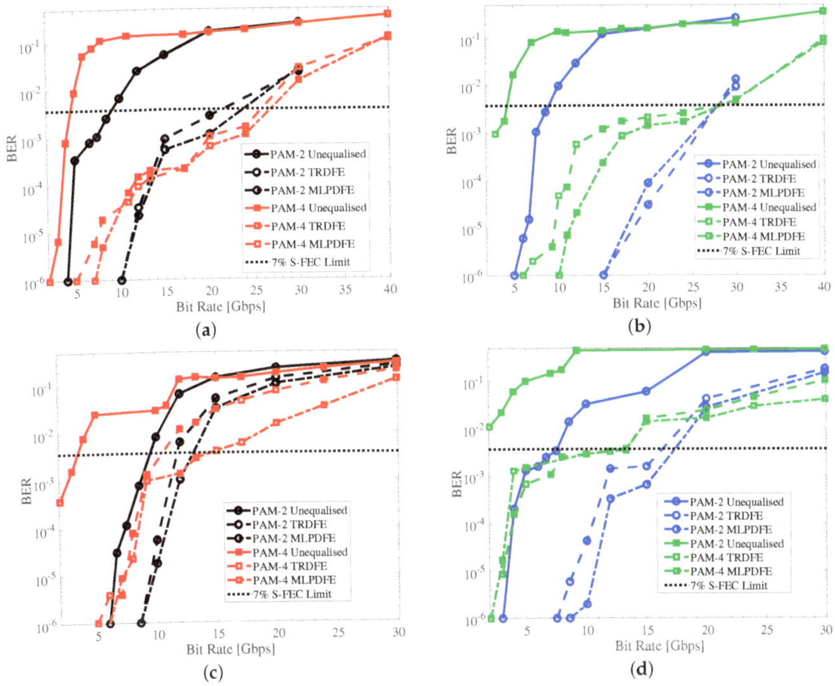

Figure 4. BER vs. bit rate plots comparing the equalisers' performance with PAM-2 and PAM-4 for: (**a**) SM-VCSEL with Setup-I, (**b**) SM-VCSEL with Setup-II, (**c**) MM-VCSEL with Setup-I, and (**d**) MM-VCSEL with Setup-II.

As with SM-VCSEL, equalisation significantly improves the performance of MM-VCSEL links. However, due to higher RIN from the MM-VCSEL, its achievable bit rates are considerably lower than those of SM-VCSEL at the 7% S-FEC limit, such as 15 Gbps with PAM-4 MLPDFE in Setup-I and ∼17.5 Gbps with PAM-2 MLPDFE with Setup-II. It is also worth noting that, unlike the SM-VCSEL, the MLPDFE offers better bit rate performance than TRDFE for PAM-2 from the MM-VCSEL because the MM-VCSEL has higher non-linearity than the SM-VCSEL. Nonetheless, we will focus only on the SM-VCSEL in the following sections due to its superior bit rate performance compared to the MM-VCSEL.

Figure 5 presents the BER results for bit rates with PAM-8 and PAM-16 using Setup-I (Figure 5a) and Setup-II (Figure 5b). Significant BER challenges are evident across different modulation schemes and setups without equalisation. Firstly, with PAM-8 modulation on Setup-I, the BER surpasses 0.07 for bit rates starting from 2 Gbps and beyond. Similarly, using PAM-16 on Setup-I exacerbates the situation, with the BER exceeding 0.15 for bit rates of 2 Gbps and higher. Conversely, with Setup-II, employing PAM-8 modulation results in a BER of ∼0.02 at a bit rate of 2 Gbps, suggesting relatively improved performance compared to Setup-I. However, with PAM-16 on Setup-II, the BER exceeds 0.09 for bit rates starting from 2 Gbps. These results underscore the critical importance of effective equalisation strategies in mitigating ISI and non-linearity for OWC links utilising PAM-8 and PAM-16.

The comparison between the TRDFE and the MLPDFE underscores the latter's superior performance, particularly in Setup-II. With the TRDFE on Setup-I, the BER surpasses

0.03 for bit rates of 2 Gbps and beyond with PAM-8, escalating to over 0.09 with PAM-16. In contrast, Setup-II exhibits improved performance with the TRDFE but still encounters BER issues exceeding 0.07 for bit rates beyond 2 Gbps with PAM-16. These results demonstrate the ineffectiveness of the TRDFE to mitigate the ISI in the presence of non-linearity.

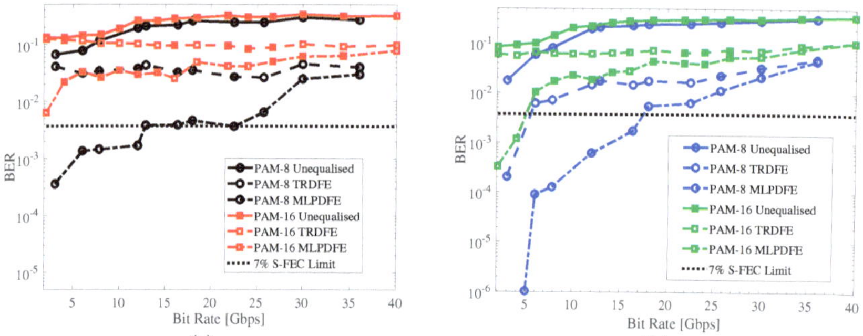

(a) (b)

Figure 5. BER vs. bit rate plots comparing the equalisers' performance for the SM-VCSEL with PAM-8 and PAM-16 for: (**a**) Setup-I and (**b**) Setup-II.

The MLPDFE, however, showcases remarkable improvements in both setups. In Setup-I, it achieves an impressive bit rate of about 22.5 Gbps at the 7% S-FEC limit with PAM-8, alongside a low BER of about 3.5×10^{-4} at a bit rate of 3 Gbps with the same modulation scheme. Similarly, in Setup-II, the MLPDFE achieves a substantial bit rate of about 17.5 Gbps with PAM-8 at the 7% S-FEC limit, coupled with an exceptionally low BER of about 10^{-6} at a bit rate of 5 Gbps. These results signify the MLPDFE's effectiveness in significantly improving transmission rates and minimising errors, which is particularly notable in Setup-II compared to the TRDFE, highlighting its superiority in optimising OWC systems with high non-linearities.

To highlight the effectiveness of the NN equaliser, eye diagrams for the output of each equaliser using PAM-8 and PAM-16 modulation schemes are presented in Figure 6. These diagrams are computed by upsampling the received PAM symbols by eight and filtering the result with a raised cosine filter with a 0.35 roll-off factor. The diagrams are presented at 12 Gbps for PAM-8 and 6 Gbps for PAM-16 in Figures 6a and 6b, respectively. Without equalisation, the PAM-8 and PAM-16 waveforms exhibit severe distortion due to ISI and non-linearities, resulting in closed eye diagrams. Implementing the TRDFE improves the situation by opening the eye diagrams; however, the levels remain unequally spaced due to incomplete mitigation of system non-linearity.

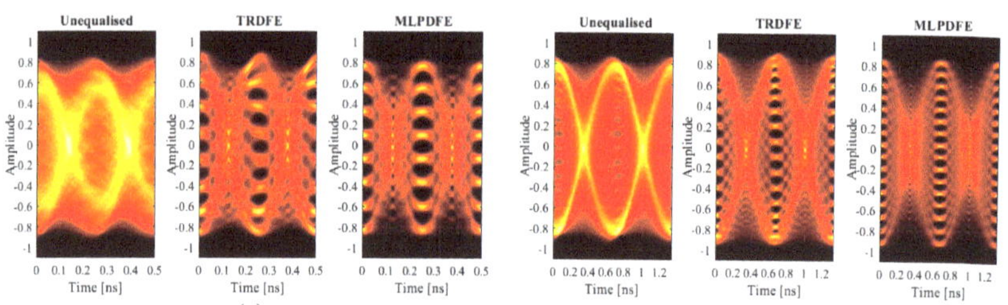

(a) (b)

Figure 6. Captured output eye diagrams with equalisers for SM-VCSEL Setup-II: (**a**) 12 Gbps with PAM-8. (**b**) 6 Gbps with PAM-16.

In contrast, the eye diagrams with the MLPDFE show more uniform spacing between levels. This uniformity is attributed to the MLPDFE's ability to compensate for the inherent non-linearity in the system, resulting in improved eye diagram characteristics. This improvement is also reflected in the BER plots in Figure 5, where the MLPDFE demonstrates superior performance compared to the TRDFE.

The results achieved with Setup-II were obtained at an ROP of ∼−3.1 dBm. However, misalignments and extensions in the OWC link can diminish the ROP, subsequently impacting the SNR and bit rate. Hence, Figure 7 is presented to illustrate the bit rate attainable at the 7% S-FEC limit for various ROPs ranging from −14 dBm to −3 dBm using PAM-2, PAM-4, PAM-8, and PAM-16 modulation schemes with both the TRDFE and MLPDFE. At lower bit rates and reduced ROPs, PAM-4 requires more power than PAM-2 to achieve similar bit rates. For instance, with the TRDFE at a bit rate of 10 Gbps, PAM-2 and PAM-4 necessitate ROPs of ∼−13.2 dBm and ∼−11.0 dBm, respectively, indicating an ROP difference of 2.2 dB in favour of PAM-2. However, at a higher bit rate of 25 Gbps, PAM-2 and PAM-4 demand ROPs of ∼−6.0 dBm and ∼−5.4 dBm, respectively, showcasing a reduced ROP difference of 0.6 dB between the two modulation schemes. PAM-4 offers a higher bit rate of ∼29 Gbps compared to PAM-2, which gives ∼28 Gbps at the high ROP of −3.1 dBm.

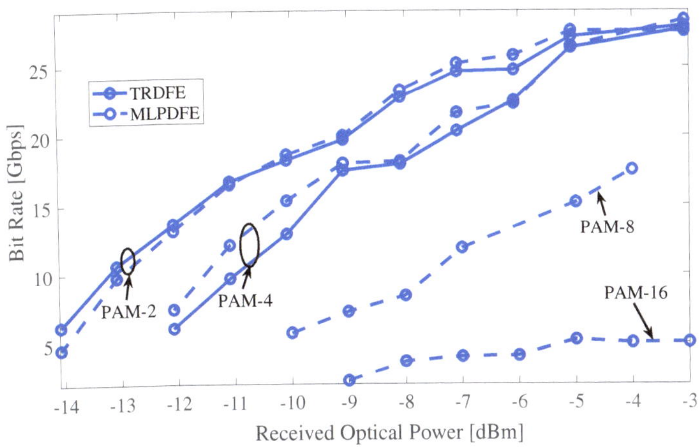

Figure 7. Plots comparing the bit rates achieved at the 7% S-FEC limit against the received optical power (ROP) with SM-VCSEL Setup-II. Solid and dashed lines denote the results from TRDFE and MLPDFE, respectively.

When considering PAM-2, the TRDFE shows a slight advantage over the MLPDFE at lower bit rates with reduced ROPs. However, as the bit rates increase with higher ROPs, the MLPDFE demonstrates superior performance over the TRDFE. For example, at a bit rate of 10 Gbps, the TRDFE and MLPDFE require ROPs of ∼−13.2 dBm and ∼−13.0 dBm, respectively, indicating a marginal ROP difference of 0.2 dB in favour of the TRDFE. However, at a higher bit rate of 25 Gbps, the TRDFE and MLPDFE demand ROPs of ∼−6.0 dBm and ∼−7.1 dBm, respectively, showcasing a notable ROP difference of 1.1 dB in favour of the MLPDFE.

With PAM-4, the MLPDFE demonstrates superior ROP performance compared to the TRDFE at lower bit rates with reduced ROPs. However, at higher bit rates with increased ROPs, the MLPDFE shows ROP requirements similar to those of the TRDFE. For instance, at a bit rate of 10 Gbps, the TRDFE and MLPDFE require ROPs of ∼−11.5 dBm and ∼−11.0 dBm, respectively, indicating a marginal ROP difference of 0.5 dB in favour of the MLPDFE. However, at a higher bit rate of 25 Gbps, both the TRDFE and MLPDFE demand an ROP of ∼−5.4 dBm, showcasing no ROP difference between the two equalisers.

However, increasing the modulation level to PAM-8 and PAM-16 does not improve the bit rate performance due to the limited SNR available in the system [29]. Consequently, no bit rates above 2 Gbps were achieved at the FEC limit with the TRDFE for these modulation levels. In contrast, the MLPDFE enables bit rates beyond 6 Gbps at ROPs greater than

−10 dBm with PAM-8. Notably, with PAM-16, the MLPDFE achieves a consistent bit rate of ∼5 Gbps at ROPs ranging from −5 dBm to −3 dBm. Compared to PAM-8, which experiences a significant drop in bit rates with a decreasing ROP, PAM-16 shows a milder reduction. This is because PAM-16 requires lower bandwidth and thus experiences lesser noise at lower ROPs for bit rates similar to that of PAM-8.

In conclusion, the MLPDFE outperforms the TRDFE when the link SNR exceeds a certain threshold. This threshold is smaller with increasing PAM levels and higher system non-linearity, as indicated by the bit rate results from the two setups (Setup-I and Setup-II) using both VCSELs.

4. Conclusions

We have conducted experimental evaluations of an OWC system that uses an eye-safe 850 nm SM-VCSEL for a 2.5 m link. The system employs PAM-M in conjunction with an NN-based DFE, also known as the MLPDFE in this study. The SM-VCSEL offers distinct advantages, including lower RIN and lesser non-linearity, allowing for higher communication speeds compared to MM-VCSELs.

We compared the BER performances of the MLPDFE against that of the TRDFE for the SM-VCSEL setup under conditions of strong and weak non-linear distortions. Our results indicate that the NN-based equaliser consistently delivers superior performance across most investigated scenarios, particularly in VCSEL-based OWC systems with high non-linear distortion and PAM constellations of eight or more levels.

Specifically, the MLPDFE outperforms the TRDFE for PAM-4 modulation at lower ROPs, whereas it excels with PAM-2 at higher ROPs. With the MLPDFE, we achieved impressive bit rates of approximately 28 Gbps, 29 Gbps, 22.5 Gbps, and 5 Gbps using multilevel PAM schemes of 2, 4, 8, and 16 levels, respectively, at the 7% second-generation super FEC (S-FEC) limit. While higher modulation levels may not always offer the best bit rate because of their dependence on the available SNR, these results underscore the efficacy of the MLPDFE in optimising the performance of SM-VCSEL-based OWC systems across various modulation schemes and non-linear distortion levels.

Author Contributions: Conceptualisation, I.N.O.O. and S.R.; methodology, I.N.O.O. and S.R.; software, I.N.O.O.; validation, I.N.O.O. and S.R.; formal analysis, I.N.O.O. and S.R.; investigation, I.N.O.O.; resources, I.K., M.M., D.P., W.M., H.H. and J.T.; data curation, I.N.O.O. and S.R.; writing—original draft preparation, I.N.O.O.; writing—review and editing, I.K., D.P., W.M., M.M., J.T., H.H. and S.R.; visualisation, I.N.O.O.; supervision, S.R., J.T. and H.H. All authors have read and agreed to the published version of the manuscript.

Funding: This work is a contribution by Project REASON, a UK government-funded project under the Future Open Networks Research Challenge (FONRC) sponsored by the Department of Science, Innovation and Technology (DSIT). The work at CSC and ICS Ltd has been partly funded by UK Research and Innovation via Project "Quantum Foundry" under the UK National Quantum Technologies Programme. This work is also supported by the UK Engineering and Physical Sciences Research Council under grant EP/Y037243/1 (TITAN Extension).

Institutional Review Board Statement: Not applicable

Informed Consent Statement: Not applicable

Data Availability Statement: The data supporting the conclusions of this article will be made available by the author (I.N.O.O.) on request.

Conflicts of Interest: Compound Semiconductor Centre Ltd. and Integrated Compound Semiconductors Ltd. had no role in the study design, in the collection, analysis or interpretation of data, in the writing of the manuscript; or in the decision to publish the results.

References

1. Matheus, L.E.M.; Vieira, A.B.; Vieira, L.F.; Vieira, M.A.; Gnawali, O. Visible light communication: Concepts, applications and challenges. *IEEE Commun. Surv. Tutor.* **2019**, *21*, 3204–3237. [CrossRef]
2. Cossu, G.; Ali, W.; Rannello, M.; Ertunc, E.; Gilli, L.; Sturniolo, A.; Messa, A.; Ciaramella, E. VCSEL-based 24 Gbit/s OWC Board-to-Board System. *IEEE Commun. Lett.* **2019**, *23*, 1564–1567. [CrossRef]
3. Loureiro, P.A.; Fernandes, G.M.; Correia, S.F.; Ferreira, R.A.; Guiomar, F.P.; Monteiro, P.P. Multi-Gigabit RGB-VLC Transmission with Jointly Optimized Lighting and Communications. *J. Light. Technol.* **2024**, *42*, 4425–4432. [CrossRef]
4. Singh, R.; Feng, F.; Hong, Y.; Faulkner, G.; Deshmukh, R.; Vercasson, G.; Bouchet, O.; Petropoulos, P.; O'Brien, D. Design and characterisation of terabit/s capable compact localisation and beam-steering terminals for fiber-wireless-fiber links. *J. Light. Technol.* **2020**, *38*, 6817–6826. [CrossRef]
5. Chun, H.; Gomez, A.; Quintana, C.; Zhang, W.; Faulkner, G.; O'Brien, D. A wide-area coverage 35 Gb/s visible light communications link for indoor wireless applications. *Sci. Rep.* **2019**, *9*, 4952. [CrossRef]
6. Michalzik, R. VCSEL fundamentals. In *VCSELs: Fundamentals, Technology and Applications of Vertical-Cavity Surface-Emitting Lasers*; Springer: Berlin/Heidelberg, Germany, 2012; pp. 19–75.
7. Othman, M.B.; Deng, L.; Pang, X.; Caminos, J.; Kozuch, W.; Prince, K.; Yu, X.; Jensen, J.B.; Monroy, I.T. MIMO-ofdm wdm pon with dm-vcsel for femtocells application. *Opt. Express* **2011**, *19*, B537–B542. [CrossRef]
8. Oh, S.; Yu, M.; Cho, S.; Noh, S.; Chun, H. Bi-LSTM-Augmented Deep Neural Network for Multi-Gbps VCSEL-Based Visible Light Communication Link. *Sensors* **2022**, *22*, 4145. [CrossRef]
9. Lu, I.C.; Yeh, C.H.; Hsu, D.Z.; Chow, C.W. Utilization of 1-GHz VCSEL for 11.1-Gbps OFDM VLC wireless communication. *IEEE Photonics J.* **2016**, *8*, 1–6. [CrossRef]
10. Wang, Z.; Zhang, L.; Wei, Z.; Wei, G.; Dong, Y.; Fu, H. 8.23 Gbps high-speed near-infrared VCSEL based facile optical wireless communication system via QAM-OFDM. In Proceedings of the P2020 Asia Communications and Photonics Conference (ACP) and International Conference on Information Photonics and Optical Communications (IPOC), Beijing, China, 24–27 October 2020; p. S4B–3.
11. *IEC 60825-1*; Safety of Laser Products-Part 1: Equipment Classification and Requirements. International Electrotechnical Commission: Geneva, Switzerland, 2007.
12. Lian, J.; Noshad, M.; Brandt-Pearce, M. Comparison of optical OFDM and M-PAM for LED-based communication systems. *IEEE Commun. Lett.* **2019**, *23*, 430–433. [CrossRef]
13. Schueppert, M.; Kruglov, R.; Loquai, S.; Bunge, C.A. Nonlinearities originated in a red RC-LED and Their impact on spectrally efficient modulation. *IEEE Photonics Technol. Lett.* **2015**, *27*, 2007–2010. [CrossRef]
14. Le Priol, R.; Hélard, M.; Haese, S.; Roy, S. Experimental comparison of PAM and CAP modulation for visible light communication under illumination constraints. *IEEE Photonics J.* **2022**, *14*, 1–11. [CrossRef]
15. Diaz, J.C.G.; Zhao, H.; Zhu, Y.; Palermo, S.; Hoyos, S. Recurrent neural network equalization for wireline communication systems. *IEEE Trans. Circuits Syst. II Express Briefs* **2022**, *69*, 2116–2120.
16. Osahon, I.N.; Rajbhandari, S.; Ihsan, A.; Tang, J.; Popoola, W.O. Multilevel PAM with ANN Equalization for an RC-LED SI-POF System. In Proceedings of the 2023 IEEE 20th Consumer Communications & Networking Conference (CCNC), Las Vegas, NV, USA, 8–11 January 2023; pp. 481–484.
17. Xu, B.; Min, T.; Yue, C.P. Design of PAM-8 VLC Transceiver System Employing Neural Network-Based FFE and Post-Equalization. *Electronics* **2022**, *11*, 3908. [CrossRef]
18. Khan, Z.; Chorchos, L.; Chang, Y.H.; Ledenstov, N.; Huang, Y.Y.; Zhao, Y.C.; Ledentsov, N.; Shi, J.W. High-Power, Low-Noise, and High-Speed 850 nm VCSEL Arrays with for Optical Wireless Transmission. In Proceedings of the 2021 Optical Fiber Communications Conference and Exhibition (OFC), San Francisco, CA, USA, 6–10 June 2021; p. Tu5C–5.
19. Lu, X.; Lu, C.; Yu, W.; Qiao, L.; Liang, S.; Lau, A.P.T.; Chi, N. Memory-controlled deep LSTM neural network post-equalizer used in high-speed PAM VLC system. *Opt. Express* **2019**, *27*, 7822–7833. [CrossRef]
20. Lu, X.; Li, Y.; Chen, X.; Li, Y.; Liu, Y. Discrete wavelet transform assisted convolutional neural network equalizer for PAM VLC system. *Opt. Express* **2024**, *32*, 10429–10443. [CrossRef]
21. Osahon, I.N.; Rajbhandari, S.; Kostakis, I.; Ihsan, A.; Powell, D.; Meredith, W.; Missous, M.; Haas, H.; Tang, J. Experimental Demonstration of 38 Gbps over 2.5 m OWC Systems with Eye-safe 850 nm SM-VCSELs. *IEEE Photonics Technol. Lett.* **2023**, *36*, 139–142. [CrossRef]
22. Ying, K.; Yu, Z.; Baxley, R.J.; Qian, H.; Chang, G.K.; Zhou, G.T. Nonlinear distortion mitigation in visible light communications. *IEEE Wirel. Commun.* **2015**, *22*, 36–45. [CrossRef]
23. Qian, H.; Yao, S.; Cai, S.; Zhou, T. Adaptive postdistortion for nonlinear LEDs in visible light communications. *IEEE Photonics J.* **2014**, *6*, 1–8. [CrossRef]
24. Nair, A.R.; Kirthiga, S. Impact of total harmonic distortion in SWIPT enabled wireless communication networks. In Proceedings of the 2021 Smart Technologies, Communication and Robotics (STCR), Sathyamangalam, India, 9–10 October 2021; pp. 1–5.
25. Burse, K.; Yadav, R.; Shrivastava, S. Channel equalisation using neural networks: A review. *IEEE Trans. Syst. Man Cybern. Part C (Appl. Rev.)* **2010**, *40*, 352–357. [CrossRef]
26. Haykin, S.O. *Neural Networks and Learning Machines*; Pearson: Upper Saddle River, NJ, USA, 2009; Volume 3.

27. Sidelnikov, O.; Redyuk, A.; Sygletos, S. Equalization performance and complexity analysis of dynamic deep neural networks in long haul transmission systems. *Opt. Express* **2018**, *26*, 32765–32776. [CrossRef]
28. Tychopoulos, A.; Koufopavlou, O.; Tomkos, I. FEC in optical communications-A tutorial overview on the evolution of architectures and the future prospects of outband and inband FEC for optical communications. *IEEE Circuits Devices Mag.* **2006**, *22*, 79–86. [CrossRef]
29. Chun, H.; Rajbhandari, S.; Faulkner, G.; Xie, E.; McKendry, J.J.; Gu, E.; Dawson, M.D.; O'Brien, D. Optimum device and modulation scheme selection for optical wireless communications. *J. Light. Technol.* **2021**, *39*, 2281–2287. [CrossRef]

photonics

MDPI

Article

Impact of Optical-to-Electrical Conversion on the Design of an End-to-End Learning RGB-LED-Based Visible Light Communication System

Jose Martin Luna-Rivera [1,*], Jose Rabadan [2], Julio Rufo [3], Carlos A. Gutierrez [1], Victor Guerra [4] and Rafael Perez-Jimenez [2]

1 Faculty of Sciences, Universidad Autonoma de San Luis Potosi, San Luis Potosi 78295, Mexico; cagutierrez@ieee.org
2 Instituto para el Desarrollo Tecnológico y la Innovación en Comunicaciones (IDeTIC), Universidad de Las Palmas de Gran Canaria, PCT Tafira, 35017 Las Palmas, Spain; jose.rabadan@ulpgc.es (J.R.); rafael.perez@ulpgc.es (R.P.-J.)
3 Departamento en Ingeniería Industrial, Universidad de La Laguna, 38200 San Cristobal de la Laguna, Spain; jrufotor@ull.edu.es
4 Pi Lighting, 1950 Sion, Switzerland; victor.guerra@pi-lighting.com
* Correspondence: mlr@uaslp.mx; Tel.: +52-444-8262300 (ext. 5641)

Abstract: Visible Light Communication (VLC) is emerging as a promising technology to meet the demands of fifth-generation (5G) networks and the Internet of Things (IoT). This study introduces a novel RGB-LED-based VLC system design that leverages autoencoders, addressing the often overlooked impact of optical-to-electrical (O/E) conversion efficiency. Unlike traditional methods, our autoencoder-based system not only improves communication performance but also mitigates the negative effects of O/E conversion. Through comprehensive simulations, we show that the proposed autoencoder structure enhances system robustness, achieving superior performance compared to traditional VLC systems. By quantitatively assessing the impact of O/E conversion—a critical aspect previously overlooked in the literature—our work bridges a crucial gap in VLC research. This contribution not only advances the understanding of VLC systems but also provides a strong foundation for future enhancements in 5G and IoT connectivity.

Keywords: visible light communication (VLC); autoencoder (AE); photodetector (PD); color-shift keying (CSK); optical wireless communication (OWC)

1. Introduction

The fifth-generation (5G) networks aim to revolutionize the Internet of Things (IoT) with enhanced speed, connectivity, and capacity. A potential technology for this transformation is Optical Wireless Communication (OWC) [1], specifically Visible Light Communication (VLC), which utilizes ubiquitous light-emitting diodes (LEDs) for dual purposes: illumination and high-speed data transmission [2]. Operating in the 400–800 THz spectrum, VLC faces challenges in signal communication at these frequencies, particularly affecting device performance at the transmitting and receiving ends.

In addressing VLC's challenges within 5G networks, our research adopts an end-to-end learning approach using neural networks to optimize transmitter and receiver structures [3,4]. This innovative method aligns with the IEEE 802.15.7 standard, which describes various modulation schemes, including CSK [5]. CSK, pivotal in our study, modulates data through color changes in RGB LEDs, balancing data transmission with illumination needs [6,7]. Performance in CSK-based VLC systems relies on maximizing the Euclidean distance between constellation symbols and maintaining desired color tones [8–10]. The VLC standard includes two receiver detectors: photodetectors and image sensors [11]. However, despite the breadth of studies on VLC [12–14], existing research overlooks hardware imperfections and component non-linearities, which are critical to system performance. Addressing these elements is vital for practical system design.

Traditionally, communication system designs have been based on mathematical models and standard optimization techniques, which focus on individual components within the system [15]. This methodology has made communication systems a complex and well-established engineering field with numerous distinct areas of investigation [16–18]. However, optimizing these components as a whole often proves complex and computationally demanding. Recently, machine learning, particularly deep learning, has emerged as a solution, treating the system as an autoencoder (AE) model [19–21]. This novel approach involves training the AE to optimize the structure of the transceiver for specific performance goals. Notable studies in this domain include Lee et al.'s multi-color VLC AE with dimming control [22], Pepe et al.'s heuristic machine learning algorithms for CSK signal classification [23], and Zhang et al.'s AE for multi-color VLC, addressing chromaticity and signal constraints [24]. The integration of deep learning into VLC systems is further explored in the works of Zou et al. [25], Ulkar et al. [26], and Shrivastava et al. [27], expanding the application of these advanced technologies in communication systems.

Integration of VLC with IoT offers a wide range of applications and opportunities. Although studies like those by Mitra et al. [28], Laakso et al. [29], and Lizarraga et al. [30] have examined LEDs' non-linear characteristics, research into the efficiency of electro-optic conversions in VLC systems remains limited. Moreover, while the responsivity of photodetectors can appear linear within certain operational ranges, it is subject to non-linear behaviors outside these limits because of saturation at high light intensities and noise at low intensities. Additionally, temperature changes and wavelength variations can further complicate this relationship. These nuances underscore the need for careful consideration in system design to maintain signal integrity and system reliability in VLC technologies.

When considering signal conversions in VLC systems, it is important to acknowledge that the process is not simply linear. Non-linearities, spectral dependencies, and the potential for signal distortion introduce complexities that demand sophisticated approaches to optimize the system. These factors can result in degraded signal integrity, higher error rates, and reduced system reliability, presenting significant challenges in the practical implementation of VLC technologies. To tackle these challenges, we are investigating the optical-to-electrical (O/E) conversion effects in CSK-based VLC systems by introducing a novel autoencoder architecture. This architecture, which highlights photoelectric conversion imperfections, is trained using stochastic gradient descent to optimize transmitter and receiver functions in RGB-LED-based VLC systems. Our approach, evaluated through symbol error rate (SER) analysis, demonstrates the potential to simplify connectivity between IoT devices and wireless networks.

The primary motivation for employing AEs in our work is their ability to learn complex, non-linear relationships directly from data without requiring explicit mathematical modeling of the system's non-linearities. This capability is particularly advantageous in the context of O/E conversion in VLC systems. By training on a diverse dataset, AEs can map the input to the output accurately, even in the presence of complex, non-linear distortions, allowing for a more robust and adaptive system design. AEs provide an end-to-end learning framework, optimizing the entire transceiver system simultaneously, which is often more efficient and effective than optimizing individual components in isolation.

The main contributions of this paper are: (1) providing new insights into the often overlooked aspect of O/E conversion in CSK-based VLC systems, addressing a gap in current VLC research; (2) introducing a novel, data-driven autoencoder architecture designed to optimize RGB-LED-based VLC systems, focusing on the nuances of O/E conversion; (3) developing a CSK-focused VLC system equipped with a single-photodiode receiver to reduce system complexity and energy consumption, thus enhancing IoT integration capabilities; and (4) conducting a thorough SER analysis to quantitatively evaluate the impact of O/E conversion on the system's receiver performance, demonstrating the effectiveness of our proposed solution in mitigating conversion-related performance degradation.

2. Signal Conversions and System Design in VLC

2.1. Signal Conversions

In VLC systems, the process of transitioning signals between electrical and optical domains presents a range of challenges that extends beyond just energy consumption. It includes considerations such as performance, system complexity, and signal accuracy. Converting electrical signals to optical signals is crucial for enabling data transmission, impacting both energy efficiency and the system's operational bandwidth and latency. On the receiving end, transforming optical signals back to electrical signals using photodetectors or image sensors is equally important. This step is essential for accurately converting transmitted photon streams into electrical signals, directly affecting the overall system performance.

The efficiency of the O/E conversion process, as quantified by the photodetector's responsivity, emerges as a crucial factor in determining the effectiveness of the VLC system. Responsivity, defined as the photocurrent produced per milliwatt of optical power, is mathematically expressed as

$$R = \frac{I_p}{P(\lambda)} \approx Q.E.\frac{\lambda}{1.24} \quad (A/W), \tag{1}$$

where I_p is the output photocurrent (in amperes) and $P(\lambda)$ represents the radiant energy (in watts). The relationship between responsivity and wavelength, particularly for standard silicon photodiodes, is vital [31].

Recognizing and addressing these multifaceted challenges, our research aims to advance the design and implementation of an end-to-end learning VLC system. For simplicity and without losing generality in demonstrating the functionality of our autoencoder strategy, we focus exclusively on the O/E conversion processes at the receiver. This focus is aimed at enhancing not only energy efficiency but also system performance and reliability. However, this approach can be extended to include the conversion of electrical to optical signals at the transmitter as well. In doing so, our goal is to bridge the gap between the theoretical potential and the practical usability of VLC technologies, ensuring that our solution is conceptually sound and viable for real-world applications.

2.2. CSK-Based System Design

The traditional VLC system, employing CSK modulation, is illustrated in Figure 1. We focus on a direct point-to-point optical communication link, using an RGB LED to transmit a signal vector, denoted as $\mathbf{s}$, from a predefined set $\mathcal{S} = \{\mathbf{s}_1, \mathbf{s}_2, \cdots, \mathbf{s}_M\}$. This vector, derived from the M different data symbols d in the set $\mathcal{M} = \{1, 2, \cdots, M\}$, is received and processed through optical filtering to separate the RGB channels, followed by conversion into electrical currents for each channel. The signal model for the received vector $\mathbf{r} \in \mathbb{R}^{3\times1}$ is given by

$$\mathbf{r} = \mathbf{CHs} + \mathbf{n}, \tag{2}$$

where $\mathbf{H} \in \mathbb{R}^{3\times3}$ represents the VLC channel matrix, $\mathbf{s} \in \mathbb{R}^{3\times1}$ transmitted signals, $\mathbf{n} \in \mathbb{R}^{3\times1}$ the noise vector, and $\mathbf{C} \in \mathbb{R}^{3\times3}$ the responsivity matrix.

Our study uses CSK modulation for its complexity and sophistication as a key technique in the VLC standard IEEE 802.15.7 [5]. CSK operates by transmitting $\log_2(M)$ bits per symbol, where M indicates the size of the constellation. This transmission is achieved by varying the color intensities of an RGB LED. A crucial aspect of CSK is the design of the constellation's symbols, which requires a meticulous balance between brightness and chromaticity to ensure efficient communication. The transmitted RGB optical power vector, denoted $\mathbf{s} = [P_r, P_g, P_b]^T$, corresponds to a specific chromaticity coordinate (x, y). This vector is calculated using the following equation, as referenced in [32]:

$$\begin{bmatrix} P_r \\ P_g \\ P_b \end{bmatrix} = \begin{bmatrix} \frac{x_r}{y_r} & \frac{x_g}{y_g} & \frac{x_b}{y_b} \\ 1 & 1 & 1 \\ \frac{1-x_r-y_r}{y_r} & \frac{1-x_g-y_g}{y_g} & \frac{1-x_b-y_b}{y_b} \end{bmatrix}^{-1} \begin{bmatrix} \frac{x}{y} \\ 1 \\ \frac{1-x-y}{y} \end{bmatrix}. \tag{3}$$

In this equation, (x_r, y_r), (x_g, y_g), and (x_b, y_b) denote the chromaticity coordinates of red, green, and blue light sources. The transmitted signal must meet two key lighting criteria: adhering to a target color, typically white, verified via a chromaticity diagram to ensure that CSK constellation symbols match this color benchmark, and maintaining within the RGB LED's dynamic range. The total optical power, the sum of the powers of the red (P_r), green (P_g), and blue (P_b) LEDs, must not exceed the limit of the dynamic range, mathematically $P_r + P_g + P_b \leq P_T$, where P_T is the maximum allowable average power.

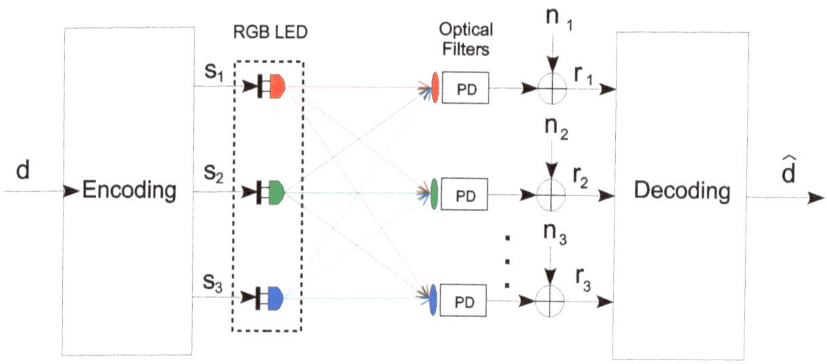

Figure 1. Traditional color space-based VLC system model.

In the exploration of VLC systems, particular attention is given to the receiver-side O/E conversion process. This conversion is critical in the transformation of optical signals into electrical signals, a process that is fundamentally influenced by the characteristics of the photodetector used. The analysis begins with a conventional photodetector array configuration, represented in Figure 1, which is standard in VLC systems. This configuration, while effective, presents an inherent challenge in managing the non-linear response characteristics induced by various factors such as quantum efficiency (*Q.E.*), wavelength (λ), and ambient temperature. To provide a comprehensive understanding of these dynamics, Figure 2 delves into the operational principles of a photodetector. It illustrates how incident light is converted into an electrical signal, with the photodetector's responsivity (R) playing a central role. The figure further demonstrates the non-linear response of the photodetector to temperature variations and wavelength changes, which can significantly impact the system's performance.

An alternative detection method employing a single photodetector array is also introduced in Figure 3. This approach aims to simplify the system's design, reducing both complexity and cost and facilitating better integration with IoT technologies. However, simplifying the design also accentuates the challenge of accurately managing the non-linear O/E conversion process. Therefore, it prioritizes implementation efficiency, offering substantial advantages in deployment and integration with IoT technologies, with careful consideration required for the conversion intricacies that could affect system performance.

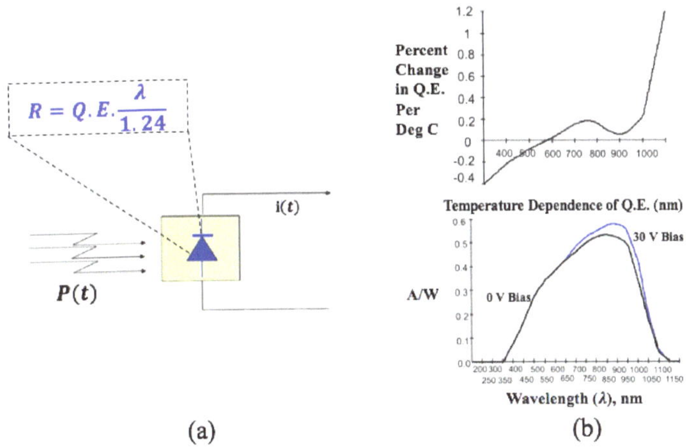

Figure 2. Photodetector response characteristics in VLC Systems: (**a**) schematically represents the photodetector converting light power into an electrical current with the responsivity equation highlighted to account for material properties; (**b**) displays the temperature dependence of quantum efficiency (*Q.E.*), illustrating the percent change in responsivity per degree Celsius and the wavelength dependence of responsivity under different bias voltages.

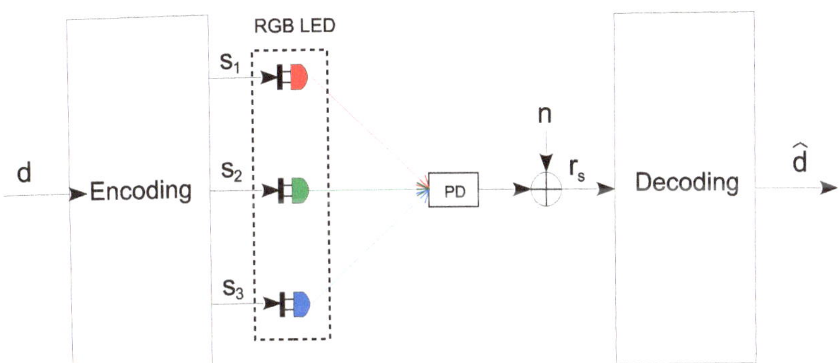

Figure 3. Single photodetector-based VLC color space-based VLC system model.

In terms of enhancing the performance of the symbol error probability, the maximum-likelihood (ML) method is used with the photodetector array. The ML method aims to minimize the Euclidean distance between the received signal vector **r** (as defined in (2)) and all possible transmitted signals. The optimal function of the ML detector can be formulated as follows:

$$\hat{d} = \arg\min_{s \in \mathcal{S}} \|\mathbf{r} - \mathbf{CHs}\|_2^2, \tag{4}$$

where $\hat{d}$ represents the estimated value of d, which is mapped to one of the signal vectors in the set $\mathcal{S}$, and $\|\cdot\|_2$ denotes the 2-norm.

An alternative to the photodetector array to detect RGB light signals is the use of a single photodiode (SPD) [33]. This approach, illustrated in Figure 3, simplifies the receiver design by using just one photodetector and omitting any optical filters. The fundamental principle of the SPD is based on the wavelength-dependent responsivity of the photodiode. This feature allows for the mapping of received optical signals to distinct scalar values, each corresponding to the wavelengths of the individual RGB LEDs. With the SPD configuration, the received signal model, initially formulated in (2), is adapted to the following:

$$r_s = \langle \boldsymbol{\rho}, \mathbf{Hs} \rangle + n, \tag{5}$$

where $\boldsymbol{\rho} = [R(\lambda_r), R(\lambda_g), R(\lambda_b)]^T$ represents the photodetector's responsivity at the peak wavelengths for the red, green, and blue components. The symbol $\langle \cdot \rangle$ signifies the dot product operation, and the photodetector output is a combination of the influence of the channel matrix and a scalar representation of noise. Similarly to the approach in (4), the SPD receiver determines the transmitted symbol by employing an optimal detection strategy. This is expressed as follows.

$$\hat{d} = \arg\min_{s \in \mathcal{S}} |r_s - \langle \boldsymbol{\rho}, \mathbf{Hs} \rangle|, \tag{6}$$

where $|\cdot|$ indicates the absolute value. This method allows the SPD receiver to effectively identify the transmitted symbol, optimizing detection in the VLC system.

3. CSK-Based VLC System Design Using Neural Networks

In this section, we delve into the design of our CSK-based VLC system, which utilizes an innovative autoencoder-based approach. Initially, we detail the architecture of our autoencoder model, focusing on its core components, including the crucial hidden layers that play a vital role in the system's learning and adaptation capabilities. We elaborate on the meticulous design of the training framework, which is tailored to optimize the weights and biases within these layers, effectively minimizing the cost function for enhanced performance.

3.1. Autoencoder Design

In our study, we conceptualize the VLC system within an autoencoder framework consisting of three primary components: the encoder, the code, and the decoder. The transmitter and receiver are represented as two separate parametric neural networks, envisioned, respectively, as the encoder and decoder. These networks are jointly optimized to meet specific performance criteria, with the Additive White Gaussian Noise (AWGN) channel representing the code.

3.1.1. Hyperparameters

Our autoencoder's efficiency is significantly influenced by the selection of optimal hyperparameters. The key parameters that we fine-tuned during the optimization process are detailed in Table 1.

Table 1. Optimized hyperparameters for autoencoder efficiency.

Encoding Dimension	Set at $N = 3$, corresponding to the number of transmitting LEDs in our CSK-based VLC system.
Hidden Layers	Two layers were determined to best balance performance and complexity.
Nodes in Hidden Layers	Each layer has 4 nodes, except the final transmitter layer with 3 nodes, as established through iterative testing.
Activation Function	The encoder employs the Exponential Linear Unit (ELU), and the decoder uses the Softmax function, suited for our multi-class scenario.
Optimizer	The Adam optimizer was chosen for its adaptive learning rate capabilities.
Epochs	Training iterations were consistently set to 10 for all combination of parameters.
Batch Size	A mini-batch size of 100 ensured fast and efficient model convergence.

3.1.2. Cost Function

Our goal is to jointly optimize the transmitter and receiver network parameters. This is achieved using the following cost function:

$$C = L_1 + \lambda L_2. \tag{7}$$

L_1 is defined using the categorical cross-entropy metric to measure the discrepancy between the input symbols and their predicted probabilities:

$$L_1 = -\sum_{m=1}^{M} d_m \log(p_m), \tag{8}$$

where d_m is the true value for the m-th symbol, and p_m is the predicted probability for the m-th symbol. The categorical cross-entropy metric (L_1) plays a pivotal role in the cost function, not only measuring the accuracy discrepancy between input values and their predicted counterparts but also implicitly favoring the maximization of the minimum Euclidean distance between constellation symbols. Well-separated symbols naturally lead to a lower cross-entropy value, which correlates with a more accurate and reliable system performance. The integration of this metric thus subtly but effectively ensures that symbols are sufficiently spaced in the constellation without necessitating an additional explicit term in the loss function for the Euclidean distance.

The second term, L_2, incorporates the chromaticity constraint for illumination, integrating the RGB constraint in the CIE chromaticity diagram:

$$L_2 = \frac{2}{(1 + e^{-|T_c - CCT|/1000}) - 1}, \tag{9}$$

where CCT represents the desired color temperature and T_c is the color temperature resulting from the RGB lights produced by the transmitter. The computation of T_c is based on McCamy's approximation [34]:

$$T_c = 437n^3 + 3601n^2 + 6831n + 5517, \tag{10}$$

where $n = (x - 0.3320)/(0.1858 - y)$, given the measured chromaticity coordinates (x, y). Lastly, the parameter λ controls the relative contribution of the penalty term provided in (9).

The balance within the cost function, as reflected in Equation (7), between the accuracy-focused L_1 and the chromaticity constraint L_2, has been carefully calibrated to support robust performance while maintaining compliance with illumination requirements.

3.2. Autoencoder Simulation Framework

In this section, we outline the simulation framework used to model the VLC system using neural networks. The proposed simulation structure can be seen in Figure 4.

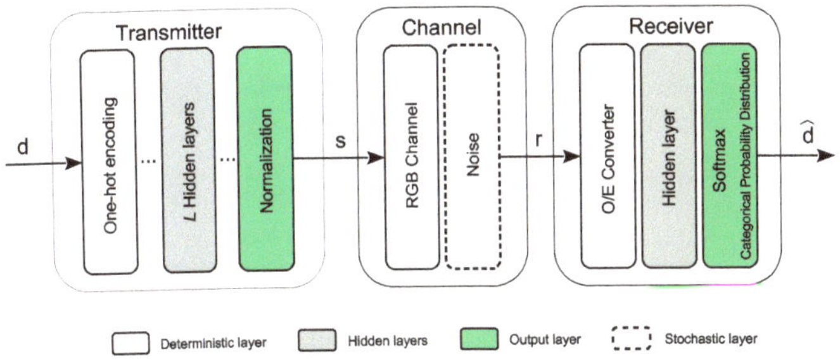

Figure 4. Proposed end-to-end VLC structure.

3.2.1. Transmitter

The transmitter begins by mapping each training input symbol $d(i)$ from the set $\mathcal{D}$ and ranging from $i = 1, 2, \cdots, I$ to a one-hot vector $\mathbf{e}(i)$ in the set $\mathcal{E}$. The set $\mathcal{E}$ comprises distinct one-hot vectors, defined as:

$$\mathcal{E} = \left\{ \mathbf{e}(i) \in \{0,1\}^M : \sum_{m=1}^{M} e_m(i) = 1 \right\}. \tag{11}$$

This vector is then passed through L fully-connected hidden layers, each with M neurons, except the last layer, which has N neurons. The output from the l-th layer, where l ranges from 1 to L, is given by:

$$\mathbf{h}_l^t(i) = \Phi_l^t(\mathbf{W}_l^t \mathbf{x}_l^t(i) + \mathbf{b}_l^t), \tag{12}$$

where $\mathbf{W}_l^t \in \mathbb{R}^{M \times M}$, $\mathbf{x}_l^t(i) \in \mathbb{R}^{M \times 1}$, and $\mathbf{b}_l^t \in \mathbb{R}^{M \times 1}$ represent the weights matrix, input vector, and bias vector, respectively, with $\Phi_l^t(\cdot)$ as the activation function. Notice that for $l = 1$, $\mathbf{x}_1^t(i) = \mathbf{e}(i)$, and for $l = L$, $\mathbf{W}_L^t \in \mathbb{R}^{N \times N}$, $\mathbf{x}_L^t(i) \in \mathbb{R}^{N \times 1}$, and $\mathbf{b}_L^t \in \mathbb{R}^{N \times 1}$. The final output from the transmitter, symbolizing the CSK constellation symbol for the i-th training symbol, is normalized to uphold power constraints:

$$\mathbf{s}(i) = \mathbf{h}_L^t(i) \circ \mathbf{v}, \tag{13}$$

where $\circ$ denotes the Hadamard product, and $\mathbf{v} = [v_1, v_2, v_3]^T$ is the normalization vector ensuring $\max(\mathbf{s}(i)) = [1,1,1]^T$.

3.2.2. Channel

Our novel autoencoder architecture is initially analyzed using an AWGN channel for simplicity and to gain basic insights before addressing more complex scenarios. This is apt for VLC, where non-line of sight reflections are minimal, aligning with the AWGN model's characteristics.

In the stochastic channel layer of the model, the transmitted optical signal vector $\mathbf{s}(i)$ is first multiplied by the identity matrix $\mathbf{H} = \mathbf{I} \in \mathbb{Z}^{3 \times 3}$. We then introduce a noise vector $\mathbf{n}(i)$, adhering to the normal distribution $\mathcal{N}(0, \sigma^2 \mathbf{I})$, to this product. Consequently, the output from the channel layer for the i-th training symbol is formulated as:

$$\mathbf{r}(i) = \mathbf{C}\mathbf{s}(i) + \mathbf{n}(i), \tag{14}$$

with $\mathbf{r}(i)$ is $\in \mathbb{R}^{N \times 1}$. It is important to note that although the noise samples for each training symbol are independent and identically distributed (i.i.d.), their specific statistics are not discernible during the training phase. This consideration is crucial for accurately modeling the stochastic nature of the channel within our autoencoder framework.

3.2.3. Receiver

The receiver's primary function in our VLC system is to decode the channel output, $\mathbf{r}(i)$, to accurately approximate the original transmitted symbol. This process begins with the output being fed through a deterministic O / E conversion layer, which is subsequently passed through the J hidden neural network layers. The output from each hidden layer, j, where j ranges from 1 to J, is formulated as:

$$\mathbf{h}_j^r(i) = \Phi_j^r\left(\mathbf{W}_j^r \mathbf{x}_j^r(i) + \mathbf{b}_j^r\right), \tag{15}$$

in this expression, $\mathbf{W}_j^r \in \mathbb{R}^{M \times M}$, $\mathbf{x}_j^r(i) \in \mathbb{R}^{M \times 1}$, and $\mathbf{b}_j^r \in \mathbb{R}^{M \times 1}$ represent the weight matrix, input vector, and bias vector for each layer, respectively. Each layer utilizes the activation function $\Phi_j^r(\cdot)$. In particular, the first layer's input is specifically $\mathbf{x}_1^r(i) = \mathbf{r}(i)$.

Following the hidden layers, the output layer uses the softmax activation function to calculate the probabilities corresponding to each predicted class. These probabilities are determined by:

$$p_m(i) = \frac{e^{h^r_J(i,m)}}{\sum_{m=1}^{M} e^{h^r_J(i,m)}}, \tag{16}$$

where $h^r_J(i,m)$ is the m-th element of the output layer. The final step in the decoding process involves determining the estimated training symbol, $\hat{d}(i)$, by identifying the element with the highest probability:

$$\hat{d}(i) = \arg\max_{m \in \mathcal{M}} \left(p_m(i) \right). \tag{17}$$

4. System Configuration and Evaluation Criteria

Following the autoencoder design outlined in Section 3, this section elaborates on the specific configurations of our VLC system for numerical evaluation. We detail the setup processes, performance metrics, and computational complexity assessments involved, offering a framework essential for understanding the numerical results discussed in Section 5.

4.1. Setup

Our approach is designed to develop an optimal CSK-based modulation set, achieved by training the autoencoder parameters in a noise-free environment. In this approach, the model acclimates to the modulation symbols derived from the training data without the influence of noise factors. The crux of this method lies in maximizing the minimum Euclidean distance among the modulation symbols. This is in contrast to noise-aware models, which require a prior estimation of noise statistics and necessitate updates to the modulation set upon any changes in these statistics. A key advantage of our noise-free approach is the rapid training process that leads to optimal results. This efficiency is due to the fact that the goal of maximizing Euclidean distance between modulation symbols is not impacted by noise, making it a robust strategy regardless of noise conditions.

Table 2 presents the detailed structure of our proposed autoencoder for the VLC network. The transmitter consists of a one-hot mapping layer, two hidden layers, and a normalization layer, with the maximum transmitted optical power for each color channel normalized to a value of one (as specified in Section 3.2.1). Conversely, the receiver integrates two fully connected hidden layers, each containing M neurons. We employed silicon photodiodes as optical detectors to investigate the nuances of the O/E conversion response. The specifications of the light sources and optical receivers used in our VLC system are listed in Table 3. Notably, the responsivity vector ρ is derived based on the documented response in [31], with defined wavelengths of $\lambda_b = 470$ nm, $\lambda_g = 530$ nm and $\lambda_r = 645$ nm.

Table 2. Autoencoder structure.

Block	Layer Type	Outputs	Activation Function
Transmitter	One-Hot Mapping	M	NA
	Hidden	M	None
	Hidden	N	ELU
	Normalization	N	NA
RGB Channel	Noise	N	NA
Receiver	Input	N	None
	Hidden	M	None
	Hidden	M	Softmax
	Arg. Max	M	NA

Table 3. Electric and optical parameters.

	Chromaticity Coordinates	PD Responsivity ρ (A/W)
Red ($\lambda_r = 645$ nm)	$(0.7006, 0.2993)$	0.42
Green ($\lambda_g = 530$ nm)	$(0.1547, 0.8059)$	0.32
Blue ($\lambda_b = 470$ nm)	$(0.1440, 0.0297)$	0.22
	CCT = 6500 K	
CIE 1931 (x,y)	$(0.3127, 0.3290)$	

We assess the performance of the VLC system across four distinct scenarios:

1. CSK-VLC-AE (baseline): This scenario serves as a baseline, employing a maximum-likelihood detector to identify data symbols in the RGB channels without considering optical-to-electrical (O/E) conversion effects. This ideal case helps establish a performance benchmark under optimal conditions.
2. CSK-VLC-AE with O/E: This scenario is similar to the baseline but includes the impact of O/E conversion at the receiver, using the same ML detection method. This comparison highlights the effects of O/E conversion on system performance.
3. CSK-VLC-AE with SPD: Here, we implement a system using a single-photodiode (SPD) detector, which inherently includes O/E conversion. This setup is crucial for understanding the performance trade-offs when employing a cost-effective, simplified detector configuration.
4. CSK-VLC with O/E (traditional non-autoencoder system): This scenario compares our autoencoder approach with a traditional VLC system using CSK modulation and ML detection that includes O/E conversion. This comparison serves as a critical benchmark to evaluate the advantages of integrating autoencoders into VLC systems.

The network model is trained with 6×10^5 samples on 10 epochs, using a mini-batch size of $100 \times N$ and Adam optimizer with a learning rate of $\rho = 0.001$. The effectiveness of these schemes is tested using 4×10^5 data symbols, analyzing the SER outcomes. During our training phase, we used both the transmitter and receiver networks to evaluate the average SER performance. For the testing phase, we analyzed 4×10^5 data symbols to compare SER across the three outlined schemes.

4.2. Performance Metrics

Evaluating the performance and complexity of our neural network-based VLC system is crucial. This subsection details the metrics we have selected for this evaluation, enabling a comprehensive comparison of the different schemes under study.

Symbol Error Rate: A key metric in assessing the efficiency of digital communication systems is the SER in relation to the SNR. To estimate the SER, we conducted simulations using the learned transmitter and receiver with a substantial number of symbols. We calculate the SER by comparing the ratio of erroneously received symbols to the total number of transmitted symbols, defined as:

$$\text{SER} = \frac{\text{number of symbols in error}}{\text{total number of transmitted symbols}}. \tag{18}$$

Minimum Euclidean distance: Another critical metric is the minimum Euclidean distance between symbols in a modulation scheme. This distance is indicative of a communication system's performance, with a larger Euclidean distance generally implying better performance. To assess this, we evaluate the minimum Euclidean distance among the signals received by our VLC receiver, calculated as:

$$d_{min} = \min_{m,k,m \neq k} ||\mathbf{s}_m - \mathbf{s}_k||_2^2, \tag{19}$$

where $\mathbf{s}_m$ and $\mathbf{s}_k$ are elements of the set $\mathcal{S}$ for $m, k = 1, 2, \cdots, M$.

4.3. Processing Complexity

In assessing the efficiency of our proposed model, it is important to consider the processing complexity of the autoencoder architecture. To this end, the computational complexities of the individual layers are detailed in Table 4, offering a granular view of the computational demands of the model. Our analysis includes a representative complexity function for the entire architecture, formulated as $c(N, M) = 3M^2 + N^2 + 2N + 5M + 4$, which simplifies to a complexity of $O(M^2 + N^2)$. Moreover, the adoption of parallel processing, a common practice in neural network operations, can significantly mitigate these computational requirements.

Table 4. Number of mathematical operations in the end-to-end autoencoder model.

Layer (Output Dimensions)	Multiplications	Additions & Divisions	Function Exp $(\cdot)$
	Transmitter		
Noise (N)	-	-	-
One-Hot Mapping (M)	-	-	-
Hidden (M)	$M \cdot M$	$M + 1$	-
Hidden (N)	$N \cdot N$	$N + 1$	N
Normalization (N)	-	N	-
	RGB Chanel		
Noise (N)	-	-	-
	Receiver		
Noise (N)	-	-	-
Input (N)	-	-	-
Hidden (M)	$M \cdot M$	$2M + 1$	-
Hidden (M)	$M \cdot M$	$2M + 1$	M
Arg. Max (M)	-	-	-

Comparing this to conventional systems, the autoencoder stands out for its balanced computational profile. Traditional VLC systems often have fluctuating computational needs based on their detection methods. For instance, optimal ML detectors with a photodetector array can have a complexity of $O(M^N)$, while a single-photodiode detector simplifies this to $O(M)$.

5. Results

This section discusses the impact of O/E conversion on our CSK-modulated autoencoder VLC system. We initially trained the CSK-VLC-AE network to incorporate the autoencoder's explicit structure and the effects of O/E conversion. The learning curves of the dataset indicated a rapid and effective acquisition of transmitter and receiver functions, with a training accuracy that quickly reached and stabilized nearly 100% after the initial epochs. The loss, combining the terms L_1 and L_2 of our methodology, showed rapid stabilization, maintaining a consistent average in the later training stages. Specifically, after the fifth epoch, the loss value leveled at around 0.42, indicating that the model had successfully converged. The Adam optimizer was particularly effective in later learning stages, even with suboptimal hyperparameters. Following a similar approach, we trained and tested two other scenarios: CSK-VLC-AE without O/E and CSK-VLC-AE with SPD. Both yielded results consistent with our initial network model.

Figure 5 illustrates the average SER performance in relation to SNR for the VLC schemes evaluated. Incorporating O/E conversion at the receiver notably degrades SER performance. Specifically, the CSK-VLC-AE scheme shows a performance drop of approximately 10 dB at an average SER of 10^{-3}, and an additional 8 dB decrease is observed in the CSK-VLC-AE with SPD scheme. To provide a benchmark, the results of a traditional non-autoencoder VLC system, CSK-VLC with O/E, which incorporates O/E effects and utilizes a maximum-likelihood detector, are also presented (see Figure 1). Despite utilizing well-established detection methods, this traditional VLC system is consistently outperformed by the CSK-VLC-AE schemes, particularly at SNRs above 12 dB and SERs below 2×10^{-1}. This highlights the autoencoder's ability to enhance SER performance amidst the challenges posed by O/E conversion. As noise levels increase, surpassing signal levels, all schemes exhibit an uptick in error rates. However, the resilience of the CSK-VLC-AE approaches underscores their potential advantages in practical VLC system applications.

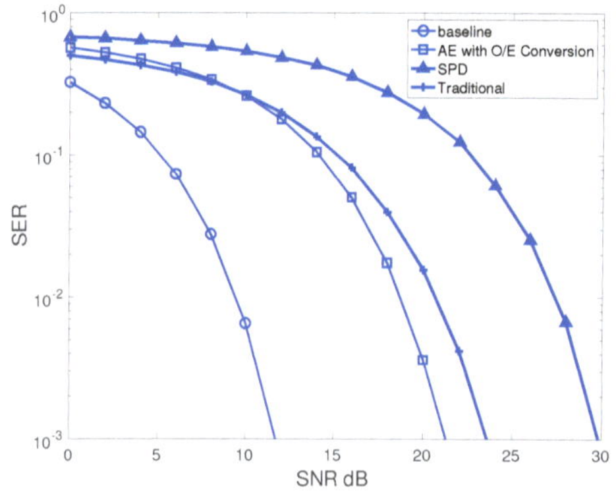

Figure 5. SER performance results of the CSK-based autoencoder VLC system according to the baseline scheme, the cases with O/E conversion and SPD detection, and an additional curve representing a traditional VLC system employing maximum-likelihood detection for comparison.

In Figures 6–8, the 4-CSK mapping constellations for the autoencoder-based schemes are displayed in the CIE 1931 color space for the respective examined configurations. The central wavelengths of RGB LEDs are indicated by circles, with the optimal color coordinates for each constellation symbol shown as squares, and the average color tonality is represented by a red cross, with an aim to reproduce the D65 color standard. Figure 6 represents the standard CSK-VLC-AE scheme, Figure 7 illustrates the CSK-VLC-AE scheme with O/E conversion, and Figure 8 shows the CSK-VLC-AE scheme with SPD. All schemes strive for D65 white tonality. Additionally, the MacAdam ellipse on the CIE 1931 xy chromaticity diagram is included to emphasize colors that are indistinguishable from the ellipse's center to the average human eye.

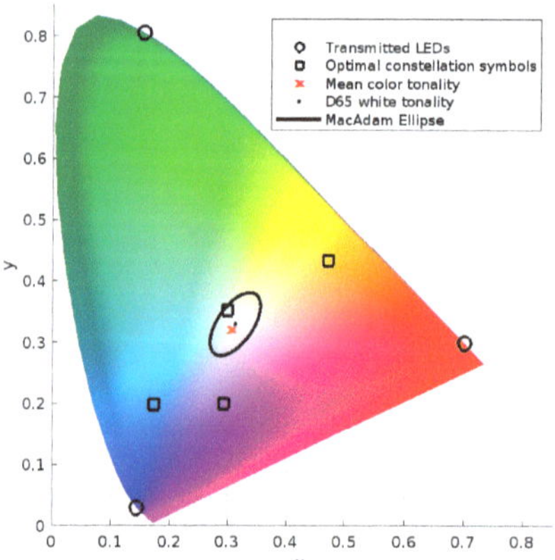

Figure 6. Resulting optimal constellation symbols (N = 3) for scheme CSK-VLC-AE for $M = 4$ and using D65 white tonality as target.

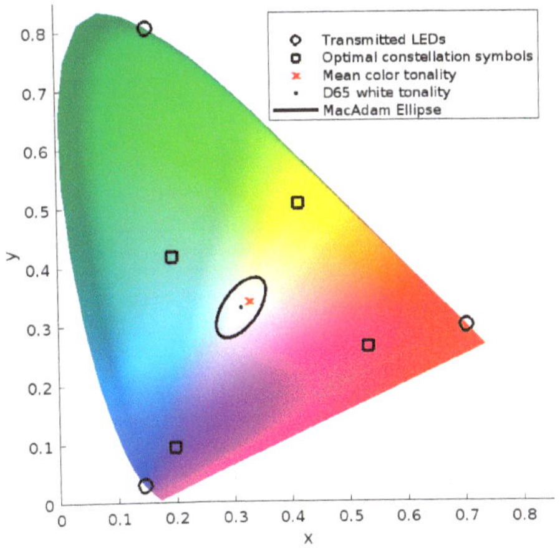

Figure 7. Resulting optimal constellation symbols (N = 3) for scheme CSK-VLC-AE with O/E for $M = 4$ and using D65 white tonality as target.

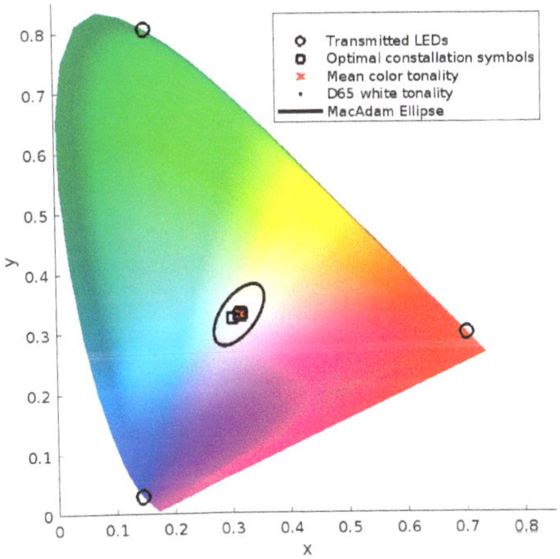

Figure 8. Resulting optimal constellation symbols (N = 3) for scheme CSK-VLC-AE with SPD for $M = 4$ and using D65 white tonality as target color.

Table 5 details the constellation sets for each case study, focusing on the minimum distance at the transmitter d_{min}^{tx} for each scheme. Evaluating the minimum distance between constellation symbols at the receiver d_{min}^{rx} reveals the effect of photoelectric conversion, with d_{min}^{rx} decreasing significantly due to this impact. The SPD-structured system further reduces this minimum Euclidean distance, as the transmitted RGB vector aligns with the sensitivity response.

The Euclidean distance at the receiver plays a crucial role in error probability. Figure 9 contrasts the constellations at the receiver for the standard CSK-VLC-AE scheme with those impacted by O/E conversion and SPD scenarios. The constellation points in the O/E and SPD scenarios are noticeably closer due to photoelectric conversion, reducing their effectiveness in minimizing error probability compared to the baseline configuration.

Table 5. CSK constellations learned for $M = 4$ and $CCT = 6500$ K.

Method	CSK Constellation Sets	d_{min}^{tx}	d_{min}^{rx}
CSK-VLC-AE	$\{[0.1227, 0.0624, 0.2000]^T$, $[0.0879, 0.3982, 1.0000]^T$, $[1.0000, 0.5765, 0.1243]^T$, $[0.8067, 1.0000, 0.8565]^T\}$	0.8676	0.8676
CSK-VLC-AE with O/E Conversion Loss	$\{[0.9998, 0.0905, 0.2493]^T$, $[0.1512, 0.0912, 1.000]^T$, $[1.0000, 1.0000, 0.1135]^T$, $[0.1599, 0.9999, 0.6370]^T\}$	0.9213	0.3265
CSK-VLC-AE with SPD	$\{[1.0000, 1.0000, 1.0000]^T$, $[0.3588, 0.3479, 0.3478]^T$, $[0.0195, 0.0210, 0.0219]^T$, $[0.6644, 0.6626, 0.6448]^T\}$	0.5297	0.2944

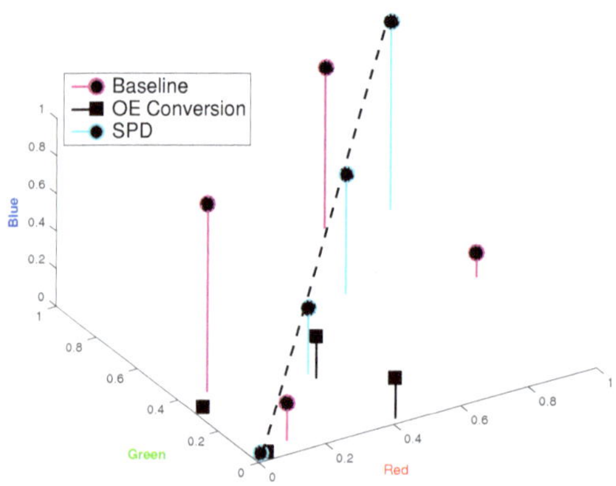

Figure 9. Receiver constellations for standard CSK-VLC-AE (baseline) and the impact of O/E conversion and SPD, showing reduced error minimization effectiveness.

6. Conclusions

In this study, we have applied the use of an autoencoder network for the design of a CSK-based VLC system, meticulously incorporating the nuances of the O/E conversion at the receiver. This end-to-end learning approach represents a notable shift from conventional mathematical models, showcasing the potential to craft multi-colored VLC transceivers leveraging sophisticated neural network algorithms. By optimizing constellation symbols with a custom cost function that combines categorical cross-entropy with chromaticity considerations, our methodology has yielded significant performance enhancements.

A critical insight from our research is the substantial influence of O/E conversion efficiency on the minimum distance between constellation symbols, which in turn, crucially affects SER performance. Specifically, our simulations show that including the O/E conversion process leads to a performance degradation of approximately 10 dB at an SER of 10^{-3}. However, our proposed autoencoder design mitigates this degradation, improving performance by 2 dB compared to a traditional CSK-VLC system without the autoencoder. Furthermore, our findings underline the adaptability of an autoencoder framework equipped with a single-photodiode receiver as an effective solution for creating cost-efficient and energy-saving IoT deployments. Despite observed performance dips in the single-photodiode receiver configuration due to altered constellation points, potential remedies such as the augmentation of emission power present a viable countermeasure, particularly in IoT networks where power limitations may be more flexible.

By delving into the impacts of O/E conversion and introducing a learning-driven design philosophy for VLC systems, our work furthers the integration of this technology with IoT infrastructures. This research not only deepens the theoretical understanding of VLC but also extends its practical reach, setting the stage for VLC technology to thrive in diverse real-world applications. Additionally, utilizing autoencoders paves the way for future research opportunities to investigate other processes that impact both the transmitter and receiver. Although this study primarily focuses on the optical-to-electrical conversion at the receiver, the framework developed can be expanded to address other overlooked processes that influence the overall system performance and quality. Subsequent studies will further confirm the effectiveness of autoencoders in optimizing VLC systems and overcoming a wider range of challenges. Potential future applications for this technology include smart homes and offices, healthcare facilities, public transportation systems, industrial automation, AR/VR applications, and retail environments. These applications showcase the adaptability and extensive usability of our proposed VLC system design in enhancing Internet of Things (IoT) connectivity and data transmission.

Author Contributions: Conceptualization, J.M.L.-R., J.R. (Jose Rabadan) and R.P.-J.; Methodology, J.M.L.-R., J.R. (Jose Rabadan) and J.R. (Julio Rufo); Software, J.M.L.-R., J.R. (Julio Rufo) and V.G.; Validation, J.M.L.-R. and V.G.; Formal analysis, J.M.L.-R., C.A.G. and R.P.-J.; Writing—original draft, J.M.L.-R., C.A.G. and J.R. (Jose Rabadan); Writing—review & editing, R.P.-J.; Funding acquisition, J.M.L.-R., J.R. (Jose Rabadan), J.R. (Julio Rufo) and R.P.-J. All authors have read and agreed to the published version of the manuscript.

Funding: This research was funded by the Spanish State Research Agency through grants PID2020-114561RB-I00 and TED2021-130049B-C21/22, MCIN/AEI/10.13039/501100011033, and the European Union NextGenerationEU/PRTR with grant TED2021-130049A-C22. Additional funding was provided by the María Zambrano Grant from the Universidad de Las Palmas de Gran Canaria and the State Government of San Luis Potosí in Mexico with grant FME-2021-01-03, Fideicomiso 23871.

Institutional Review Board Statement: Not applicable.

Informed Consent Statement: Not applicable.

Data Availability Statement: Data is contained within the article.

Conflicts of Interest: The authors declare no conflicts of interest.

References

1. Khalighi, M.A.; Uysal, M. Survey on Free Space Optical Communication: A Communication Theory Perspective. *IEEE Commun. Surv. Tutor.* **2014**, *16*, 2231–2258. [CrossRef]
2. Pathak, P.H.; Feng, X.; Hu, P.; Mohapatra, P. Visible Light Communication, Networking, and Sensing: A Survey, Potential and Challenges. *IEEE Commun. Surv. Tutor.* **2015**, *17*, 2047–2077. [CrossRef]
3. Qin, Z.; Ye, H.; Li, G.Y.; Juang, B.-H.F. Deep Learning in Physical Layer Communications. *IEEE Wirel. Commun.* **2019**, *26*, 93–99. [CrossRef]
4. An, Y.; Wang, S.; Zhao, L.; Ji, Z.; Ganchev, I. A Learning-Based End-to-End Wireless Communication System Utilizing a Deep Neural Network Channel Module. *IEEE Access* **2023**, *11*, 17441–17453. [CrossRef]
5. *IEEE Std 802.15.7-2018 (Revision of IEEE Std 802.15.7-2011)*; IEEE Standard for Local and Metropolitan Area Networks—Part 15.7: Short-Range Optical Wireless Communications. IEEE: Piscataway, NJ, USA, 2019; pp. 1–407. [CrossRef]
6. Das, P.; Kim, B.-Y.; Park, Y.; Kim, K.-D. A New Color Space Based Constellation Diagram and Modulation Scheme for Color Independent VLC. *Adv. Electr. Comput. Eng.* **2012**, *12*, 11–18. [CrossRef]
7. Drost, R.J.; Sadler, B.M. Constellation Design for Color-Shift Keying Using Billiards Algorithms. In Proceedings of the 2010 IEEE Globecom Workshops, Miami, FL, USA, 6 December 2010; pp. 980–984. [CrossRef]
8. Monteiro, E.; Hranilovic, S. Design and Implementation of Color-Shift Keying for Visible Light Communications. *J. Light. Technol.* **2014**, *32*, 2053–2060. [CrossRef]
9. Monteiro, E.; Hranilovic, S. Constellation Design for Color-Shift Keying Using Interior Point Methods. In Proceedings of the 2012 IEEE Globecom Workshops, Anaheim, CA, USA, 3–7 December 2012; pp. 1224–1228. [CrossRef]
10. Liang, X.; Yuan, M.; Wang, J.; Ding, Z.; Jiang, M.; Zhao, C. Constellation Design Enhancement for Color-Shift Keying Modulation of Quadrichromatic LEDs in Visible Light Communications. *J. Light. Technol.* **2017**, *35*, 3650–3663. [CrossRef]
11. IEEE 802.15.7-2011 and IEEE802.15.7r1, "IEEE 802.15 WPANTM". Available online: https://www.ieee802.org/15/pub/IEEE%20802_15%20WPAN%2015_7%20Revision1%20Task%20Group.htm (accessed on 9 November 2022).

12. Koonen, T. Indoor Optical Wireless Systems: Technology, Trends, and Applications. *J. Light. Technol.* **2018**, *36*, 1459–1467. [CrossRef]
13. Matheus, L.E.M.; Vieira, A.B.; Vieira, L.F.M.; Vieira, M.A.M.; Gnawali, O. Visible Light Communication: Concepts, Applications and Challenges. *IEEE Commun. Surv. Tutor.* **2019**, *21*, 3204–3237. [CrossRef]
14. Căilean, A.-M.; Dimian, M. Current Challenges for Visible Light Communications Usage in Vehicle Applications: A Survey. *IEEE Commun. Surv. Tutor.* **2017**, *19*, 2681–2703. [CrossRef]
15. Sklar, B.; Harris, F. *Digital Communications: Fundamentals and Applications*, 3rd ed.; Pearson: London, UK, 2020.
16. Rappaport, T.S. *Wireless Communications: Principles and Practice*, 2nd ed.; Prentice-Hall: Upper Saddle River, NJ, USA, 2002.
17. Djordjevic, I.B. *Advanced Optical and Wireless Communications Systems*; Springer: Cham, Switzerland, 2018.
18. Gagliardi, R.M.; Karp, S. *Optical Communications*, 2nd ed.; Wiley: New York, NY, USA, 1995.
19. Li, Z.; Shi, J.; Zhao, Y.; Li, G.; Chen, J.; Zhang, J.; Chi, N. Deep Learning Based End-to-End Visible Light Communication with an In-Band Channel Modeling Strategy. *Opt. Express* **2022**, *30*, 28905–28921. [CrossRef] [PubMed]
20. O'Shea, T.; Hoydis, J. An Introduction to Deep Learning for the Physical Layer. *IEEE Trans. Cogn. Commun. Netw.* **2017**, *3*, 563–575. [CrossRef]
21. Chi, N.; Jia, J.; Hu, F.; Zhao, Y.; Zou, P. Challenges and Prospects of Machine Learning in Visible Light Communication. *J. Commun. Inf. Netw.* **2020**, *5*, 302–309. [CrossRef]
22. Lee, H.; Lee, I.; Lee, S.H. Deep Learning Based Transceiver Design for Multi-Colored VLC Systems. *Opt. Express* **2018**, *26*, 6222–6238. [CrossRef] [PubMed]
23. Pepe, A.; Wei, Z.; Fu, H.Y. Heuristic, Machine Learning Approach to 8-CSK Decision Regions in RGB-LED Visible Light Communication. *OSA Contin.* **2020**, *3*, 473–482. [CrossRef]
24. Zhang, D.-F.; Yu, H.-Y.; Zhu, Y.-J.; Zhu, Z.-R. A Transceiver Design Based on an Autoencoder Network for Multi-Color VLC Systems. *IEEE Photonics J.* **2020**, *12*, 7902716. [CrossRef]
25. Zou, C.; Yang, F. Dimming-Aware Deep Learning Approach for OOK-Based Visible Light Communication. *J. Light. Technol.* **2020**, *38*, 5733–5742. [CrossRef]
26. Ulkar, M.G.; Baykas, T.; Pusane, A.E. VLCnet: Deep Learning Based End-to-End Visible Light Communication System. *J. Light. Technol.* **2020**, *38*, 5937–5948. [CrossRef]
27. Shrivastava, S.; Chen, B.; Chen, C.; Wang, H.; Dai, M. Deep Q-Network Learning Based Downlink Resource Allocation for Hybrid RF/VLC Systems. *IEEE Access* **2020**, *8*, 149412–149434. [CrossRef]
28. Mitra, R.; Bhatia, V. Chebyshev Polynomial-Based Adaptive Predistorter for Nonlinear LED Compensation in VLC. *IEEE Photonics Technol. Lett.* **2016**, *28*, 1053–1056. [CrossRef]
29. Laakso, M.; Dowhuszko, A.A.; Wichman, R. Predistortion of OFDM Signals for VLC Systems Using Phosphor-Converted LEDs. In Proceedings of the 2022 IEEE 23rd International Workshop on Signal Processing Advances in Wireless Communication (SPAWC), Oulu, Finland, 4–6 July 2022; pp. 1–5. [CrossRef]
30. Lizarraga, E.M.; Wayar, J.; Linares, M.F.; Dowhuszko, A.A. Effects of LED Non-linear Response and Waveform Settings on Multi-Cell VLC System Deployments. In Proceedings of the 2021 IEEE Latin-American Conference on Communications (LATINCOM), Santo Domingo, Dominican Republic, 17–19 November 2021; pp. 1–5. [CrossRef]
31. Hamamatsu. Si PIN Photodiodes (S1223 Series). Available online: https://www.hamamatsu.com/eu/en/product/optical-sensors/photodiodes/si-photodiodes/S1223-01.html (accessed on 1 December 2023).
32. Wyszecki, G.; Stiles, W.S. *Color Science*; Wiley: New York, NY, USA, 1982; Volume 8.
33. Luna-Rivera, J.M.; Suarez-Rodriguez, C.; Guerra, V.; Perez-Jimenez, R.; Rabadan, J.; Rufo, J. Low-Complexity CSK-based Visible Light Communications System. *IET Optoelectron.* **2015**, *9*, 191–198. [CrossRef]
34. McCamy, C.S. Correlated Color Temperature as an Explicit Function of Chromaticity Coordinates. *Color Res. Appl.* **1992**, *17*, 142–144. [CrossRef]

photonics

MDPI

Article

Regeneration of 200 Gbit/s PAM4 Signal Produced by Silicon Microring Modulator (SiMRM) Using Mach–Zehnder Interferometer (MZI)-Based Optical Neural Network (ONN)

Tun-Yao Hung [1], David W. U Chan [2], Ching-Wei Peng [1], Chi-Wai Chow [1,*] and Hon Ki Tsang [2]

1 Department of Photonics & Graduate Institute of Electro-Optical Engineering, College of Electrical and Computer Engineering, National Yang Ming Chiao Tung University, Hsinchu 30010, Taiwan
2 Department of Electronic Engineering, The Chinese University of Hong Kong, Shatin, Hong Kong
* Correspondence: cwchow@nycu.edu.tw

Abstract: We propose and demonstrate a Mach–Zehnder Interferometer (MZI)-based optical neural network (ONN) to classify and regenerate a four-level pulse-amplitude modulation (PAM4) signal with high inter-symbol interference (ISI) generated experimentally by a silicon microing modulator (SiMRM). The proposed ONN has a multiple MZI configuration achieving a transmission matrix that resembles a fully connected (FC) layer in a neural network. The PAM4 signals at data rates from 160 Gbit/s to 240 Gbit/s (i.e., 80 GBaud to 120 GBaud) were experimentally generated by a SiMRM. As the SiMRM has a limited 3-dB modulation bandwidth of ~67 GHz, the generated PAM4 optical signal suffers from severe ISI. The results show that soft-decision (SD) forward-error-correction (FEC) requirement (i.e., bit error rate, BER < 2.4×10^{-2}) can be achieved at 200 Gbit/s transmission, and the proposed ONN has nearly the same performance as an artificial neural network (ANN) implemented using traditional computer simulation.

Keywords: silicon photonics (SiPh); silicon-on-insulator (SOI); pulse amplitude modulation (PAM); silicon microring modulator (SiMRM); optical neural network (ONN)

1. Introduction

From streaming 4 K/8 K videos to accessing cloud-based Internet services, the need for high-speed and reliable Internet connectivity is on the rise. To satisfy these bandwidth demands, high-capacity optical transmission technologies are required. Recently, 800 Gbit/s systems were proposed utilizing eight lanes of 50 Gbaud four-level pulse amplitude modulation (PAM4) (i.e., 8 × 100 Gbit/s/λ) or by utilizing four lanes of 100 Gbaud PAM4 (i.e., 4 × 200 Gbit/s/λ) [1,2]. It was also reported that an aggregate data rate of 1.6 Tbit/s transceiver (TRx) was realized by utilizing eight lanes of 200 Gbit/s [3]. For beyond 1 Tbit/s transmission [4], a single-lane data rate at or beyond 200 Gbit/s is required with improved power and space efficiencies [5]. Nowadays, silicon photonics (SiPh) is widely considered as one of the important optical integration technologies for the next generation data center optical networks and optical interconnects [6–11]. SiPh devices consume less power and produce less heat than conventional electronic circuits, offering great advantages of energy-efficient bandwidth upgrade. In addition, SiPh is compatible with the mature, complementary metal–oxide–semiconductor (CMOS) fabrication technologies, which potentially allow integration of photonic and electronic devices at mass volume cost effectively. Recently, different high-speed SiPh modulators have been reported [12]. Although SiPh-based modulators provide many merits, such as low power consumption and a small footprint, there are still many challenges for data center interconnect applications [13]. One is the limited electrical-to-optical (EO) bandwidth (i.e., 50~60 Gbaud) and limited extinction ratio (ER) of the SiPh modulators. Hence, different digital signal processing (DSP) techniques are employed to further enhance the data rates, such as Volterra equalization [14], feed-forward equalization (FFE), and decision feedback equalization (DFE) [15], as well as machine learning approaches, including long short-term memory neural network (LSTMNN) [16], recurrent neural network (RNN) [17], etc.

As discussed before, machine learning approaches have been successfully applied in optical communications and networking [18,19]. Neuromorphics is an attempt to migrate the elements in machine learning algorithms to a hardware platform [20]. This could lead to much faster and more energy efficient data processing [21]. Thanks to the advancements in photonics technologies, bringing together neuromorphics and photonics could offer a high-bandwidth and low-power-consumption operation when compared with electronics [22]. An optical neural network (ONN) enables the running of machine learning algorithms more efficiently [23]. Once an ONN is trained, its architecture could be passive, and the computation using optical signals will be operated without the need of additional power consumption. ONNs can be implemented using free-space optics, which can provide the advantages of negligible crosstalk with lower losses [24]. Recently, many researchers have explored ONNs using an integrated approach with programmable silicon interferometers for matrix and vector multiplications [25,26]. This enables chip-scale parameter calculations in neural networks. The basic component is the Mach–Zehnder Interferometer (MZI), which is utilized to manipulate both power coupling ratio and phase. The multiple MZI configuration can achieve a transmission matrix that resembles a fully connected layer in a neural network. Besides the MZI-based ONN, microring-based ONN [27] and phase change material-based ONN [28] are also promising.

In this work, we propose and demonstrate an ONN to regenerate the four-level pulse amplitude modulation (PAM4) signal with high inter-symbol interference (ISI) generated experimentally by a silicon microring modulator (SiMRM). The proposed ONN has a multiple MZI configuration achieving a transmission matrix that resembles a fully connected layer in a neural network. Here, the PAM4 signals at data rates from 160 Gbit/s to 240 Gbit/s (i.e., 80 GBaud to 120 GBaud) were experimentally generated using a silicon microring modulator (SiMRM) [29]. It is also worth mentioning that the PAM4 signal can be generated by other schemes, such as injection-locked vertical-cavity surface-emitting lasers (VCSELs) [30,31]. As the SiMRM has a 3-dB modulation bandwidth of ~67 GHz, the expected PAM4 data rate is ~134 Gbit/s (i.e., 2 bit/symbol $\times$ 67 Gbaud). When the data rate is operated at >200 Gbit/s, the generated PAM4 optical signal suffers from severe ISI. After the utilization of the proposed MZI-based ONN, the result shows that soft-decision (SD) forward-error-correction (FEC) requirement (i.e., bit error rate, BER < 2.4×10^{-2}) can be achieved at 200 Gbit/s transmission, and the proposed ONN has nearly the same performance with the artificial neural network (ANN) implemented using computer software.

2. Theory of the MZI-Based ONN

The proposed ONN has a multiple MZI configuration achieving a transmission matrix resembles a fully connected layer in a neural network. Figure 1 shows a typical 2×2 MZI, which is composed of two 3-dB couplers, a phase shifter θ situated on the top arm inside the MZI, and a phase shifter φ situated at the MZI output. The phase shifter θ controls the MZI output power, while the phase shifter φ determines the phase of the MZI outputs. This configuration permits adaptable rotation within the unitary matrix, thus contributing to its versatility. Equation (1) shows the transformation matrix of MZI, where θ and φ represent the internal and external phase shift values, respectively.

$$S_{MZI} = je^{j\left(\frac{\theta}{2}\right)} \begin{bmatrix} e^{j\varphi}\sin\left(\frac{\theta}{2}\right) & e^{j\varphi}\cos\left(\frac{\theta}{2}\right) \\ \cos\left(\frac{\theta}{2}\right) & -\sin\left(\frac{\theta}{2}\right) \end{bmatrix} \quad (1)$$

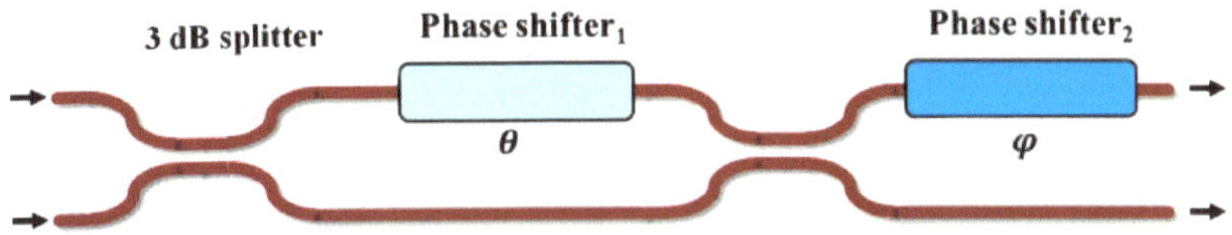

Figure 1. A typical 2 × 2 MZI used in the ONN. It consists of two 3-dB couplers, a phase shifter θ=, and a phase shifter φ.

Figure 2 shows the architecture of the ONN utilized for the classification of ISI distorted PAM4 signals. This MZI network architecture is known as Reck mesh architecture [32]. The number of MZIs in a $N \times N$ Reck mesh is $\frac{N(N-1)}{2}$, where N represents the number of input ports and output ports. These MZIs are organized in $(N-1)$ rows, with the count of MZIs in each row decreasing from $(N-1)$ to 1 from top to bottom. The first port is for receiving the PAM4 data, while the second part is for optical pumping. This will be discussed in detail in a later section.

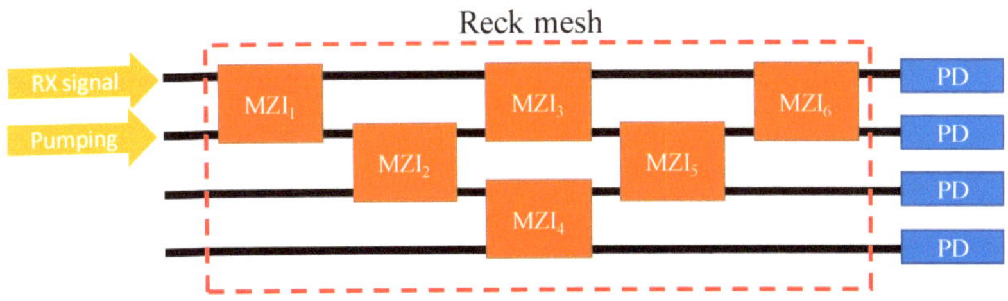

Figure 2. The architecture of MZI-based ONN in Reck mesh architecture.

The transformation matrix of each MZI in the mesh can be expanded to a $N \times N$ dimensional Hilbert space. Take the 4 × 4 Reck mesh for example, the 4 × 4 dimensional Hilbert space of each MZI is shown in Equations (2)–(4).

$$D_n = \begin{bmatrix} \multicolumn{2}{c}{S_{MZI_n}} & 0 & 0 \\ & & 0 & 0 \\ 0 & 0 & 1 & 0 \\ 0 & 0 & 0 & 1 \end{bmatrix} \quad n = 1,\ 3,\ 6 \tag{2}$$

$$D_n = \begin{bmatrix} 1 & 0 & 0 & 0 \\ 0 & \multicolumn{2}{c}{S_{MZI_n}} & 0 \\ 0 & & & 0 \\ 0 & 0 & 0 & 1 \end{bmatrix} \quad n = 2,\ 5 \tag{3}$$

$$D_n = \begin{bmatrix} 1 & 0 & 0 & 0 \\ 0 & 1 & 0 & 0 \\ 0 & 0 & \multicolumn{2}{c}{S_{MZI_n}} \\ 0 & 0 & & \end{bmatrix} \quad n = 4 \tag{4}$$

The S_{MZI_n} in the equations is the nth MZI transformation matrix as shown in Equation (1). The entire Hilbert space of the network system is derived from the inner product of $\boldsymbol{D_n}$. Therefore, the entire Hilbert space in the Reck mesh can be written as Equation (5). Hence, the input-output relationship of the MZI network can be expressed as Equation (6), where $\boldsymbol{Y}$ represents the output optical field matrix, $\boldsymbol{X}$ is the input optical field matrix, and $\boldsymbol{H}$ denotes the Hilbert space matrix. This operation is like the fully connected layer shown in Figure 3.

$$\boldsymbol{H} = \boldsymbol{D_6} \cdot \boldsymbol{D_5} \cdot \boldsymbol{D_4} \cdot \boldsymbol{D_3} \cdot \boldsymbol{D_2} \cdot \boldsymbol{D_1} \tag{5}$$

$$Y = X \cdot H \tag{6}$$

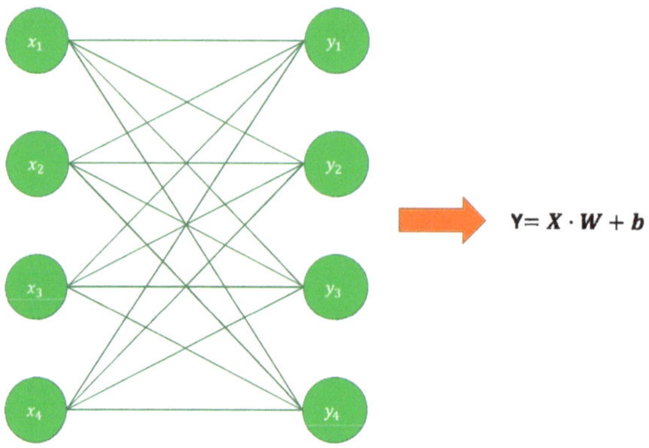

Figure 3. The 4 × 4 fully connected layer neural network operation.

In a fully connected layer, each connection line from x_i to y_j can be written as $x_i\, w_{i,j} + b_{i,j}$, where $w_{i,j}$ and $b_{i,j}$ are the weight and bias value at connect line, respectively. The relationship between x_i and y_j is illustrated in Equation (7). Using a matrix to express this relationship, we can obtain Equation (8), where $\boldsymbol{Y}$ is output matrix, $\boldsymbol{X}$ is input matrix, $\boldsymbol{W}$ is weight matrix, and $\boldsymbol{b}$ is the bias matrix. Comparing Equation (8) with Equation (6), it can be observed that they are very similar.

$$y_j = \sum_{i=1}^{i=n} x_i w_{i,j} + b_{i,j} \tag{7}$$

$$Y = X \cdot W + b \tag{8}$$

Therefore, we can use same way in a neural network like a back-propagation algorithm to optimize $\boldsymbol{H}$ matrix value in the lower loss function value as shown in Equation (9),

$$\boldsymbol{H}_{t+1} = \boldsymbol{H}_t - \alpha \cdot \nabla_{\boldsymbol{H}_t} L \tag{9}$$

where α is the learning rate, ∇ is the gradient operator, L is the loss function value, and t is the current epoch. Due to the unitary property inherent in linear transformation matrices, the inverse matrix $[\mathbf{S_{MZI}}]^{-1}$ of each MZI is equal to its conjugate transpose as Equation (10)

$$S_{MZI}{}^{-1} = -je^{-j\left(\frac{\theta}{2}\right)} \begin{bmatrix} e^{-j\varphi}\sin\left(\frac{\theta}{2}\right) & \cos\left(\frac{\theta}{2}\right) \\ e^{-j\varphi}\cos\left(\frac{\theta}{2}\right) & -\sin\left(\frac{\theta}{2}\right) \end{bmatrix} \tag{10}$$

Hence, the decomposition of $\boldsymbol{H}$ is equivalent to the reverse arrangement of MZIs. This leads to successive products culminating in the eventual formation of the identity matrix as shown in Equation (11). Through the sequential multiplication of $\boldsymbol{H}$ by $[\boldsymbol{D_n}]^{-1}$ in a defined order, the off-diagonal elements in both the upper and lower triangles of the matrix would eventually become 0. Subsequently, Gaussian elimination can be applied to determine the phase shift values φ and θ at each phase shifter.

$$\boldsymbol{H} \cdot [\boldsymbol{D_1}]^{-1} \cdot [\boldsymbol{D_2}]^{-1} \cdot [\boldsymbol{D_3}]^{-1} \cdot [\boldsymbol{D_4}]^{-1} \cdot [\boldsymbol{D_5}]^{-1} \cdot [\boldsymbol{D_6}]^{-1} = \begin{bmatrix} 1 & 0 & 0 & 0 \\ 0 & 1 & 0 & 0 \\ 0 & 0 & 1 & 0 \\ 0 & 0 & 0 & 1 \end{bmatrix} \tag{11}$$

When the MZI-based ONN has been trained, it can be operated as a PAM4 signal classifier as illustrated in Figure 4. It shows that after the trained MZI-based ONN, different photodiodes (PDs) will be detected corresponding to different levels in the PAM4 input data. However, this part is only the linear operation, and nonlinear activation is needed to handle more complicated scenario.

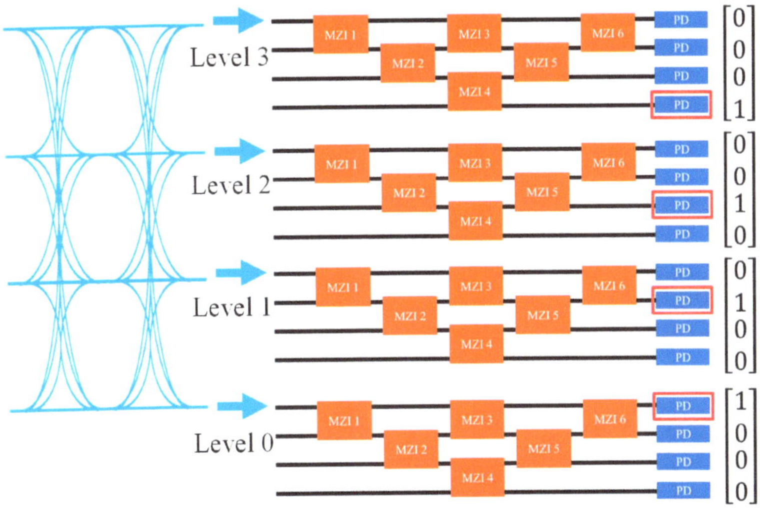

Figure 4. After proper training, the MZI-based ONN is acted as a PAM4 signal classifier.

The nonlinear activation function plays a pivotal role in the functionality of a neural network. In the ONN, one way to achieve nonlinear activation is use the structure shown in Figure 5, which is known as the electro-optic nonlinear activation function [24]. As illustrated in Figure 5, the electro-optic nonlinear activation function structure consists of a directional coupler (DC), a PD, an electric amplifier, and a MZI. In the proposed work, the electrical amplifier is implemented off chip. The DC splitter divides the light into two paths. One pathway receives a fraction α of the input light power, which is then sent to the PD for conversion into an electric signal. In contrast, the remaining fraction of the input light power, which is $1-\alpha$, is directed to the MZI after an appropriate time delay. The PD output voltage will be amplified by the electric amplifier and combined with a proper voltage V_b to input to the MZI phase shift. The operation of electro-optic nonlinear activation function is illustrated in Equation (12), with the two internal components defined in Equations (13) and (14).

$$f(z) = j\sqrt{1-\alpha}e^{-j\left(\frac{g_\varphi|z|^2}{2}+\frac{\varphi_b}{2}\right)} \cdot \cos\left(\frac{g_\varphi|z|^2}{2}+\frac{\varphi_b}{2}\right)z \quad (12)$$

$$\varphi_b = \pi\frac{V_b}{V_\pi} \quad (13)$$

$$g_\varphi = \pi\frac{\alpha GR}{V_\pi} \quad (14)$$

Above, z is the input light field, α is the DC split power ratio, V_π is the voltage of the MZI phase shift π, G is the gain of the electric amplifier, and R is the responsivity. Hence, by controlling the V_b, we can conveniently modify Equation (13) to a different nonlinear activation function. By connecting the electro-optic nonlinear activation function in series after the MZI network mesh, a neural network with an activation function can be realized.

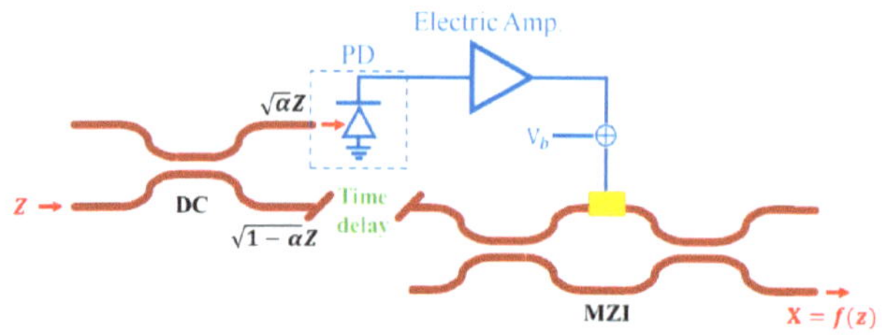

Figure 5. The structure of electro-optic nonlinear activation functions. MZI: Mach–Zehnder Interferometer; DC: directional coupler; PD: photodetector.

3. Experimental Setup

Figure 6 illustrates the experimental setup to obtain the PAM4 optical signal. At the transmitter (Tx) side, a 1550 nm wavelength distributed feedback (DFB) laser with an output power of 6 dBm is launched into a silicon photonic (SiPh) chip with an SiMRM. The SiMRM was fabricated by the multi-project wafer (MPW) scheme in CUMEC. The electrical PAM4 signal is generated by an arbitrary waveform generator (AWG, Keysight M8194A) with 45 GHz analog bandwidth. Subsequently, the signal is amplified by a 60 GHz radio-frequency (RF) amplifier. The Tx digital signal processing (DSP) includes PAM4 symbol mapping, pre-distortion, upsampling, channel estimation, and pre-emphasis. The pre-distortion and pre-emphasis serve to alleviate non-linear distortion and tackle issues related to high-frequency roll-off, stemming from the limited bandwidth of the AWG. The optical PAM4 signal is produced via a SiMRM with a bandwidth ~67 GHz and operated at −3 V bias, measured by a lightwave component analyzer (LCA; Keysight N4373D). At the receiver (Rx) side, the optical PAM4 signal is detected by a 70 GHz bandwidth PD connected to a real-time oscilloscope (RTO, Keysight UXR0802A) with 80 GHz bandwidth and 256 GSa/s sampling rate. To evaluate transmission performance related to different received optical powers, a variable optical attenuator (VOA) is employed. The Rx DSP invovles time synchronization for ensuring proper alignment of the received signal with the transmitted signal, resampling to adjust the signal sampling rate to match with the neural network, the proposed ONN processing, symbol demapping, and BER evaluation. Inset of Figure 5 shows the photo of the SiMRM with diameter of ~10 μm. It was fabricated on a silicon-on-insulator (SOI) platform with a staring wafer of 220 nm silicon layer and 2 μm buried oxide layer (BOX). The SiMRM has a loaded Q of ~3000.

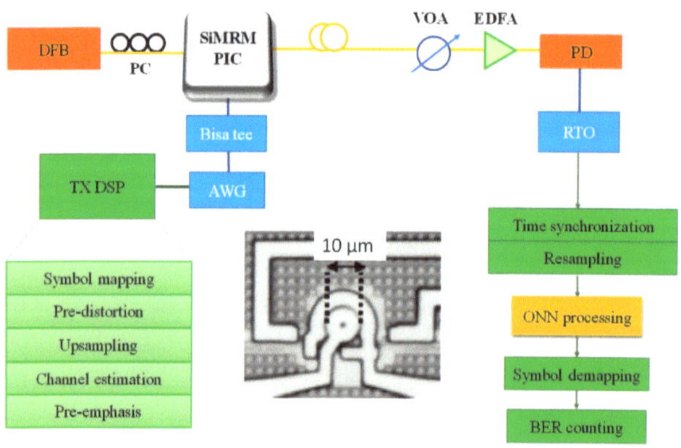

Figure 6. The experimental setup to obtain the PAM4 optical signal. AWG: arbitrary waveform generator; DFB: distributed feedback laser diodes; PC: polarization controller; EDFA: erbium-doped fiber amplifier; VOA: variable optical attenuator; PD: photodetector; RTO: real-time oscilloscope. Inset: photo of the SiMRM.

4. Result and Discussion

In this work, we use Neuroptica [33,34], which is a customized ONN simulator programmed in Python to simulate the PAM4 signal classify by ONN processing. As discussed above, Figure 2 shows the architecture of a Reck-based ONN to classify the experimentally obtained PAM4 signal. We only use two ports for the classification of the distorted PAM4 signal as indicated in Figure 2. The first port is for receiving the PAM4 data, while the second part is for optical pumping. In this work, the optical pumping is needed to increase signal resolvability and provide additional optical power to amplify the PAM4 data. Similar to the case of coherent detection, the pumping light can amplify the optical signal like the local oscillator (LO) light. Here, we did not consider the additional noise of pumping light in our simulation. However, the influence of additional noise from pumping light on the system will be similar to that of a coherent transmission system. To simulate the PD, a square law detection is implemented at the output ports. The classification result depends on the maximum element in the output matrix. Therefore, the target data should be processed by one-hot encode. To update the ONN parameters, cross-entropy loss function is employed, and the optimizer is the Adam. In order to evaluate the performance of proposed ONN, a fully connected ANN using traditional computer simulation is also performed for comparison. This ANN has a four by four fully connected layer with the ReLU activation function. As the ANN is used to compare with the proposed ONN, it has the same number of neurons as the ONN. Hence, it will theoretically have the same performance as the ONN. The dataset used is experimental data obtained from our previous work in [29]. The received waveforms are adjusted by resampling so that there is one sample per symbol. The data length of each transmission data rate experiment is 2^{17} bauds. We use 20% data for training and 80% for testing. In the proof-of-concept demonstration illustrated in Figure 6, the input data are experimentally generated by a bandwidth-limited SiMRM chip. This experimental ISI-distorted optical PAM4 signal will be detected by a separated PD, and a RTO will store the electrical PAM4 signal as shown in Figure 6. Hence, this stored electrical PAM4 signal can be used for the ONN simulation. In the future ONN chip implementation, the ISI distorted optical PAM4 signal can be directly launched into the ONN chip "RX signal" port as shown in Figure 2; hence, no additional OE conversion by the PD is needed. In this case, four on-chip PDs on the ONN chip are used as shown in Figure 2. The optical amplification can be realized by the pumping light as discussed before; hence, VOA and EDFA may not be necessary. Figure 7 shows the accuracy and loss curves for the proposed ONN. It is evident from the results that the ONN exhibits convergence at approximately 100 epochs.

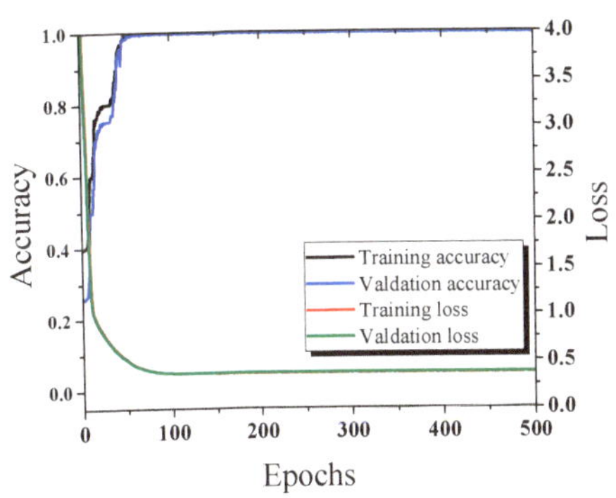

Figure 7. The accuracy and loss curves for the proposed ONN.

Figure 8 illustrates the BER performance of PAM4 signals utilizing both the proposed ONN and ANN. The ONN can recover and classify distorted PAM4 signals within the range of 160 Gbit/s to 240 Gbit/s (i.e., 80 GBaud to 120 GBaud). The data rate achieving the SD-FEC threshold (i.e., BER = 2.4×10^{-2}) can be up to 200 Gbit/s.

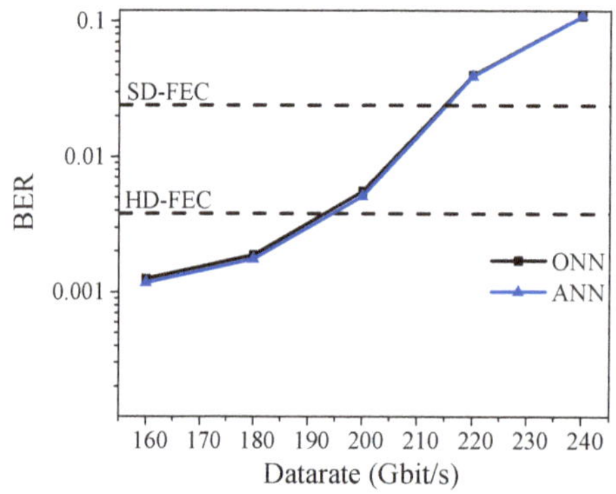

Figure 8. BER performances of ONN and ANN used for classifying the distorted PAM4 signal without the activation function.

It is worth noting that the proposed ONN without an activation function is particularly sensitive to signal power variations. When the signal power is low, the accuracy of the model tends to decrease significantly. Figure 9 illustrates the accuracy and loss performance of different normalized input signal amplitudes. For better understanding, here, the normalized signal amplitude represents the first level of the PAM4 signal, and the four levels in the PAM4 have the same separation. Taking the signal amplitude of 0.8 as an example, the PAM4 values would be 0.8, 1.6, 2.4 and 3.2. We can observe from Figure 9 that the accuracy and loss performance are poor when the normalized input signal amplitude is lower than 0.6. At the normalized input signal of 0.1, the model accuracy falls below 50%. According to our simulation results, the ONN accuracy reduces when the signal amplitude is less than 0.4. This happens because when the signal amplitude is too low, the ASE noise from the EDFA and the thermal and shot noises from the PD become dominant, causing the ONN to fail in performing classification and prediction. When the signal amplitude is larger than 0.4, the ASE and PD noises will not be the dominating factors, and we can observe that the ONN accuracy is ~1 when signal amplitude is between 0.6 and 1.0. To solve this issue, the electro-optic nonlinear activation function discussed in Figure 5 above is included into the ONN model. This enhances the capability of the ONN model to handle nonlinear problems.

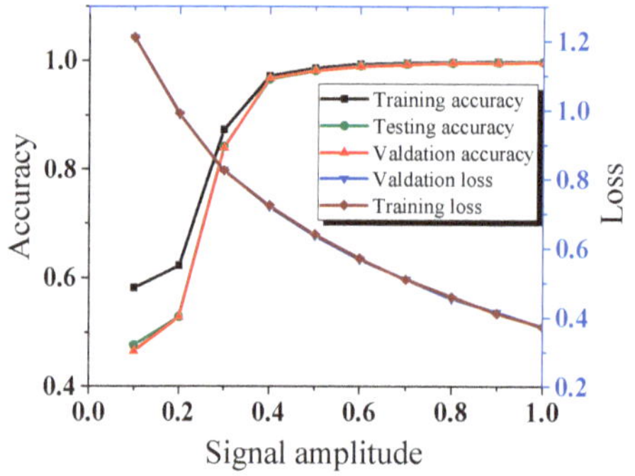

Figure 9. Accuracy and loss performance of different normalized input signal amplitudes without activation function.

Figure 10 shows the modified ONN model with electro-optic nonlinear activation functions. In this architecture, each output port of the first Reck mesh will be connected to an electro-optic nonlinear activation function. The output of the electro-optic nonlinear activation function will then be connected to the input port of the second Reck mesh, and subsequently will be connected to a PD. Furthermore, the fusion of the activation function and the fully connected layer can be considered as a two-layer fully connected ONN, interconnected through activation functions

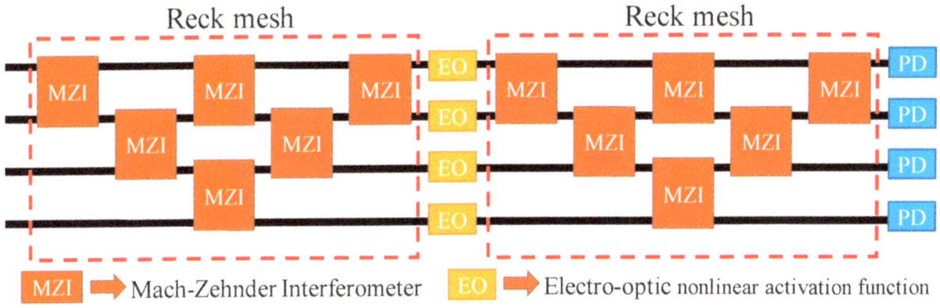

Figure 10. Modified ONN model with electro-optic nonlinear activation functions. MZI: Mach–Zehnder Interferometer; EO: electro-optic nonlinear activation function; PD: photodetector.

In the modified ONN, the parameters of the electro-optic nonlinear activation function as optimized. The α is set to be 0.1, V_π of the MZI phase shift is 5 V, the V_b is set to be -5 V, G is set to be 20, and the responsivity R is set to be 1. Therefore, φ_b is set to be -π, and g_φ is set to be 0.4π. Figure 11 shows the transmission coefficient (i.e., $\frac{|f(z)|^2}{|z|^2}$) of the electro-optic nonlinear activation function with normalized input field Z. We can observe that the electro-optic nonlinear activation function defined exhibits similarities to the sigmoid function but shifted towards the positive x-axis. In the simulation work here, the $\alpha = 0.1$ is used for reducing the loss for electro-optic nonlinear activation function. The electro-optic nonlinear activation function will have different characteristics under different φ_b and g_φ. Here, we found that the nonlinear activation function as illustrated in Figure 11 has a better performance in our model. Therefore, φ_b is set to be $-\pi$, and g_φ is set to be 0.4π.

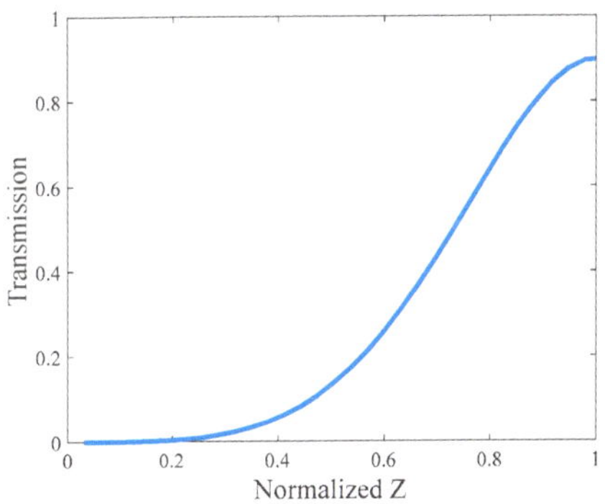

Figure 11. Modified ONN model with electro-optic nonlinear activation functions.

Figure 12 illustrates the accuracy and loss performance of different normalized input signal amplitudes with the electro-optic nonlinear activation function. Comparing the results to the ONN model without an electro-optic nonlinear activation function shown in Figure 10, the accuracy and loss performance in Figure 12 have been significantly improved, particularly at low input signal powers. We can observe that even when the normalized input signal amplitude is as low as 0.1, the accuracy remains at an impressive value of 99.7%.

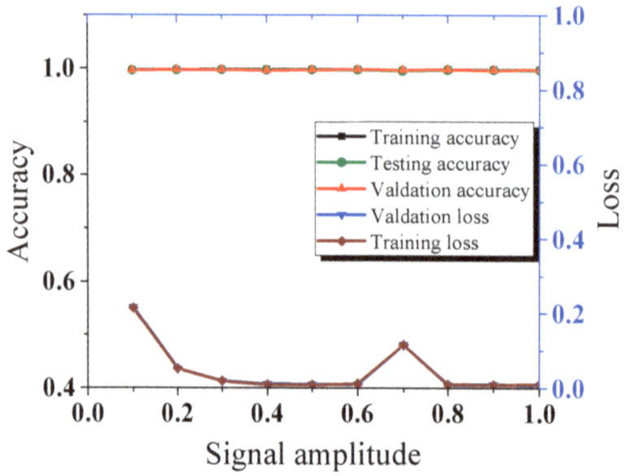

Figure 12. Accuracy and loss performance of different normalized input signal amplitudes with an activation function.

Analyzing the BER performance of PAM4 signals involves using the modified ONN with an electro-optic nonlinear activation function. It can be observed that the BER performance of the modified ONN model with the electro-optic nonlinear activation function is nearly the same as that without the activation function illustrated in Figure 8. The data rate achieving the SD-FEC threshold (i.e., BER = 2.4×10^{-2}) can be up to 200 Gbit/s. This reveals that when the input signal power is high enough, no additional bit error will be introduced for the ONN without the electro-optic nonlinear activation function. However, the introduction of activation function increases the robustness of the proposed ONN. We analyze the impact of the phase shift error on MZI ONN performance. To simulate the phase error of phase shift, we introduce a random normal distribution $N\left(0,\sigma^2\right)$ and add it to the final training results of the phase shift value for each phase shifter in the MZIs. Here, σ is the standard deviation of the phase error. Therefore, the θ and φ in Equation (1) are now written as $\hat{\theta}$ and $\hat{\varphi}$ as shown in Equations (15) and (16).

$$\hat{\theta} = \theta + N\left(0,\sigma^2\right) \tag{15}$$

$$\hat{\varphi} = \varphi + N\left(0,\sigma^2\right) \tag{16}$$

Then, we analyze the impact of the phase error on the ONN. Figure 13 shows the BER performance under various standard deviation phase errors at a data rate of 160 Gbit/s. Here, each BER point is obtained by averaging 1000 BER calculations to ensure the randomness. By analyzing phase errors from 0° to 1.5°, we can observe that the BER performance remains within the SD-FEC threshold when the standard deviation of phase errors is up to 1°. In Figure 13, we also compare the BER performance of the ONN model with and without electro-optic nonlinear activation function under different standard deviation phase errors. Under 1° phase error, the ONN model with electro-optic nonlinear activation function achieves a slightly lower Bit Error Rate (BER) compared to the standard deviation phase errors. This shows the ONN model with the electro-optic nonlinear activation function possesses a higher tolerance for phase errors, providing a more stable and reliable performance under 1° of phase error.

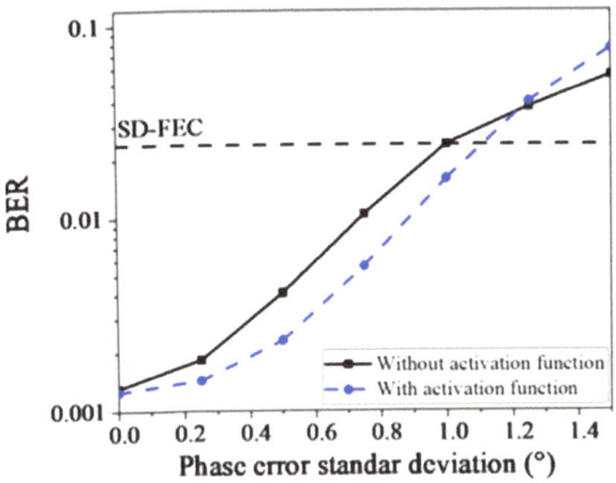

Figure 13. BER performance under various standard deviation phase errors at a data rate of 160 Gbit/s.

5. Conclusions

We proposed and demonstrated an ONN to regenerate PAM4 signal with high ISI generated experimentally by a SiMRM. As the SiMRM has a 3-dB modulation bandwidth of ~67 GHz, the expected PAM4 data rate is ~134 Gbit/s (i.e., 2 bit/symbol × 67 Gbaud). When the data rate is operated at >200 Gbit/s, the generated PAM4 optical signal suffers from severe ISI. The proposed ONN has a multiple MZI configuration achieving a transmission matrix that resembled a fully connected layer in a neural network. The PAM4 signals at data rates from 160 Gbit/s to 240 Gbit/s (i.e., 80 GBaud to 120 GBaud) were experimentally generated using a SiMRM with limited modulation bandwidth of ~67 GHz. The proposed ONN is performed via Neuroptica, which is a customized ONN simulator programmed in Python. Results showed that SD-FEC requirement (i.e., BER < 2.4×10^{-2}) can be achieved at 200 Gbit/s transmission, and the proposed ONN has nearly the same performance with ANN implemented using traditional computer simulation. Moreover, we also discussed the effect of electro-optic nonlinear activation function on the ONN model. By comparing the ONN model with and without electro-optic nonlinear activation function in different input signal amplitudes, it can be observed that the accuracy and loss can be significantly improved at low input signal amplitudes. Even at the normalized input signal amplitude of 0.1, the accuracy can still achieve 99.7%. Furthermore, we analyzed the impact of the phase shift error of MZI to the ONN model. Both ONN model with and without electro-optic nonlinear activation function can still achieve SD-FEC threshold under a 1° phase shift error.

Author Contributions: All authors contributed to the study conception and design. Data collection and analysis were performed by T.-Y.H., D.W.U.C. and C.-W.P. The first draft of the manuscript was written by T.-Y.H. All authors have read and agreed to the published version of the manuscript.

Funding: This work was supported by National Science and Technology Council, Taiwan, under Grant NSTC-112-2221-EA49-102-MY3, NSTC-112-2218-E-011-006, NSC-112-3111-E-A49-001, NSTC-110-2221-E-A49-057-MY3.

Institutional Review Board Statement: Not applicable.

Informed Consent Statement: Not applicable.

Data Availability Statement: The data presented in this study are available on request from the corresponding author.

Conflicts of Interest: The authors declare no conflicts of interest.

References

1. Soref, R.; Dong, P.; Chen, J.; Melikyan, A.; Fan, T.; Fryett, T.; Li, C.; Chen, J.; Koeppen, C. Silicon photonics for 800G and beyond. In *Proceedings of the Optical Fiber Communication Conference*; Optica Publishing Group: San Diego, CA, USA, 2022; p. M4H-1.
2. 800G Pluggable MSA Group. 2022. Available online: https://www.800gmsa.com/ (accessed on 7 April 2024).
3. Available online: https://www.marvell.com/company/newsroom/marvell-extends-connectivity-leadership-with-industrys-first-pam4-optical-dsp.html (accessed on 7 April 2024).
4. Zhou, X.; Urata, R.; Liu, H. Beyond 1 Tb/s intra-data center interconnect technology: IM-DD OR coherent? *J. Light. Technol.* **2020**, *38*, 475–484. [CrossRef]
5. Ozolins, O.; Joharifar, M.; Salgals, T.; Louchet, H.; Schatz, R.; Gruen, M.; Dippon, T.; Kruger, B.; Pittala, F.; Che, D.; et al. Optical amplification-free high baudrate links for intra-data center communications. *J. Light. Technol.* **2023**, *41*, 1200–1206. [CrossRef]
6. Malik, A.; Liu, S.; Timurdogan, E.; Harrington, M.; Netherton, A.; Saeidi, M.; Blumenthal, D.J.; Theogarajan, L.; Watts, M.; Bowers, J.E. Low power consumption silicon photonics datacenter interconnects enabled by a parallel architecture. In *Proceedings of the Optical Fiber Communication Conference*; Optica Publishing Group: San Francisco, CA, USA, 2021; p. W6A-3.
7. Zhou, J.; Wang, J.; Zhu, L.; Zhang, Q. Silicon photonics for 100Gbaud. *J. Light. Technol.* **2021**, *39*, 857–867. [CrossRef]
8. Ahmed, A.H.; Sharkia, A.; Casper, B.; Mirabbasi, S.; Shekhar, S. Silicon-photonics microring links for datacenters—Challenges and opportunities. *IEEE J. Sel. Top. Quant. Electron.* **2016**, *22*, 194–203. [CrossRef]
9. Peng, C.W.; Chow, C.W.; Kuo, P.C.; Chen, G.H.; Yeh, C.H.; Chen, J.; Lai, Y. DP-QPSK coherent detection using 2D grating coupled silicon based receiver. *IEEE Photonics J.* **2021**, *13*, 7900105. [CrossRef]
10. Rahim, A.; Hermans, A.; Wohlfeil, B.; Petousi, D.; Kuyken, B.; Van Thourhout, D.; Baets, R.G. Taking silicon photonics modulators to a higher performance level: State-of-the-art and a review of new technologies. *Adv. Photonics* **2021**, *3*, 024003. [CrossRef]
11. Luo, L.W.; Ophir, N.; Chen, C.P.; Gabrielli, L.H.; Poitras, C.B.; Bergmen, K.; Lipson, M. WDM-compatible mode-division multiplexing on a silicon chip. *Nat. Comm.* **2014**, *5*, 3069. [CrossRef]
12. Zhang, F.; Zhang, L.; Ruan, X.; Yang, F.; Ming, H.; Li, Y. High baud rate transmission with silicon photonic modulators. *IEEE J. Sel. Top. Quantum Electron.* **2021**, *27*, 8300709. [CrossRef]
13. Dourado, D.M.; de Farias, G.B.; Gounella, R.H.; Rocha, M.D.L.; Carmo, J.P. Challenges in silicon photonics modulators for data center interconnect applications. *Opt. Laser Technol.* **2021**, *144*, 107376. [CrossRef]
14. Hsu, Y.; Tzu, T.C.; Lin, T.C.; Chuang, C.Y.; Wu, X.; Chen, J.; Yeh, C.H.; Tsang, H.K.; Chow, C.W. 64-Gbit/s PAM-4 20-km transmission using silicon micro-ring modulator for optical access networks. In *Proceedings of the Optical Fiber Communication Conference*; Optica Publishing Group: Los Angeles, CA, USA, 2017; p. M3H.2.
15. Chan, D.W.U.; Wu, X.; Lu, C.; Lau, A.P.T.; Tsang, H.K. Efficient 330-Gb/s PAM-8 modulation using silicon microring modulators. *Opt. Lett.* **2023**, *48*, 1036–1039. [CrossRef]
16. Hung, T.Y.; Chan, D.W.U.; Peng, C.W.; Chow, C.W.; Tsang, H.K. 300-Gbit/s/λ 8-Level pulse-amplitude-modulation (PAM8) with a silicon microring modulator utilizing long short term memory regression and deep neural network classification. *Opt. Laser Technol.* **2024**, *171*, 110379. [CrossRef]
17. Deligiannidis, S.; Mesaritakis, C.; Bogris, A. Performance and complexity evaluation of recurrent neural network models for fibre nonlinear equalization in digital coherent systems. In Proceedings of the 2020 European Conference on Optical Communications (ECOC), Brussels, Belgium, 6–10 December 2020; pp. 1–4. [CrossRef]
18. Hung, N.T.; Stainton, S.; Le, S.T.; Haigh, P.A.; Tien, H.P.; Vien, N.D.N.; Tuan, N.V. High-speed PAM4 transmission using directly modulated laser and artificial neural network nonlinear equalizer. *Opt. Laser Technol.* **2023**, *157*, 108642. [CrossRef]
19. Wang, C.; Du, J.; Chen, G.; Wang, H.; Sun, L.; Xu, K.; Liu, B.; He, Z. QAM classification methods by SVM machine learning for improved optical interconnection. *Opt. Comm.* **2019**, *444*, 1–8. [CrossRef]
20. Shastri, B.J.; Tait, A.N.; de Lima, T.F.; Pernice, W.H.P.; Bhaskaran, H.; Wright, C.D.; Prucnal, P.R. Photonics for artificial intelligence and neuromorphic computing. *Nat. Photonics* **2021**, *15*, 102–114. [CrossRef]
21. Tait, A.N.; de Lima, T.F.; Zhou, E.; Wu, A.X.; Nahmias, M.A.; Shastri, B.J.; Prucnal, P.R. Neuromorphic photonic networks using silicon photonic weight banks. *Sci. Rep.* **2017**, *7*, 7430. [CrossRef]
22. Liao, K.; Dai, T.; Yan, Q.; Hu, X.; Gong, Q. Integrated photonic neural networks: Opportunities and challenges. *ACS Photonics* **2023**, *10*, 2001–2010. [CrossRef]
23. Zhang, D.; Tan, Z. A review of optical neural networks. *App. Sci.* **2022**, *12*, 5338. [CrossRef]
24. Zhou, T.; Lin, X.; Wu, J.; Chen, Y.; Xie, H.; Li, Y.; Wu, H.; Fang, L.; Dai, Q. Large-scale neuromorphic optoelectronic computing with a reconfigurable diffractive processing unit. *Nat. Photonics* **2021**, *15*, 367–373. [CrossRef]
25. Shokraneh, F.; Geoffroy-Gagnon, S.; Nezami, M.S.; Liboiron-Ladouceur, O. A single layer neural network implemented by a 4×4 MZI-based optical processor. *IEEE Photonics J.* **2019**, *11*, 4501612. [CrossRef]
26. Mojaver, K.R.; Zhao, B.; Leung, E.; Safaee, S.M.R.; Liboiron-Ladouceur, O. Addressing the programming challenges of practical interferometric mesh based optical processors. *Opt. Exp.* **2023**, *31*, 23851–23866. [CrossRef]
27. Ma, X.; Peserico, N.; Shastri, B.J.; Sorger, V.J. Design and testing of a Silicon Photonic Tensor Core with integrated lasers. In Proceedings of the 2023 IEEE Silicon Photonics Conference (SiPhotonics), Washington, DC, USA, 4–7 April 2023; pp. 1–2. [CrossRef]
28. Teo, T.Y.; Ma, X.; Pastor, E.; Wang, H.; George, J.K.; Yang, J.K.W.; Wall, S.; Miscuglio, M.; Simpson, R.E.; Sorger, V.J. Programmable chalcogenide-based all-optical deep neural networks. *Nanophotonics* **2022**, *11*, 4073–4088. [CrossRef]

29. Chan, D.W.U.; Wu, X.; Zhang, Z.; Lu, C.; Lau, A.P.T.; Tsang, H.K. C-band 67 GHz silicon photonic microring modulator for dispersion-uncompensated 100 Gbaud PAM-4. *Opt. Lett.* **2022**, *47*, 2935–2938. [CrossRef]
30. Wu, H.W.; Lu, H.H.; Tsai, W.S.; Huang, Y.C.; Xie, J.Y.; Huang, Q.P.; Tu, S.C. A 448-Gb/s PAM4 FSO communication with polarization-multiplexing injection-locked VCSELs through 600 m free-space link. *IEEE Access* **2020**, *8*, 28859–28866. [CrossRef]
31. Tsai, W.S.; Li, C.Y.; Lu, H.H.; Lu, Y.F.; Tu, S.C.; Huang, Y.C. 256 Gb/s four-channel SDM-based PAM4 FSO-UWOC convergent system. *IEEE Photon. J.* **2019**, *11*, 7902008. [CrossRef]
32. Reck, M.; Zeilinger, A.; Bernstein, H.J.; Bertani, P. Experimental realization of any discrete unitary operator. *Phys. Rev. Lett.* **1994**, *73*, 58. [CrossRef]
33. Williamson, I.A.D.; Hughes, T.W.; Minkov, M.; Bartlett, B.; Pai, S.; Fan, S. Reprogrammable electro-optic nonlinear activation functions for optical neural networks. *IEEE J. Sel. Top. Quantum Electron.* **2020**, *26*, 7700412. [CrossRef]
34. Bartlett, B.; Minkov, M.; Hughes, T.; Williamson, I.A.D. Neuroptica: Flexible Simulation Package for Optical Neural Networks, GitHub Repository. 2019. Available online: https://github.com/fancompute/neuroptica (accessed on 7 April 2024).

photonics

Article

Exploration of Four-Channel Coherent Optical Chaotic Secure Communication with the Rate of 400 Gb/s Using Photonic Reservoir Computing Based on Quantum Dot Spin-VCSELs

Dongzhou Zhong [1,*], Tiankai Wang [1], Yujun Chen [1], Qingfan Wu [1], Chenghao Qiu [1], Hongen Zeng [1], Youmeng Wang [1] and Jiangtao Xi [1,2]

1 School of Electronics and Information Engineering, Wuyi University, Jiangmen 529020, China; 2112204006@wyu.edu.cn (T.W.); 2112204025@wyu.edu.cn (Y.C.); 2112204070@wyu.edu.cn (Q.W.); 2112304007@wyu.edu.cn (C.Q.); 2112304088@wyu.edu.cn (H.Z.); 2112304080@wyu.edu.cn (Y.W.); jiangtao@uow.edu.au (J.X.)

2 School of Electrical, Computer, Telecommunications Engineering, University of Wollongong, Wollongong 2522, Australia

* Correspondence: zhongdz@wyu.edu.cn

Abstract: In this work, we present a novel four-channel coherent optical chaotic secure communication (COCSC) system, incorporating four simultaneous photonic reservoir computers in tandem with four coherent demodulation units. We employ a quartet of photonic reservoirs that capture the chaotic dynamics of four polarization components (PCs) emitted by a driving QD spin-VCSEL. These reservoirs are realized utilizing four PCs of a corresponding reservoir QD spin-VCSEL. Through these four concurrent photonic reservoir structures, we facilitate high-quality wideband-chaos synchronization across four pairs of PCs. Leveraging wideband chaos synchronization, our COCSC system boasts a substantial 4×100 GHz capacity. High-quality synchronization is pivotal for the precise demasking or decoding of four distinct signal types, QPSK, 4QAM, 8QAM and 16QAM, which are concealed within disparate chaotic PCs. After initial demodulation via correlation techniques and subsequent refinement through a variety of digital signal processing methods, we successfully reconstruct four unique baseband signals that conform to the QPSK, 4QAM, 8QAM and 16QAM specifications. Careful examination of the eye diagrams, bit error rates, and temporal trajectories of the coherently demodulated baseband signals indicates that each set of baseband signals is flawlessly retrieved. This is underscored by the pronounced eye openings in the eye diagrams and a negligible bit error rate for each channel of baseband signals. Our results suggest that delay-based optical reservoir computing employing a QD spin-VCSEL is a potent approach for achieving multi-channel coherent optical secure communication with optimal performance and enhanced security.

Keywords: quantum-dot (QD) spin-vertical-cavity surface-emitting laser; photonic reservoir computing; chaotic synchronization; coherent optical chaos secure communication

1. Introduction

As is well known, there are several methods for optical communication multiplexing, including wavelength division multiplexing (WDM), optical time division multiplexing (OTDM), and polarization multiplexing (PM). Coherent optical communication based on polarization-multiplexing was extensively studied in the 1980s due to the high sensitivity of coherent receivers, which could enhance unrepeated transmission distance [1]. However, related research and development were interrupted in the 1990s due to the rapid advances in high-capacity WDM systems. In 2005, the demonstration of digital carrier-phase estimation in coherent receivers sparked renewed interest in coherent optical communications [2,3]. This was because the digital coherent receiver allowed for a variety of spectrally efficient modulation formats, such as M-ary phase-shift keying and quadrature-amplitude modulation (QAM), which rely upon stable carrier-phase estimation in the digital domain. Additionally, linear transmission impairments, such as group-velocity dispersion (GVD) and polarization-mode dispersion (PMD) of transmission fibers, can be addressed via digital signal processing (DSP). These advantages of the born-again coherent receiver afford considerable potential for innovating existing optical communication systems. Recently, 100-Gb/s transmission systems, which employ QPSK modulation, polarization-division multiplexing, and phase diversity homodyne detection assisted with high-speed DSP at a symbol rate of 25 GBd, have been developed and introduced into commercial networks [4]. Worldwide efforts are now underway to develop coherent receivers that can handle a bit rate of over 400 Gb/s per WDM channel.

In recent decades, there has been a growing focus on enhancing the security of fiber-optic communication through the use of optical chaotic secure communications that employ various devices [5]. As these methods have become increasingly capable of high-speed and high-capacity data transmission, most of the current studies are focused on multi-channel optical chaotic secure communications, including WDM, OTDM and PM chaotic secure communications. Efforts in this area aim to develop secure communication systems that can operate over multiple channels simultaneously, with the goal of improving both the speed and security of fiber-optic communication. Several researchers have already demonstrated successful implementations of WDM and OTDM chaotic secure communications. Furthermore, as digital signal processing (DSP) becomes increasingly integrated with coherent optical communication, high-speed coherent optical transmission systems are poised to play a more significant role in the global optical network infrastructure. The ongoing evolution of fiber-optic communication promises significant enhancements in both capacity and security. As a result, coherent optical chaotic secure communication (COCSC) has generated considerable interest from researchers and industry experts who are working to explore and develop this promising technology. However, it is worth noting that to date, COCSC has not been widely reported and there are several new challenges in key areas of the technology that will need to be addressed moving forward. These new challenges include the following: first, knowing how to realize multi-channel COCSCs with high-speed and high-capacity, and second, knowing how to achieve high-quality chaotic synchronization and coherent demodulation.

It is anticipated that quantum dot spin vertical-cavity surface-emitting lasers (QD-Spin-VCSELs) can be employed to implement high-speed and high-capacity multi-channel coherent optical chaotic secure communications (COCSCs). QD-Spin-VCSELs possess femtosecond dynamic characteristics [6], temperature stability [7], lower lasing current [8], ultra-large bandwidth [9], and independent control of output polarization [10–12], making them well-suited for the realization of multi-channel COCSCs with high-speed and high-capacity. Furthermore, these lasers can achieve ultrafast operation from both their ground and excited states, presenting promising opportunities for ultrafast dual-wavelength laser modules that emit ultrafast dynamics. Each beam of light emitted from the ground state (GS) and excited state (ES) includes components with right circular polarization (RCP) and left circular polarization (LCP). The utilization of ultrafast chaotic RCPs and LCPs from the ground and excited states holds significant potential for realizing a four-channel COCSC system with high speed and high capacity. However, one of the challenges in such a COCSC system pertains to achieving high-quality chaotic synchronization and coherent demodulation. Traditional chaotic synchronization methods, such as leading synchronization and lagging synchronization, are limited by the symmetry between the driving laser and the response laser, as well as the need for a perfect match of their parameters. However, recently developed photon reservoir computing (RC) systems have demonstrated promising performances in chaotic synchronization prediction and chaotic signal separation. These RC systems are expected to alleviate the challenges faced in high-speed COCSC. In particular, a QD spin-VCSEL can generate four polarization components (PCs) from the GS and ES emissions. Four parallel RCs system are constructed by using the four PCs from its GS and ES emissions, where the spacing between two nonlinear nodes is very short. These four parallel RCs are potentially applied to address the challenge of high-quality chaos synchronizations.

Photon reservoir computing systems utilize the nonlinear dynamics of chaotic lasers to process and predict information [13,14]. They consist of a chaotic laser, which acts as a "reservoir" of nonlinear dynamics, and a readout layer that learns to map the reservoir dynamics to the desired output. This enables the system to capture and utilize the complex dynamics of chaotic signals for various applications [15–17], including chaos synchronization and prediction. The advantage of photon RC systems lies in their ability to effectively handle the mismatch between the driving laser and the response laser, as well as the variability in their parameters. By utilizing the reservoir dynamics, these systems can adapt and learn from the input chaotic signals, allowing for robust synchronization and separation even in the presence of imperfections and parameter mismatches. In the context of high-speed COCSC, photon RC systems hold great potential for enhancing the synchronization performance and enabling coherent demodulation in multi-channel communication systems. By leveraging the capabilities of photon RC systems, it is expected that the challenges associated with achieving high-quality chaotic synchronization and coherent demodulation can be effectively addressed.

Recently, there have been several works proposing a delay-based photon RC system based on electronically pumped spin-VCSELs [15,18]. This RC system utilizes the nonlinear dynamical x polarization component (X-PC) and Y-PC from the VCSEL output to perform two parallel reservoir computers, which are capable of predicting two independent optical chaotic time-series simultaneously and their synchronizations. The output X-PC and Y-PC from the electronically pumped spin-VCSEL can be interchanged continuously under external perturbations and optical feedback, which can affect the predictive performance of the two parallel RCs. Compared to an electrically-pumped VCSEL, a QD spin-VCSEL offers flexible spin control of the lasing output and provides more control parameters [19,20]. This enables better controllability for polarization switching and weakly correlated GS and ES dynamics [21,22]. These advantages allow for the realization of four parallel RCs using the four PCs from the GS and ES emissions of the QD-spin-VCSEL. Additionally, a QD-spin-VCSEL can generate ultrafast chaotic dynamics when subjected to short feedback delays, resulting in very short spacing between two virtual nodes with sufficient nodes. Therefore, four RCs using the four PCs from the ground state and excited state emissions can effectively handle four high-speed chaotic time-series in parallel and their synchronizations.

In this study, we introduce a unique four-channel COCSC system that uses four concurrent photonic reservoir computers coupled with a coherent demodulation device. Within this system, a QD-spin-VCSEL is employed as the driving laser, and a separate QD-spin-VCSEL serves as the reservoir laser. We individually modulate four distinct encoded messages (QPSK, 4QAM, 8QAM and 16QAM) to four PCs, originating from the GS and ES emissions in the drive laser QD-spin-VCSEL. Additionally, we build four parallel photonic reservoirs using four PCs, sourced from the GS and ES of the reservoir QD-spin-VCSEL, maintaining a minimal distance between two non-linear nodes. By leveraging a concurrent simulation of Matlab and VPI [23], these four photonic RCs help us overcome the obstacle of chaos synchronization for four pairs of PCs generated by the drive and reservoir QD-spin-VCSELs. We exhibit a four-channel COCSC with a 4×100 GHz capacity using chaos synchronizations founded on these quartet parallel photonic reservoirs. Once the output weights are trained within the nonlinear node states, the four parallel reservoirs can be employed for synchronization and decryption. Further, we coherently demodulate four channels of baseband signals (or bit sequence signals) hidden in modulation messages through a polarization diversity digital coherent receiver (PDDCR) and a variety of DSP methods. We examine the impact of the sampling period and the interval of the virtual nodes on training errors. We approximate the effects of the injection and feedback strengths on chaotic synchronizations. Conclusively, we evaluate the transmission performances of the four-channel baseband signals within this COCSC system, analyzing elements such as bit error rates and eye diagrams.

2. Theoretical Framework and Simulation Experiment Setup

Figure 1 displays the fundamental block diagram of a quad-channel COCSC system, built on four concurrent photonic reservoir computers. This intricate system is composed of the transmitter module (TM), the reservoir computing module (RCM), and the coherent demodulation module (CDM). Within the TM, the ground state of the QD-spin-VSEL generates the chaotic X-PC and Y-PC, marked as GS-PC$_x$ and GS-PC$_y$, respectively. Interestingly, its excited state yields two additional photonic currents recognizable as ES-PC$_x$ and ES-PC$_y$. Each of the QPSK, 4QAM, 8QAM and 16QAM is IQ modulated with a group of bit sequences (baseband signal). In this scheme, there are four distinct groups of bit sequence signals, as depicted as b^1–b^4, individually. For the convenience of discussion, the temporal dynamics of the QPSK, 4QAM, 8QAM and 16QAM are described by $S_1(t)$, $S_2(t)$, $S_3(t)$ and $S_4(t)$, respectively. The QPSK, 4QAM, 8QAM and 16QAM are masked within the chaotic GS-PC$_x$, GS-PC$_y$, ES-PC$_x$ and ES-PC$_y$, respectively. These four channels of chaotic masked signals are integrated into a single optical fiber utilizing a wavelength division multiplexer (WDM Mux). In the RCM, after fiber transmission, the combined signals are partitioned into four-channel chaotic masked signals via a wavelength division demultiplexer (WDM DeMux). Each channel of chaotic masked signal is subsequently bisected into dual beams. A singular beam of chaotic masked signal is introduced to a photonic RC. Here, the predicted outputs from the RC$_1$–RC$_4$ are denoted as the GS-PC$^{'}_x$, GS-PC$^{'}_y$, ES-PC$^{'}_x$ and ES-PC$^{'}_y$, respectively. Once output weights are precisely trained within the non-linear node states of each photonic RC, the GS-PC$_x$, GS-PC$_y$, ES-PC$_x$ and ES-PC$_y$ can be perfectly synchronized with GS-PC$^{'}_x$, GS-PC$^{'}_y$, ES-PC$^{'}_x$ and ES-PC$^{'}_y$ correspondingly. In this scenario, signal types QPSK, 4QAM, 8QAM and 16QAM can be demodulated by applying synchronous subtraction between the chaotic masked signal and each RC's predicted output. These demodulated messages, noted as $S^{'}_1(t)$, $S^{'}_2(t)$, $S^{'}_3(t)$ and $S^{'}_4(t)$, are then channeled into their respective coherent demodulation units (CDUs with the subscripts of 1–4). Post coherent demodulation and DSP, four sets of bit sequence signals are further decoded. These reinstated signal bits are referred to as $b^{'1}$–$b^{'4}$, respectively.

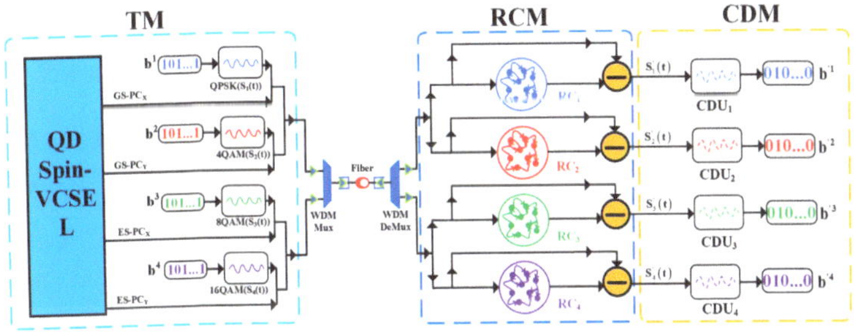

Figure 1. Principle block diagram of four-channel coherent optical chaotic secure communication based on four parallel photonic reservoir computers. Here, TM: transmitter module; RCM: reservoir computing module; CDM: coherent demodulation module; CDU: coherent demodulation unit; b^1–b^4: baseband signals (bit sequence signals); $b^{'1}$–$b^{'4}$: demodulation baseband signals; GS-PC$_x$ and GS-PC$_y$: X-PC and Y-PC from the ground state emission of the QD-spin-VCSEL, respectively; WDM Mux: wavelength division multiplexer; WDM DeMux: wavelength division demultiplexer; and ES-PC$_x$ and ES-PC$_y$: X-PC and Y-PC from the excited state emission of the QD-spin-VCSEL, respectively.

Following the principal block diagram displayed in Figure 1, Figure 2a,b illustrate the simulation experiment setup for a four-channel COCSC system. In this configuration, the QD spin-VCSEL marked with subscript 1 functions as the driving laser, while the QD spin-VCSEL designated by subscript 2 serves as the reserve laser. The CWs, labelled from

1–8, represent the continuous wave lasers. Optical Isolators (ISs, with subscripts from 1–12) are put into service to prevent optical feedback. The neutral density filters (NDFs, labelled from 1–10) are employed to regulate light intensity. The QPSK transmitter (QPSKT), 4QAM transmitter (4QAMT), 8QAM transmitter (8QAMT) and 16QAM transmitter (16QAMT) generate QPSK, 4QAM, 8QAM and 16QAM signals, respectively. The fiber polarization beam splitters (FPBS), carrying subscripts 1–6, are used to partition the light into two distinct polarization components. Bidirectional ports (BPs, labelled 1–4) combine two bidirectional ports into a singular bidirectional multiport of width 2. Lastly, photodiodes (PDs), labeled from 1–12, are designated to convert light waves into corresponding current signals.

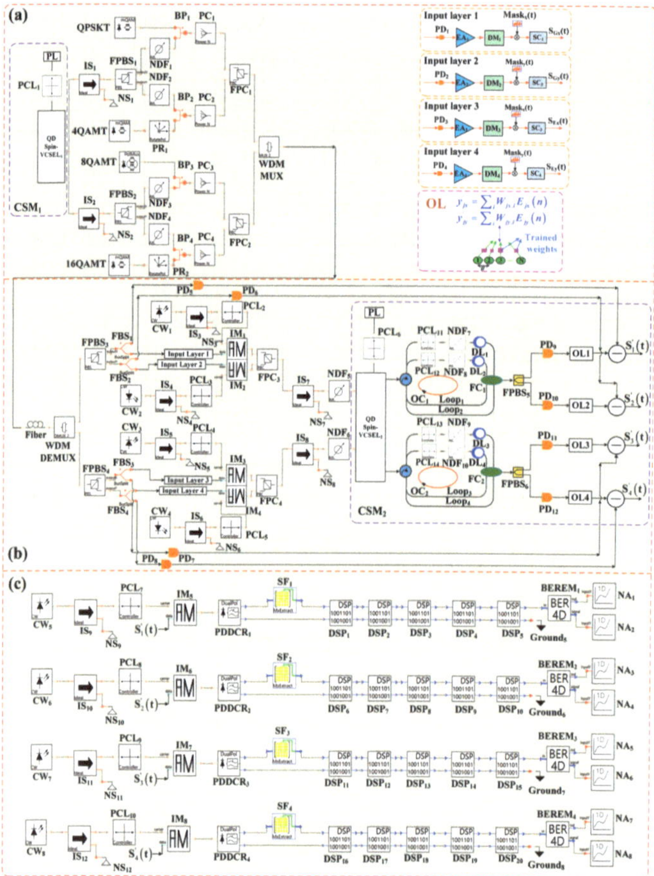

Figure 2. Simulation experiment setup of a four-channel COCSC system, founded on four parallel reservoirs. Here, (**a**) Transmitter; (**b**) Chaos-synchronization prediction and demodulation using reservoirs; (**c**) Coherent demodulation and DSP processing; PL: pumped light; PCL: polarization controller; IS: isolator; FPBS: fiber-optic polarization splitter; QPSKT: QPSK transmitter; 4QAMT: 4QAM transmitter; 8QAMT: 8QAM transmitter; 16QAMT: 16QAM transmitter; NS: empty source; BP: bidirectional ports; PC: power combiner; FPC: fiber polarization coupler; WDM Mux: wavelength division multiplexer; WDM DeMux: wavelength division demultiplexer; CW: continuous wave laser; NDF: the neutral density filter; PD: photodetector; AM: amplitude modulator; DL: delay line; FC: fiber coupler; OL: output layer; CSM: co-simulation module; EA: Electrical amplifier; DM: discrete module; SC: proportional operation circuit; Mask: masked signal; PDDCR: polarization-diversity digital coherent receiver; DSP: digital signal processor; SF: submatrix finder; BEREM: bit error rate estimation module; and NA: numerical analyzer.

In Figure 2a,b , in the QPSKT, the QPSK modulation scheme utilizes a baseband signal b^1 for its in-phase (I) and quadrature (Q) components. Similarly, in the 4QAMT, for 4QAM modulation, the I and Q components are modulated using the odd and even parts of the signal b^2. In the 8QAMT, for 8QAM modulation, the odd and even parts of the signal b^3 are used as the respective signals for the I and Q components. Lastly, for 16QAM modulation, the I and Q components are modulated using the odd and even parts of the signal b^4. In these four modules (QPSKT, 4QAMT, 8QAMT and 16QAMT), QPSK, 4QAM, 8QAM and 16QAM signals are optically modulated using continuous wave lasers and then converted into optical signals at the output ports of these modules. In the QD spin-VCSEL labeled with subscript 1, the light emitted from its ground state (GS) is divided into two chaotic polarization components (GS-PC$_x$ and GS-PC$_y$) using the FPBS$_1$, with their amplitudes represented as $E_{Gx}(t)$ and $E_{Gy}(t)$, respectively. Likewise, the light emitted from its excited state (ES) is separated into two chaotic polarization components (ES-PC$_x$ and ES-PC$_y$) using the FPBS$_2$, and their amplitudes are indicated by $E_{Ex}(t)$ and $E_{Ey}(t)$, respectively. The QPSK and 4QAM signals are concealed within the chaotic GS-PC$_x$ and GS-PC$_y$ using the power combiners 1 and 2 (PC$_1$ and PC$_2$), respectively. These two chaotic hidden signals can be described as ($E_{Gx}(t)$ + $S_1(t)$) and ($E_{Gy}(t)$ + $S_2(t)$), respectively, and are combined into a single beam through the fiber polarization coupler 1 (FPC$_1$). The 8QAM and 16QAM signals are masked within the ES-PC$_x$ and ES-PC$_y$ using the PC$_3$ and PC$_4$, respectively. These two chaotic masked signals are represented as ($E_{Ex}(t)$ + $S_3(t)$) and ($E_{Ey}(t)$ + $S_4(t)$), respectively, and merged into a single beam via FPC$_2$. The mixed light-waves from FPC$_1$ and FPC$_2$ are coupled into an optical fiber through the WDM Mux. After fiber transmission, the multiplexed light-waves are split into two beams with different wavelengths via the WDM DeMux. One beam of light from the WDM DeMux is divided into GS-PC$_x$ and GS-PC$_y$, which contain hidden messages, via the FPBS$_3$. The GS-PC$_x$, carrying the QPSK signal, is further split into two parts using the fiber beam splitter 1 (FBS$_1$). One part is injected into input layer 1, and the other is converted into a current signal by the PD$_5$. The GS-PC$_y$ with 4QAM, ES-PC$_x$ with 8QAM and ES-PC$_y$ with 16QAM are processed similarly.

The input layers provide the connections to the reservoirs. Initially, in input layers 1 and 2, the GS-PC$_x$, including QPSK and the GS-PC$_y$ with 4QAM, are transformed into two distinct current signals via the PD$_1$ and PD$_2$, amplified using electric amplifiers EA$_1$ and EA$_2$, and eventually sampled as separate input data series through the discrete modules DM$_1$ and DM$_2$, respectively. These data series are designated as $u_{Gx}(n-L_{Gx})$ and $u_{Gy}(n-L_{Gy})$. Moreover, the sampled time series of the QPSK, 4QAM, 8QAM and 16QAM are respectively described as $I_1(n)$, $I_2(n)$, $I_3(n)$ and $I_4(n)$, where $I_1(n) = |S_1(n)|^2$, $I_2(n) = |S_2(n)|^2$, $I_3(n) = |S_3(n)|^2$ and $I_4(n) = |S_4(n)|^2$. As a result, $u_{Gx}(n-L_{Gx}) = (C_{Gx}(n-L_{Gx}) + I_1(n-L_{Gx}))$, $u_{Gy}(n-L_{Gy}) = (C_{Gy}(n-L_{Gy}) + I_2(n-L_{Gy}))$, where $C_{Gx}(n-L_{Gx}) = |E_{Gx}(n-L_{Gx})|^2$ and $C_{Gy}(n-L_{Gy}) = |E_{Gy}(n-L_{Gy})|^2$. The term n denotes the discrete time index, while L_{Gx} and L_{Gy} signify the discrete channel delay lengths for GS-PC$_x$ and GS-PC$_y$, respectively. Input layers 3 and 4 process ES-PC$_x$ containing 8QAM and ES-PC$_y$ carrying 16QAM in a similar manner, yielding respective input data as $u_{Ex}(n-L_{Ex})$ and $u_{Ey}(n-L_{Ey})$. Here, $u_{Ex}(n-L_{Ex})$ equals ($C_{Ex}(n-L_{Ex})$ + $I_3(n-L_{Ex})$) and $u_{Ey}(n-L_{Ey})$ corresponds to ($C_{Ey}(n-L_{Ey}) + I_4(n-L_{Ey})$), where $C_{Ex}(n-L_{Ex}) = |E_{Ex}(n-L_{Ex})|^2$ and $C_{Ey}(n-L_{Ey}) = |E_{Ey}(n-L_{Ey})|^2$. L_{Ex} represents the discrete channel delay length for the ES-PC$_x$ and L_{Ey} illustrates that of the ES-PC$_y$. Importantly, $C_{Gx}(n-L_{Gx})$, $C_{Gy}(n-L_{Gy})$, $C_{Ex}(n-L_{Ex})$ and $C_{Ey}(n-L_{Ey})$ are considered four distinct prediction targets. The sampled data, $u_{Gx}(n-L_{Gx})$ and $u_{Ex}(n-L_{Ex})$, are multiplied by the mask signal, Mask$_x$, while $u_{Gy}(n-L_{Gy})$ and $u_{Ey}(n-L_{Ey})$ are multiplied by Mask$_y$. Both Mask$_x$ and Mask$_y$ are chaotic signals, as illustrated in [24]. Post scaling with a scaling factor γ through the scaling operation circuits (SC$_1$–SC$_4$), the four input layers yield output signals denoted as $S_{Gx}(n)$, $S_{Gy}(n)$, $S_{Ex}(n)$ and $S_{Ey}(n)$, respectively. These are respectively modulated with the optical-field phases of CW$_1$–CW$_4$. The FPC$_3$ first couples the modulated $S_{Gx}(n)$ and $S_{Gy}(n)$ into a single beam, which is then injected into the ground state of the reservoir QD spin-VCSEL. Similarly, the FPC$_4$ couples the modulated $S_{Ex}(n)$ and $S_{Ey}(n)$ into a single beam, which is subsequently injected into the excited state of the reservoir QD spin-VCSEL.

Within the reservoir, the GS and ES of the QD spin-VCSEL are both influenced by dual feedback mechanisms. The feedback loops for the GS are denoted by subscripts 1 and 2, while those for the ES are indicated by subscripts 3 and 4. Each loop employs a NDF and a PCL to adjust the feedback intensity and polarization direction of the feedback beam, respectively. The delay time established by the delay lines (DL_1–DL_4) is defined as τ. In the output layers (OLs), the GS-PC_x' and GS-PC_y' emissions from the QD spin-VCSEL are bifurcated using $FPBS_5$. Similarly, the ES-PC_x' and ES-PC_y' emissions are split through $FPBS_6$. The intensity values of GS-PC_x', GS-PC_y', ES-PC_x' and ES-PC_y' are sampled at intervals of θ and are considered as virtual nodes. Accordingly, the total number N of virtual nodes along each delay line is determined by the ratio $N = \tau/\theta$. The states of the N virtual nodes along the DL_1–DL_4 are weighted and linearly summed up. The combined weighted states from the DL_1 and DL_2 are represented as $y'_{Gx}(n)$ and $y'_{Gy}(n)$, respectively, while those from the DL_3 and DL_4 are signified as $y'_{Ex}(n)$ and $y'_{Ey}(n)$. In this setup, by calibrating the output weights, $y'_{Gx}(n)$ and $y'_{Gy}(n)$ can achieve synchronization with $C_{Gx}(n-L_{Gx})$ and $C_{Gy}(n-L_{Gy})$, respectively. Likewise, $y'_{Ex}(n)$ and $y'_{Ey}(n)$ can be attuned to synchronize with $C_{Ex}(n-L_{Ex})$ and $C_{Ey}(n-L_{Ey})$. Under these synchronization conditions, the concealed messages QPSK and 4QAM are decoded by the synchronous subtraction of $y'_{Gx}(n)$ from $C_{Gx}(n-L_{Gx})$ and $y'_{Gy}(n)$ from $C_{Gy}(n-L_{Gy})$, with the retrieved messages designated as $S_1'(t)$ and $S_2'(t)$, respectively. In a similar fashion, the messages 8QAM and 16QAM are decoded by the synchronous subtraction of $y'_{Ex}(n)$ from $C_{Ex}(n-L_{Ex})$ and $y'_{Ey}(n)$ from $C_{Ey}(n-L_{Ey})$, with their decoded equivalents presented as $S_3'(t)$ and $S_4'(t)$, correspondingly.

As illustrated in Figure 2c, the decoded messages, $S_1'(t)$, $S_2'(t)$, $S_3'(n)$ and $S_4'(n)$, are initially modulated with the optical field phases of the CW_5–CW_8 using intensity modulators (IM_1–IM_4) and then each injected into its corresponding coherent demodulation unit (CDU). Each CDU comprises a polarization-diversity digital coherent receiver (PDDCR), a submatrix finder (SF), a set of five digital signal processors (DSPs), and a bit error rate estimation module (BEREM). The PDDCR, depicted in VPI [23], models an optical coherent quadrature receiver that encompasses a local oscillator, optical hybrids, post-detection electrical filters, and analog-to-digital converters. The SF is used to extract specified elements of the input matrix. The DSPs with subscripts 1, 6, 11 and 16 address the compensation of group velocity dispersion and nonlinear effects within the optical fiber, whereas the DSPs labeled with subscripts 2, 7, 12 and 17 are designated to down-sample the in-phase and quadrature signals to match the baud rate. The DSPs marked with subscripts 3, 8, 13 and 18 are dedicated to estimating and correcting frequency discrepancies between the received optical signal and the local oscillator. The DSPs inscribed with subscripts 4, 9, 14 and 19 adjust and align the clock phase of both transmitter and receiver. The DSPs tagged with the subscripts 5, 10, 15 and 20 are dedicated to estimating and correcting phase discrepancies between the received optical signal and the local oscillator. The BEREMs labeled with subscripts 1-4, as four-dimensional bit error rate modules, are capable of generating BERs for the baseband signals and facilitating their demodulation. After processing through the four CDUs, four sets of baseband signals (or bit streams) encapsulated within the decoded modulation messages, $S_1'(t)$, $S_2'(t)$, $S_3'(t)$ and $S_4'(t)$ are effectively reconstructed. These recovered bit streams are denoted as b'^1-b'^4, respectively.

Drawing on the spin-flip model (SFM) of vertical-cavity surface-emitting lasers (VCSELs) put forth by Miguel et al. [25], the interconnected rate equations characterizing the QD spin-VCSEL$_1$ (which serves as the driving QD spin-VCSEL) are delineated as follows [19,26]:

$$\begin{aligned}\frac{dn^{\pm}_{D,WL}}{dt} =&\frac{h_{D,2}}{eN_{QD}}[\eta^{\pm}(I_{E,th}-I_{G,th})+I_{G,th}]-\gamma_{D,0}n^{\pm}_{D,WL}(\frac{h_{D,2}-n^{\pm}_{D,ES}}{2h_{D,2}})\\ &-\gamma_{D,n}n^{\pm}_{D,WL}\mp\gamma_{D,s}(n^{+}_{D,WL}-n^{-}_{D,WL}),\end{aligned} \tag{1}$$

$$\begin{aligned}\frac{dn^{\pm}_{D,ES}}{dt} =&\frac{1}{4}\gamma_{D,0}n^{\pm}_{D,WL}(\frac{h_{D,2}-n^{\pm}_{D,ES}}{h_{D,2}})-\gamma_{D,n}(h_{D,2}+n^{\pm}_{D,ES})-2\gamma_{D,n}n^{\pm}_{D,ES}|E^{\pm}_{D,ES}|^2\\ &-\gamma_{D,21}(h_{D,2}+n^{\pm}_{D,ES})(\frac{h_{D,1}-n^{\pm}_{D,GS}}{2h_{D,1}})\mp\gamma_{D,s}(n^{+}_{D,ES}-n^{-}_{D,ES}),\end{aligned} \tag{2}$$

$$\begin{aligned}\frac{dn^{\pm}_{D,GS}}{dt} =&\gamma_{D,21}(\frac{h_{D,2}+n^{\pm}_{D,ES}}{h_{D,2}})(h_{D,1}-n^{\pm}_{D,GS})-\gamma_{D,n}(h_{D,1}+n^{\pm}_{D,GS})\\ &-2\gamma_{D,n}n^{\pm}_{D,GS}|E^{\pm}_{D,GS}|^2\mp\gamma_{D,s}(n^{+}_{D,GS}-n^{-}_{D,GS}),\end{aligned} \tag{3}$$

$$\frac{dE^{\pm}_{D,GS}}{dt} =k_D(n^{\pm}_{D,GS}-1)(1+i\alpha_D)E^{\pm}_{D,GS}-(\gamma_{D,a}+i\gamma_{D,p})E^{\mp}_{D,GS}+\sqrt{\beta_{sp}}\xi^{\pm}_{D,GS}, \tag{4}$$

$$\frac{dE^{\pm}_{D,ES}}{dt} =k_D(n^{\pm}_{D,ES}-1)(1+i\alpha_D)E^{\pm}_{D,ES}-(\gamma_{D,a}+i\gamma_{D,p})E^{\mp}_{D,ES}+\sqrt{\beta_{sp}}\xi^{\pm}_{D,ES}. \tag{5}$$

The interrelated rate equations governing the QD spin-VCSEL$_2$ (the reservoir QD spin-VCSEL) under the influence of optical feedback and optical injection are revised as follows [19,26]:

$$\frac{dn^{\pm}_{WL}}{dt} =\frac{h_2}{eN_{QD}}[\eta^{\pm}(I_{E,th}-I_{G,th})+I_{G,th}]-\gamma_0 n^{\pm}_{WL}(\frac{h_2-n^{\pm}_{ES}}{2h_2})-\gamma_n n^{\pm}_{WL}\mp\gamma_s(n^{+}_{WL}-n^{-}_{WL}), \tag{6}$$

$$\begin{aligned}\frac{dn^{\pm}_{ES}}{dt} =&\frac{1}{4}\gamma_0 n^{\pm}_{WL}(\frac{h_2-n^{\pm}_{ES}}{h_2})-\gamma_n(h_2+n^{\pm}_{ES})-\gamma_{21}(h_2+n^{\pm}_{ES})(\frac{h_1-n^{\pm}_{GS}}{2h_1})\\ &-2\gamma_n n^{\pm}_{ES}|E^{\pm}_{ES}|^2\mp\gamma_s(n^{+}_{ES}-n^{-}_{ES}),\end{aligned} \tag{7}$$

$$\frac{dn^{\pm}_{GS}}{dt} =\gamma_{21}(\frac{h_2+n^{\pm}_{ES}}{h_2})(h_1-n^{\pm}_{GS})-\gamma_n(h_1+n^{\pm}_{GS})-2\gamma_n n^{\pm}_{GS}|E^{\pm}_{GS}|^2\mp\gamma_s(n^{+}_{GS}-n^{-}_{GS}), \tag{8}$$

$$\begin{aligned}\frac{dE^{\pm}_{GS}}{dt} =&k(n^{\pm}_{GS}-1)(1+i\alpha)E^{\pm}_{GS}-(\gamma_a+i\gamma_p)E^{\mp}_{GS}\\ &-i\Delta\omega_G E^{\pm}_{GS}+k_{inj}E^{1,2}_{inj}+k_f E^{\pm}_{GS}(t-\tau)e^{-i\omega_G\tau}+\sqrt{\beta_{sp}}\xi^{\pm}_{GS},\end{aligned} \tag{9}$$

$$\begin{aligned}\frac{dE^{\pm}_{ES}}{dt} =&k(n^{\pm}_{ES}-1)(1+i\alpha)E^{\pm}_{ES}-(\gamma_a+i\gamma_p)E^{\mp}_{ES}\\ &-i\Delta\omega_E E^{\pm}_{ES}+k_{inj}E^{3,4}_{inj}+k_f E^{\pm}_{ES}(t-\tau)e^{-i\omega_E\tau}+\sqrt{\beta_{sp}}\xi^{\pm}_{ES}.\end{aligned} \tag{10}$$

In Equations (1)–(10), the subscript D designates the driving QD spin-VCSEL. The symbols + and − represent the right circular polarization (RCP) and left circular polarization (LCP) of the emitted light, respectively. The dynamic variables, indicated by n_{WL} and $n_{GS}(n_{ES})$, signify the normalized carrier concentrations in the Wetting Layer (WL) and at the ground (excited) state energy levels. Lasing is facilitated via the transitions from the excited state or the ground state to the valence band (VB), generating right (E^{+}_{ES}, E^{+}_{GS}) and left (E^{-}_{ES}, E^{-}_{GS}) circularly polarized light at two distinct wavelengths. The carrier injection thresholds for the excited and ground states are symbolized by $I_{E,th}$ and $I_{G,th}$, respectively. The remaining parameters for the aforementioned QD Spin-VCSELs are as follows: k and k_D are the photon decay rates; α and α_D represent the linewidth enhancement factors;

h_1 and $h_{D,1}$ are the normalized differential gain coefficients for the ground state transitions; and h_2 and $h_{D,2}$ are those for the excited state transitions. $\gamma_{D,n}$ and γ_n represent the carrier recombination rates; $\gamma_{D,21}$ and γ_{21} denote the intradot relaxation rates at which spin-polarized carriers relax from the excited state to the spin-up (down) ground state; $\gamma_{D,0}$ and γ_0 are the rates of carrier capture from the WL into the excited state; $\gamma_{D,s}$ and γ_s correspond to the spin relaxation rates; $\gamma_{D,p}$ and γ_p represent the birefringence rates; and $\gamma_{D,\gamma}$ and γ_γ are related to the dichroism rates. τ indicates the feedback time along any of the delay lines (DL_1–DL_4) shown in Figure 2; ω_G is the resonant frequency of the light emitted from the ground state; and ω_E is the resonant frequency of light emitted from the excited state. $\Delta\omega_G$ represents the frequency detuning between CW_1 (CW_2) and the ground state emission of the reservoir QD Spin-VCSEL; $\Delta\omega_E$ denotes the frequency detuning between CW_3 (CW_4) and the excited state emission of the reservoir QD Spin-VCSEL. β_{sp} is the rate of spontaneous emission, also viewed as an indicator of noise strength. The terms $\xi^{\pm}_{D,GS}$, $\xi^{\pm}_{D,ES}$, $\xi^{\pm}_{DS}$ and $\xi^{\pm}_{ES}$ embody independent Gaussian white noise sources with zero mean and unit variance. k_f is the feedback coupling strength; k_{inj} stands for the strength of optical injection. E^1_{inj} and E^2_{inj} are the slowly varying complex amplitudes of the CW_1 and CW_2, which are converted to RCP and LCP by the PCL_2 and PCL_3; E^3_{inj} and E^4_{inj} are the injected optical fields for the CW_3 and CW_4, likewise converted by the PCL_4 and PCL_5. E^1_{inj} and E^2_{inj} account for the light fields E^+_{GS} and E^-_{GS}, respectively, while E^3_{inj} and E^4_{inj} are charged with generating the optical fields E^+_{ES} and E^-_{GS}. The total pump strengths $\eta = \eta^+ + \eta^-$, η^+ and η^- are the pump intensities for the RCP and LCP components, respectively.

The left and right circular polarization components of the GS and ES emissions of the driving QD Spin-VCSEL are replaced with the orthogonal linear components as follows:

$$\begin{aligned} E_{D,Gx} &= \frac{E^+_{D,GS} + E^-_{D,GS}}{\sqrt{2}}, \quad E_{D,Gy} = -i\frac{E^+_{D,GS} - E^-_{D,GS}}{\sqrt{2}}, \\ E_{D,Ex} &= \frac{E^+_{D,ES} + E^-_{D,ES}}{\sqrt{2}}, \quad E_{D,Ey} = -i\frac{E^+_{D,ES} - E^-_{D,ES}}{\sqrt{2}}. \end{aligned} \tag{11}$$

The left and right circular polarization components of the GS and ES emissions of the reservoir QD Spin-VCSEL are rewritten in terms of the orthogonal linear components as follows:

$$\begin{aligned} E_{Gx} &= \frac{E^+_{GS} + E^-_{GS}}{\sqrt{2}}, \quad E_{Gy} = -i\frac{E^+_{GS} - E^-_{GS}}{\sqrt{2}}, \\ E_{Ex} &= \frac{E^+_{ES} + E^-_{ES}}{\sqrt{2}}, \quad E_{Ey} = -i\frac{E^+_{ES} - E^-_{ES}}{\sqrt{2}}. \end{aligned} \tag{12}$$

In the QPSKT and m-QAMT (m equals 4, 8, 16) presented in Figure 2a, the QPSK and m-QAM signals can be generated through the process of IQ modulation, where baseband signals (b^1–b^4) modulate a continuous light source. Subsequently, these modulated signals are combined via polarization beam combining techniques. The resulting QPSK and m-QAM signals are characterized by

$$S_j(t) = \frac{1}{2}E_{in,j}(t)\left[\sum_{k=1}^{L_k}\cos(\varphi^j_{1k}) + i\sum_{k=1}^{L_k}\cos(\phi^j_{2k})\right], \tag{13}$$

where the subscript $j = 1$ denotes QPSK. The subscripts $j = 2, 3, 4$ represent 4-QAM, 8-QAM, and 16-QAM, respectively. When $j = 1$ and 2, $L_k = 4$. If $j = 3$, $L_k = 8$, while $j = 4$, $L_k = 16$. The subscript k indicates the kth group of bits in the time sequence. In Equation (13), the phases φ^j_{1k} and ϕ^j_{2k} are respectively written as follows:

$$\varphi^j_{1k} = \arcsin\left[\mathrm{Re}(IQ^j_k)\right], \quad \phi^j_{2k} = \arcsin\left[\mathrm{Im}(IQ^j_k)\right], \tag{14}$$

where the terms IQ^1_k–IQ^4_k are respectively described as

$$IQ^1_{,k} = \cos\left(\frac{2\pi \cdot n^1_k}{2^m}\right) + i \cdot \sin\left(\frac{2\pi \cdot n^1_k}{2^m}\right), \tag{15}$$

$$IQ^{j_n}_{x,k} = \frac{\sum\limits_{l=1}^{m/2} 2^{m/2-l}\left(2 \cdot I^{j_n}_{(k-1)\cdot m/2+l} - 1\right) + i \cdot \sum\limits_{l=1}^{m/2} 2^{m/2-l}\left(2 \cdot Q^{j_n}_{(k-1)\cdot m/2+l} - 1\right)}{2^{m/2-1}}, \tag{16}$$

where the superscript j_n = 2, 3, 4. The variables n^1_k, I and Q can be expressed as follows: $n^1_k = \sum_m^{l=1} b^1_{(k-l)\cdot m+l} \cdot 2^{l-1}$, $I^{j_n}_\ell = b^{j_n}_{2\ell-1}$ and $Q^{j_n}_\ell = b^{j_n}_{2\ell}$, where $\ell = 1, 2, \cdots, N$.

In Equations (9) and (10), the slowly varying amplitudes E^1_{inj}–E^4_{inj} of the complex electric field can be described as [15]

$$\begin{aligned} E^1_{inj}(t) &= \sqrt{S_{Gx}(t) \cdot I_{d1}}, \quad E^2_{inj}(t) = \sqrt{S_{Gy}(t) \cdot I_{d2}}, \\ E^3_{inj}(t) &= \sqrt{S_{Ex}(t) \cdot I_{d3}}, \quad E^4_{inj}(t) = \sqrt{S_{Ey}(t) \cdot I_{d4}}, \end{aligned} \tag{17}$$

where the light intensities $I_{d1} = |E^1_{inj,0}|^2$, $I_{d2} = |E^2_{inj,0}|^2$, $I_{d3} = |E^3_{inj,0}|^2$ and $I_{d4} = |E^4_{inj,0}|^2$. The terms $E^1_{inj,0}$, $E^2_{inj,0}$, $E^3_{inj,0}$ and $E^4_{inj,0}$ are the amplitudes of the continuous-wave lasers CW_1–CW_4, respectively. The masked input signals $S_{Gx}(t)$, $S_{Gy}(t)$, $S_{Ex}(t)$ and $S_{Ey}(t)$ can be expressed as

$$\begin{aligned} S_{Gx}(t) &= \text{Mask}_x(t) \times C_{Gx}(n - L_{Gx}) \times \gamma, S_{Gy}(t) = \text{Mask}_y(t) \times C_{Gy}(n - L_{Gy}) \times \gamma, \\ S_{Ex}(t) &= \text{Mask}_x(t) \times C_{Ex}(n - L_{Ex}) \times \gamma, S_{Ey}(t) = \text{Mask}_y(t) \times C_{Ey}(n - L_{Ey}) \times \gamma, \end{aligned} \tag{18}$$

where the masked signals $\text{Mask}_x(t)$ and $\text{Mask}_y(t)$ are chaotic signals, as presented in [24]. γ is a scaling factor. The discrete channel delay lengths L_{Gx}, $L_{Ex} = \tau_x/h$, and L_{Gy}, $L_{Ey} = \tau_y/h$, where h is the step size, τ_x is the channel delay of the GS-PC_x or ES-PC_x, and τ_y is the channel delay of the GS-PC_y or ES-PC_y.

In such a system presented in Figure 2, chaos synchronization between each pair of PCs (i.e., GS-PC_x and GS-PC'_x, GS-PC_y and GS-PC'_y, ES-PC_x and ES-PC'_x, and ES-PC_y and ES-PC'_y) plays a key role in in security and encrypted message recovery. In the following, we use four parallel RCs to address chaos synchronization between each pair of PCs. According to lag chaotic synchronization theory, the lag synchronization solution is obtained as follows.

$$\begin{aligned} y'_{Gx}(n) &= C_{Gx}(n - L_{Gx}), \quad y'_{Gy}(n) = C_{Gy}(n - L_{Gy}), \\ y'_{Ex}(n) &= C_{Ex}(n - L_{Ex}), \quad y'_{Ey}(n) = C_{Ey}(n - L_{Ey}), \end{aligned} \tag{19}$$

where the time-dependent outputs y'_{Gx}, y'_{Gy}, y'_{Ex} and y'_{Ey} are respectively regarded as linear functions of the GS-PC_x, GS-PC_y, ES-PC_x and ES-PC_y such that

$$\begin{aligned} y'_{Gx}(n) &= W_{Gx,1}b_{out} + W_{Gx,2}C_{Gx}(n - L_{Gx}) + \sum_{i=1}^{N} W_{Gx,i+2}I_{Gx,i}(n), \\ y'_{Gy}(n) &= W_{Gy,1}b_{out} + W_{Gy,2}C_{Gy}(n - L_{Gy}) + \sum_{i=1}^{N} W_{Gy,i+2}I_{Gy,i}(n), \\ y'_{Ex}(n) &= W_{Ex,1}b_{out} + W_{Ex,2}C_{Ex}(n - L_{Ex}) + \sum_{i=1}^{N} W_{Ex,i+2}I_{Ex,i}(n), \\ y'_{Ey}(n) &= W_{Ey,1}b_{out} + W_{Ey,2}C_{Ey}(n - L_{Ey}) + \sum_{i=1}^{N} W_{Ey,i+2}I_{Ey,i}(n), \end{aligned} \tag{20}$$

where $\mathbf{W}_{Gx}$, $\mathbf{W}_{Gy}$, $\mathbf{W}_{Ex}$ and $\mathbf{W}_{Ey}$ represent the output weight matrix; $W_{Gx,i}$, $W_{Gy,i}$, $W_{Ex,i}$ and $W_{Ey,i}$ respectively represent the ith-element of $\mathbf{W}_{Gx}$, $\mathbf{W}_{Gy}$, $\mathbf{W}_{Ex}$ and $\mathbf{W}_{Ey}$; $I_{Gx,i}(n)$, $I_{Gy,i}(n)$,

$I_{Ex,i}(n)$ and $I_{Ey,i}(n)$ respectively represent the ith output state of the GS-PC$'_x$, GS-PC$'_y$, ES-PC$'_x$ and ES-PC$'_y$. Here, $I_{Gx,i}(n) = |E_{G_x}(i)|^2$, $I_{Gy,i}(n) = |E_{G_y}(i)|^2$, $I_{Ex,i}(n) = |E_{E_x}(i)|^2$ and $I_{Ey,i}(n) = |E_{E_y}(i)|^2$. b_{out} is a constant and equal to 1. Previous studies have shown that these output weight matrices can be analytically given by [27]

$$\begin{aligned} &\mathbf{W}_{Gx}=\mathbf{Y}_{Gx}\mathbf{X}^{\mathrm{Tr}}\left(\mathbf{X}_{Gx}\mathbf{X}_{Gx}^{\mathrm{Tr}}+\mu\mathbf{\Pi}\right)^{-1}, \mathbf{W}_{Gy}=\mathbf{Y}_{Gy}\mathbf{X}^{\mathrm{Tr}}\left(\mathbf{X}_{Gy}\mathbf{X}_{Gy}^{\mathrm{Tr}}+\mu\mathbf{\Pi}\right)^{-1}, \\ &\mathbf{W}_{Ex}=\mathbf{Y}_{Ex}\mathbf{X}^{\mathrm{Tr}}\left(\mathbf{X}_{Ex}\mathbf{X}_{Ex}^{\mathrm{Tr}}+\mu\mathbf{\Pi}\right)^{-1}, \mathbf{W}_{Ey}=\mathbf{Y}_{Ey}\mathbf{X}^{\mathrm{Tr}}\left(\mathbf{X}_{Ey}\mathbf{X}_{Ey}^{\mathrm{Tr}}+\mu\mathbf{\Pi}\right)^{-1}, \end{aligned} \tag{21}$$

where the superscript T_r represents the transpose of the matrix; $\mathbf{\Pi}$ is an identity matrix; μ is utilized to avoid overfitting the ridge regression parameter, which is set to 10^{-6}; $\mathbf{X}_{Gx}$, $\mathbf{X}_{Gy}$, $\mathbf{X}_{Ex}$ and $\mathbf{X}_{Ey}$ all are matrices and their lth columns are $[b_{out}; C_{Gx}(l-L_{Gx}); I_{Gx,i}(l)]$, $[b_{out}; C_{Gy}(l-L_{Gy}); I_{Gy,i}(l)]$, $[b_{out}; C_{Ex}(l-L_{Ex}); I_{Ex,i}(l)]$ and $[b_{out}; C_{Ey}(l-L_{Ey}); I_{Ey,i}(l)]$, respectively; $\mathbf{Y}_{Gx}$ and $\mathbf{Y}_{Ex}$ both are matrices, and their lth columns are $[C_{Gx}(l-L_{Gx}+1)]$ and $[C_{Ex}(l-L_{Ex}+1)]$, respectively; and $\mathbf{Y}_{Gy}$ and $\mathbf{Y}_{Ey}$ both are matrices, and their lth columns are $[C_{Gy}(l-L_{Gy}+1)]$ and $[C_{Ey}(l-L_{Ey}+1)]$, respectively. According to the complete lag synchronization theory (see Equations (19)) , we obtain

$$\begin{aligned} &S_1'(n) \approx S_1(n-L_{Gx}), \quad S_2'(n) \approx S_2(n-L_{Gy}), \\ &S_3'(n) \approx S_1(n-L_{Ex}), \quad S_4'(n) \approx S_2(n-L_{Ey}). \end{aligned} \tag{22}$$

3. Results and Discussions

The parameter values for the driving quantum dot (QD) spin-vertical cavity surface-emitting laser (VCSEL) are detailed in Table 1, while those for the reservoir QD spin-VCSEL are outlined in Table 2. Our initial step is to model the power spectral density (PSD) profiles and temporal samples stemming from the driving QD spin-VCSEL, employing concurrent simulations within Matlab (version R2021a) and VPI (version 11.1) software environments. Within Matlab, Equations (1)–(5) are executed via the fourth-order Runge–Kutta numerical approach, adopting a time step (h) of 0.78 ps. The sampling periods for the four distinct input data streams ($u_{Gx}(n-L_{Gx})$, $u_{Gy}(n-L_{Gy})$, $u_{Ex}(n-L_{Ex})$, $u_{Ey}(n-L_{Ey})$) are denoted by T and are uniformly set at 10 ps. The constants L_{Gx}, L_{Gy}, L_{Ex} and L_{Ey} are all given the value of 2.0513×10^4, which is based on τ_x, τ_y being 16 ns and h amounting to 0.78 ps. Concurrently, the dynamical output from the four parallel reservoirs, utilizing the reservoir QD spin-VCSEL, is also modeled with the integration of Matlab and VPI, where Equations (6)–(10) are solved through the fourth-order Runge–Kutta method with a finer time step of 0.048 ps. Within the present framework, both the encoding rate of the messages and the data processing speed of the reservoirs are influenced by the effective bandwidths of the driving and reservoir VCSELs. Figure 3a–d depict the PSD distributions of the GS-PC$_x$ and GS-PC$_y$, as well as the ES-PC$_x$ and ES-PC$_y$ emitted by the driving QD spin-VCSEL. According to the representations in Figure 3, the PSD distributions for these PCs consistently demonstrate chaotic behavior. The effective 3 dB bandwidths for both the GS-PC$_x$ and GS-PC$_y$ are calculated to be 180 GHz, whereas the ES-PC$_x$ and ES-PC$_y$ are slightly higher at 200 GHz. Correspondingly, the effective 3 dB bandwidths for the GS-PC$'_x$, GS-PC$'_y$, ES-PC$'_x$ and ES-PC$'_y$ of the reservoir system exhibit similar characteristics to those of their driving system counterparts (GS-PC$_x$, GS-PC$_y$, ES-PC$_x$ and ES-PC$_y$, respectively). These outcomes suggest that our system is capable of achieving high-speed, four-channel coherent optical chaotic secure communications.

Table 1. Parameter values of the driving QD Spin-VCSEL.

The Parameter and Symbol	Value	The Parameter and Symbol	Value
The photon decay rate κ_D	250 ns^{-1}	The capture rate $\gamma_{D,0}$	400 ns^{-1}
Linewidth enhancement factor α_D	3	Intradot relaxation rate $\gamma_{D,21}$	50 ns^{-1}
Total pump intensity η	4	Spin relaxation rate $\gamma_{D,s}$	10 ns^{-1}
Dichroism $\gamma_{D,a}$	0 ns^{-1}	Carrier recombination rate $\gamma_{D,n}$	1 ns^{-1}
Birefringence $\gamma_{D,p}$	30 ns^{-1}	Electron charge e	1.6×10^{-19} C
Quantum dot density N_{QD}	1.5×10^{17} m^{-2}	The gain coefficient $h_{D,1}$	1.1995

Table 2. Parameter values of the reservoir QD Spin-VCSEL.

The Parameter and Symbol	Value	The Parameter and Symbol	Value
The field decay rate κ	300 ns^{-1}	Central frequency detuning $\Delta\omega_E$	-20×10^9 rad/s
Line-width enhancement factor α	3	The capture rate γ_0	600 ns^{-1}
Total pump intensity η	4	Intradot relaxation rate γ_{21}	40 ns^{-1}
Dichroism γ_a	0.1 ns^{-1}	Spin relaxation rate γ_s	20 ns^{-1}
Birefringence γ_p	20 ns^{-1}	Carrier recombination rate γ_n	1 ns^{-1}
Center frequency ω_G	2×10^{14} rad/s	Injection strength k_{inj}	35 ns^{-1}
Center frequency ω_E	10^{14} rad/s	Feedback strength k_f	30 ns^{-1}
Central frequency detuning $\Delta\omega_G$	0 rad/s	The gain coefficient h_1	1.1665

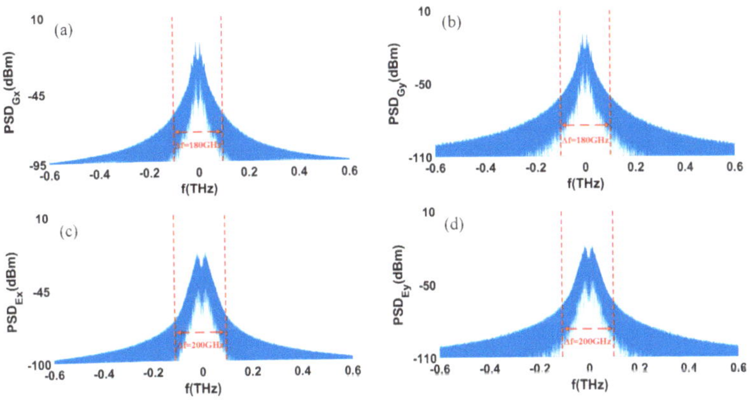

Figure 3. Power spectral density (PSD) distributions of the four polarization components GS-PC$_x$, GS-PC$_y$, ES-PC$_x$ and ES-PC$_y$ from the driving QD Spin-VCSEL output. Here, (**a**) the PSD of GS-PC$_x$ (PSD$_{Gx}$); (**b**) the PSD of the GS-PC$_y$ (PSD$_{Gy}$); (**c**) the PSD of ES-PC$_x$ (PSD$_{Ex}$); and (**d**) the PSD of the ES-PC$_y$ (PSD$_{Ey}$).

The chaotic GS-PC$'_x$, GS-PC$'_y$, ES-PC$'_x$ and ES-PC$'_y$ produced by the reservoir QD spin-VCSEL, as four parallel reservoirs, are utilized to perform the predictions of the delayed outputs GS-PC$_x$, GS-PC$_y$, ES-PC$_x$ and ES-PC$_y$, respectively. We collect 5096 samples of these delayed outputs at a sampling interval of 10 ps. After discarding the initial 1000 samples to remove transients, we allocate 2048 samples for training each of the four reservoirs, and an equivalent number of subsequent points for testing the corresponding reservoir. Moreover, the prediction performance is bolstered by implementing chaotic mask signals derived from two coupled semiconductor lasers, detailed in [24]. These mask signals are normalized with standard deviations set to 1 and mean values calibrated to 0. Each reservoir's virtual node interval, denoted by θ, is fixed at 40 fs. Here, all rates for the QPSK, 4QAM, 8QAM and 16QAM are 100 Gb/s. The input data sampling period T is maintained at 10 ps, resulting in a data processing rate of 100 Gb/s. We establish the number of virtual nodes, N, at 250, where $N = \tau/\theta$ and $\tau = T$. We maintain the scale factor γ, at a value of 1. To assess the predictions for the GS-PC$_x$, GS-PC$_y$, ES-PC$_x$ and ES-PC$_y$ made by these four parallel reservoirs using the reservoir QD spin-VCSEL, we introduce the normalized mean square error ($NMSE$) as a metric to compare the delayed predictive targets against their associated reservoir outputs, which is given as follows:

$$NMSE_{jx} = \frac{1}{L}\frac{\sum_{n=1}^{L}\left(y'_{jx}(n) - C_{jx}(n - L_{jx})\right)^2}{\mathrm{var}(y'_{jx}(n))}, (j = G, E)$$
$$NMSE_{jy} = \frac{1}{L}\frac{\sum_{n=1}^{L}\left(y'_{jy}(n) - C_{jy}(n - L_{jy})\right)^2}{\mathrm{var}(y'_{jy}(n))}, (j = G, E) \quad (23)$$

where the subscripts Gx, Ex, Gy and Ey represent GS-PC$_x$, ES-PC$_x$, GS-PC$_y$ and ES-PC$_y$, respectively. L_{Gx}, L_{Gy}, L_{Ex} and L_{Ey} are the defined lengths of the testing data set for each variable. L represents the total number of data points in the testing data set. The term "var" denotes the variance of the data. When $NMSE_{jx}$ and $NMSE_{jy}$ are both 0, it means that the outputs of the reservoirs (GS-PC$_x$, GS-PC$_y$, ES-PC$_x$ and ES-PC$_y$) perfectly match with their corresponding predicted targets ($C_{Gx}(n-L_{Gx})$, $C_{Gy}(n-L_{Gy})$, $C_{Ex}(n-L_{Ex})$ and $C_{Ey}(n-L_{Ey})$, respectively) . On the other hand, if $NMSE_{jx}$ and $NMSE_{jy}$ both are 1, it means that the reservoir outputs are completely different from the predicted targets. When $NMSE_{jx}$ and $NMSE_{jy}$ are both less than 0.1, it indicates that each reservoir is able to accurately infer the chaotic dynamics of its corresponding predicted target, which is the PC of the driving QD Spin-VCSEL output.

To intuitively observe the ability to predict the chaotic dynamics of the GS-PC$_x$, ES-PC$_x$, GS-PC$_y$ and ES-PC$_y$ in our system, Figure 4 presents their predictive results. In this figure, T = 10 ps, θ = 40 fs, and N = 250. The samples of the delayed GS-PC$_x$, GS-PC$_y$, ES-PC$_x$ and ES-PC$_y$ from the driving QD Spin-VCSEL output are denoted as $C_{Gx}(n-L_{Gx})$, $C_{Gy}(n-L_{Gy})$, $C_{Ex}(n-L_{Ex})$ and $C_{Ey}(n-L_{Ey})$, respectively. The samples of the trained GS-PC$'_x$, GS-PC$'_y$, ES-PC$'_x$ and ES-PC$'_y$ from the reservoir QD spin-VCSEL output are denoted as $y'_{Gx}(n)$, $y'_{Gy}(n)$, $y'_{Ex}(n)$ and $y'_{Ey}(n)$, respectively. As observed from Figure 4, the chaotic trajectories of the $C_{Gx}(n-L_{Gx})$, $C_{Gy}(n-L_{Gy})$, $C_{Ex}(n-L_{Ex})$ and $C_{Ey}(n-L_{Ey})$ are almost identical to those of the $y'_{Gx}(n)$, $y'_{Gy}(n)$, $y'_{Ex}(n)$ and $y'_{Ey}(n)$, respectively. In Figure 5a, when T = 10 ps, θ = 40 fs, and N = 250, the prediction errors ($NMSE_{Gx}$ and $NMSE_{Gy}$) of the GS-PC$_x$ and GS-PC$_y$ are 0.0359 and 0.0375, respectively. The $NMSE_{Ex}$ and $NMSE_{Ey}$ for the ES-PC$_x$ and ES-PC$_y$ are 0.0995 and 0.0865, respectively. These indicate that the four parallel reservoirs based on the reservoir QD spin-VCSEL can accurately predict the chaotic dynamics of the GS-PC$_x$, GS-PC$_y$, ES-PC$_x$ and ES-PC$_y$, respectively.

To comprehensively observe the prediction abilities of the four parallel reservoirs on the chaotic dynamics of the delayed GS-PC$_x$, GS-PC$_y$, ES-PC$_x$ and ES-PC$_y$, Figure 5a illustrates the relationship between the prediction errors ($NMSE_{Gx}$, $NMSE_{Gy}$, $NMSE_{Ex}$ and $NMSE_{Ey}$) and the sampling period T when θ is 40 fs. As shown in Figure 5a, $NMSE_{Gx}$ and $NMSE_{Gy}$ exhibit an almost linear decrease from 0.0362 to 0.0350 and from 0.0376 to 0.0366, respectively, as T increases from 2 ps to 128 ps. Similarly, the $NMSE_{Ex}$ and $NMSE_{Ey}$ also reveal a linear decrease from 0.0998 to 0.0961 and from 0.0867 to 0.0836, respectively. The reason why a longer sampling period T leads to reduced training error might be explained as follows. In this work, $\theta = T/N$ is fixed at 40 fs, and a smaller N is associated with a smaller T, resulting in a lower-dimensional state space. This situation can make the training of the four parallel reservoirs based on the reservoir QD spin-VCSEL become unstable and more difficult, consequently leading to a larger $NMSE$. Additionally, when T is fixed at a certain value, the $NMSE_{Ex}$ and $NMSE_{Ey}$ are significantly larger than $NMSE_{Gx}$ and $NMSE_{Gy}$. This may be explained by the fact that ES-PC$_x$ and ES-PC$_y$ have more complex chaotic dynamics than GS-PC$_x$ and GS-PC$_y$, respectively, making the predictions of ES-PC$_x$ and ES-PC$_y$ more challenging compared to those of GS-PC$_x$ and GS-

PC$_y$. Figure 5b shows the relationship between the prediction errors ($NMSE_{Gx}$, $NMSE_{Gy}$, $NMSE_{Ex}$ and $NMSE_{Ey}$) and the virtual node interval θ when T is fixed at 10 ps. From the observations in Figure 5, it can be seen that as θ increases from 1 fs to 320 fs, the $NMSE_{Ex}$ and $NMSE_{Ey}$ slowly increase from 0.0979 to 0.0998 and from 0.0853 to 0.0868, respectively. Then, they gradually stabilize at 0.0998 and 0.0865. On the other hand, the $NMSE_{Gx}$ and $NMSE_{Gy}$ remain nearly constant at 0.0363 and 0.0376, respectively. The results indicate that when T = 10 ps, the choice of the virtual node interval θ has a slight impact on the prediction accuracy for the GS-PC$_x$ and GS-PC$_y$. However, for the ES-PC$_x$ and ES-PC$_y$, the prediction errors slightly increase with an increase in θ, suggesting a potential sensitivity to the chosen θ.

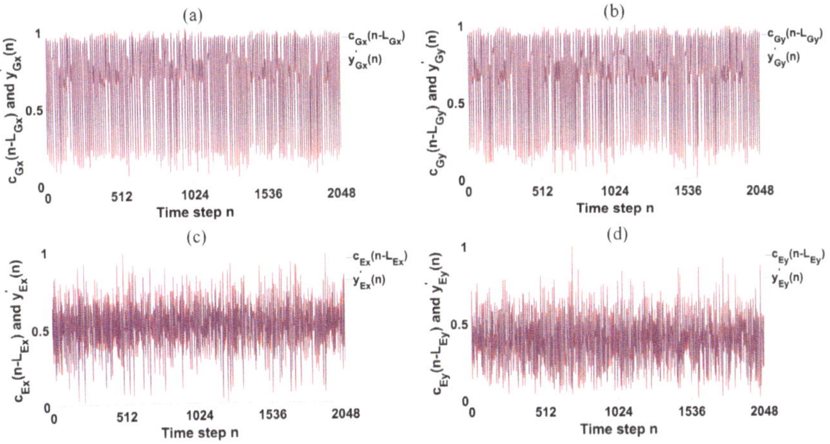

Figure 4. Samples of four delayed polarization components emitted by the driving QD spin-VCSEL (blue solid line) and the outputs of four parallel reservoir based on the reservoir QD Spin-VCSEL (red dashed line). Here, (**a**) $C_{Gx}(n-L_{Gx})$ and $y'_{Gx}(n)$; (**b**) $C_{Gy}(n-L_{Gy})$ and $y'_{Gy}(n)$; (**c**) $C_{Ex}(n-L_{Ex})$ and $y'_{Ex}(n)$; and (**d**) $C_{Ey}(n-L_{Ey})$ and $y'_{Ey}(n)$.

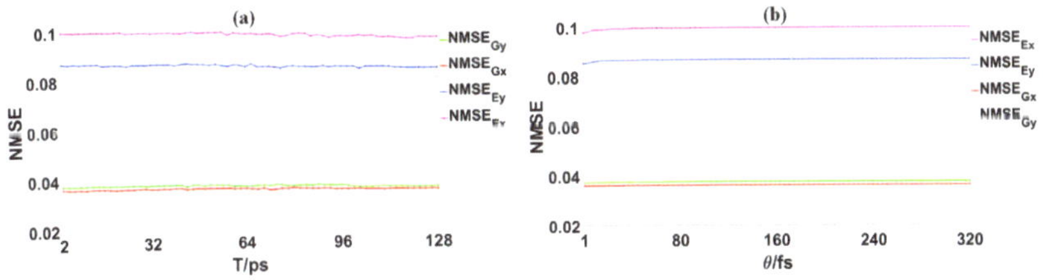

Figure 5. Dependence of the prediction errors ($NMSE_{Gx}$, $NMSE_{Gy}$, $NMSE_{Ex}$, and $NMSE_{Ey}$) on the sampling period T and the virtual node interval θ. Here, (**a**) $NMSE_{Gx}$, $NMSE_{Gy}$, $NMSE_{Ex}$, and $NMSE_{Ey}$ via T, when θ = 40 fs. (**b**) $NMSE_{Gx}$, $NMSE_{Gy}$, $NMSE_{Ex}$, and $NMSE_{Ey}$ via θ, while T = 10 ps.

The results obtained from Figures 4 and 5 demonstrate that the four parallel reservoirs, based on the reservoir QD spin-VCSEL, are capable of reproducing the chaotic dynamics of the GS-PC$_x$, GS-PC$_y$, ES-PC$_x$ and ES-PC$_y$ emitted by the driving QD spin-VCSEL. This indicates that the delayed GS-PC$_x$, GS-PC$_y$, ES-PC$_x$ and ES-PC$_y$ can successfully synchronize with the GS-PC$'_x$, GS-PC$'_y$, ES-PC$'_x$ and ES-PC$'_y$ outputs by the reservoir QD spin-VCSEL, respectively. To further analyze the qualities of their chaos synchronizations, the correlation coefficients are introduced and defined as follows.

$$\rho_{jx} = \frac{\left\langle \left[C_{jx}(n-L_{jx}) - \langle C_{jx}(n-L_{jx})\rangle\right]\left[y'_{jx}(n) - \left\langle y'_{jx}(n)\right\rangle\right]\right\rangle}{\left\langle \left[C_{jx}(n-L_{jx}) - \langle C_{jx}(n-L_{jx})\rangle\right]^2\right\rangle^{1/2}\left\langle \left[y'_{jx}(n) - \left\langle y'_{jx}(n)\right\rangle\right]^2\right\rangle^{1/2}},$$
$$\rho_{jy} = \frac{\left\langle \left[C_{jy}(n-L_{jy}) - \langle C_{jy}(n-L_{jy})\rangle\right]\left[y'_{jy}(n) - \left\langle y'_{jy}(n)\right\rangle\right]\right\rangle}{\left\langle \left[C_{jy}(n-L_{jy}) - \langle C_{jy}(n-L_{jy})\rangle\right]^2\right\rangle^{1/2}\left\langle \left[y'_{jy}(n) - \left\langle y'_{jy}(n)\right\rangle\right]^2\right\rangle^{1/2}}, \tag{24}$$

where j = G, E (the same below). The symbol $\langle\rangle$ represents the time average. ρ ranges from -1 to 1. With the bigger absolute value of ρ, the higher quality of synchronization can be obtained. When ρ equals to ±1, the in-phase and anti-phase synchronous solutions in this system exist.

In Figure 6, the correlations ρ_{Gx}, ρ_{Gy}, ρ_{Ex} and ρ_{Ey} are shown as a function of k_{inj} and k_f. It can be observed that ρ_{Gx}, ρ_{Gy}, ρ_{Ex} and ρ_{Ey} exhibit minimal changes as k_{inj} and k_f increase in the range of 0.1 ns^{-1} to 50 ns^{-1}. Within these two parameter spaces, ρ_{Gx} and ρ_{Gy} both range between 0.9849 and 0.9857, while ρ_{Ex} and ρ_{Ey} fluctuate between 0.949 and 0.96. This indicates that ρ_{Gx}, ρ_{Gy}, ρ_{Ex} and ρ_{Ey} possess strong robustness to variations in k_{inj} and k_f, which are key parameters of the reservoir QD spin-VCSEL. Furthermore, as all ρ_{Gx}, ρ_{Gy}, ρ_{Ex} and ρ_{Ey} are greater than 0.949, it can be concluded that the GS-PC$'_x$, GS-PC$'_y$, ES-PC$'_x$ and ES-PC$'_y$ can effectively synchronize with the delayed GS-PC$_x$, GS-PC$_y$, ES-PC$_x$ and ES-PC$_y$, respectively. Notably, ρ_{Gx} and ρ_{Gy} are higher than ρ_{Ex} and ρ_{Ey}, respectively. This is attributed to the fact that the $NMSE_{Gx}$ and $NMSE_{Gy}$ for the GS-PC$_x$ and GS-PC$_y$ are lower compared to the $NMSE_{Gx}$ and $NMSE_{Gy}$ for the ES-PC$_x$ and ES-PC$_y$, respectively.

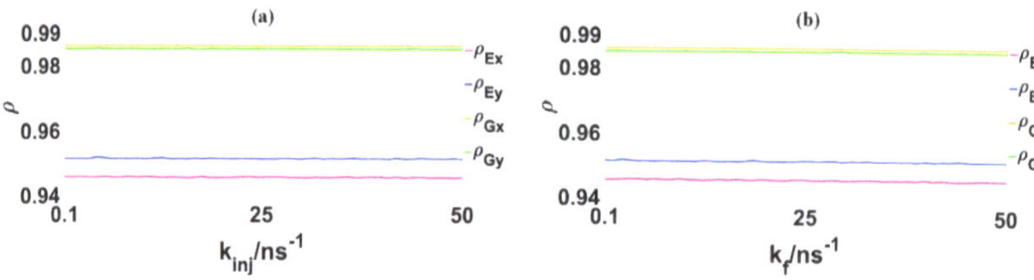

Figure 6. Dependences of the correlation coefficients (ρ_{Gx}, ρ_{Gy}, ρ_{Ex}, ρ_{Ey}) on the parameters k_{inj} and k_f when T = 10 ps and θ = 40 fs. Here, (**a**) ρ_{Gx}, ρ_{Gy}, ρ_{Ex}, $\rho_{Ey} \propto k_{inj}$; (**b**) ρ_{Gx}, ρ_{Gy}, ρ_{Ex}, $\rho_{Ey} \propto k_f$.

Here, by optimizing some key parameter values of the reservoir QD spin-VCSEL, ρ_{Gx}, ρ_{Gy}, ρ_{Ex} and ρ_{Ey} are taken as 0.9856, 0.9851, 0.9495 and 0.9581, respectively. The optimized parameters are as follows: T = 10 ps; θ = 40 fs; η = 4; k_f = 30 ns^{-1}; and k_{inj} = 30 ns^{-1}. By performing high-quality chaos synchronization between each pair of PCs (i.e., GS-PC$_x$ and GS-PC$'_x$, GS-PC$_y$ and GS-PC$'_y$, ES-PC$_x$ and ES-PC$'_x$, and ES-PC$_y$ and ES-PC$'_y$) using the reservoir QD spin-VCSEL, one of the messages QPSK, 4QAM, 8QAM and 16QAM can be decoded by synchronously dividing a reservoir-generated chaos and a delayed chaos masked message. The temporal traces of the delayed encoding message ($S_1(n-L_{Gx})$, or QPSK), the delayed chaos masked message ($U_{Gx}(n-L_{Gx})$), and the decoding message ($S'_1(n)$) are displayed in Figure 7(a$_1$–a$_3$). As observed from Figure 7(a$_1$–a$_3$), the temporal trajectory of $S_1(n-L_{Gx})$ is very similar to that of ($S'_1(n)$). Furthermore, $U_{Gx}(n-L_{Gx})$ exhibits a chaotic state. Figure 7(a$_4$–a$_6$) present the temporal trajectories of $S_2(n-L_{Gy})$ (4QAM), $U_{Gy}(n-L_{Gy})$ and ($S'_2(n)$). As seen from these figures, the temporal trajectory of $S_2(n-L_{Gy})$ is basically identical to that of ($S'_2(n)$), while $U_{Gy}(n-L_{Gy})$ shows a chaotic state.

Moreover, as displayed in Figure 7a(7–a12), the temporal trajectories of $S_3(n-L_{Ex})$ (8QAM) and $S_4(n-L_{Ey})$ (16QAM) are almost the same as those of $(S_3^{'}(n))$ and $(S_4^{'}(n))$, respectively. $U_{Ex}(n-L_{Ex})$ and $U_{Ey}(n-L_{Ey})$ both exhibit a chaotic state. Moreover, as in Figure 8, we present the eye-diagrams for these four decoded messages ($S_1^{'}(n)$, $S_2^{'}(n)$, $S_3^{'}(n)$ and $S_4^{'}(n)$). One sees from this figure that the "eyes" sizes of the eye-diagrams of these decoded messages are enough large, indicating that the decoded messages of the system have a relatively large tolerance error for noise and jitter and have good quality. However, the superposition of multiple decoded messages causes the signal line of each eye-diagram to become thicker and appear fuzzy. The reason is that very small synchronization errors may be converted into noise and superimposed on the signal line of the eye-diagram. These results indicate that the encoding messages QPSK, 4QAM, 8QAM and 16QAM can be effectively masked in a chaotic carrier and successfully recovered using reservoir computing.

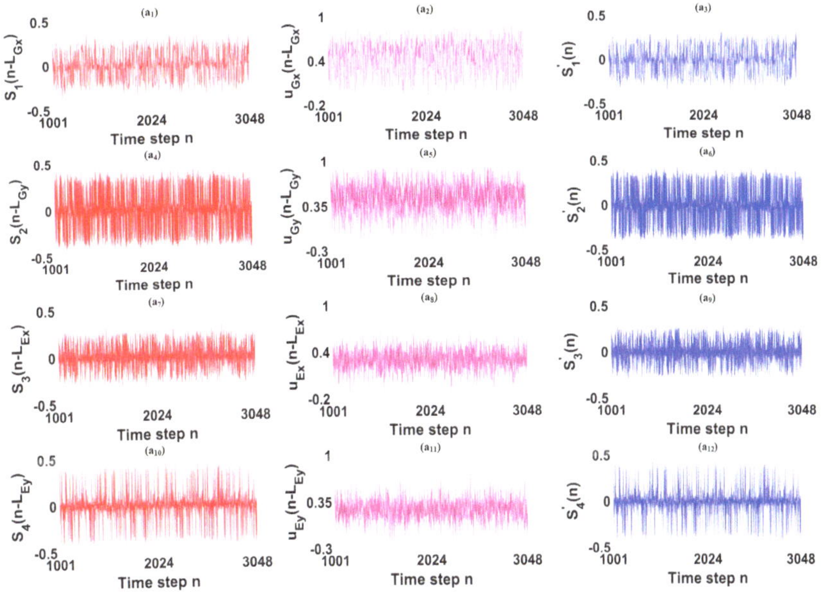

Figure 7. Temporal trajectories of the delayed encoding messages, the delayed chaos masked messages, and the decoding messages in the reservoir computing system. Here, (**a1**) the delayed encoding message $S_1(n-L_{Gx})$ via time step n; (**a2**) the delayed chaos masked message $U_{Gx}(n-L_{Gx})$ via time step n; (**a3**) the decoding message $S_1^{'}(n)$ via time step n; (**a4**) $S_2(n-L_{Gy})$ via time step n; (**a5**) $U_{Gy}(n-L_{Gy})$ via time step n; (**a6**) $S_2^{'}(n)$ via time step n; (**a7**) $S_3(n-L_{Ex})$ via time step n; (**a8**) $U_{Ex}(n-L_{Ex})$ via time step n; (**a9**) $S_3^{'}(n)$ via time step n; (**a10**) $S_4(n-L_{Ey})$ via time step n; (**a11**) $U_{Ey}(n-L_{Ey})$via time step n; and (**a12**) $S_4^{'}(n)$ via time step n.

The bit error rate (BER) is a commonly utilized metric to gauge the quality of data transmission in optical chaos-based secure communication systems [15]. The BER is defined as the ratio of the number of errored bits to the overall number of bits transmitted. Figure 9 showcases the dependences of the BERs for the decoded messages ($S_1^{'}(t)$, $S_2^{'}(t)$, $S_3^{'}(t)$ and $S_4^{'}(t)$) and their associated baseband signals ($b_1^{'}$, $b_2^{'}$, $b_3^{'}$ and $b_4^{'}$) on two key parameters (k_{inj} and k_f). As evidenced by Figure 9(a1,a2), the BERs for $S_1^{'}(t)$, $S_2^{'}(t)$, $S_3^{'}(t)$ and $S_4^{'}(t)$ exhibit oscillatory behavior as k_{inj} is adjusted within the range of 0.1 ns^{-1} to 50 ns^{-1}. Their BER values, respectively, fluctuate within the following ranges: from 1.02×10^{-2} to 1.22×10^{-2} for $S_1^{'}(t)$, from 6.1×10^{-3} to 7.5×10^{-3} for $S_2^{'}(t)$, from 3.4×10^{-3} to 6.1×10^{-3} for $S_3^{'}(t)$, and from 7.1×10^{-3} to 9.2×10^{-3} for $S_4^{'}(t)$. Within this k_{inj} range, all four decoded messages demonstrate minor oscillatory fluctuations in their *BER*s. The *BER*s cap at 1.5×10^{-2} for $S_1^{'}(t)$ and at 8.7×10^{-3} for $S_2^{'}(t)$, while those for $S_3^{'}(t)$ and $S_4^{'}(t)$ do not surpass 3.4×10^{-2} and 9.5×10^{-3}, respectively. Based on findings from earlier studies [28–30], a *BER* that closes at or below 0.01 is indicative of potentially high-quality data transmission within an optical chaos communication framework. As depicted in Figure 2c, when demodulated through correlation and refined by various digital signal processing methods, four distinct baseband signal sets (or bitstreams) encapsulated within the decoded messages $S_1^{'}(t)$, $S_2^{'}(t)$, $S_3^{'}(t)$ and $S_4^{'}(t)$ are successfully reconstructed. Consequently, the *BER* ranges for these retrieved baseband signals ($b_1^{'}$, $b_2^{'}$, $b_3^{'}$ and $b_4^{'}$) remain constant and effectively zero, irrespective of k_{inj} and k_f variations. Figure 10 delves into the performance of the four retrieved baseband signals by presenting their temporal trajectories and eye-diagrams alongside those of the original baseband signals b_1–b_4. An inspection of Figure 10 reveals a striking similarity between the temporal profiles of the original signals b_1, b_2, b_3 and b_4 and their retrieved counterparts $b_1^{'}$, $b_2^{'}$, $b_3^{'}$ and $b_4^{'}$, respectively. The eye-diagrams corresponding to the original and retrieved baseband signals also correspond closely, with b_1, b_2, b_3 and b_4, showing a remarkable resemblance to $b_1^{'}$, $b_2^{'}$, $b_3^{'}$ and $b_4^{'}$. Notably, the eye openings in the eye-diagrams for $b_1^{'}$, $b_2^{'}$, $b_3^{'}$ and $b_4^{'}$ are sufficiently large, which is an important indicator of signal integrity. The insights gathered from Figures 9 and 10 strongly support the effectiveness of our proposed coherent optical chaotic communication system in delivering secure and high-quality data transmission.

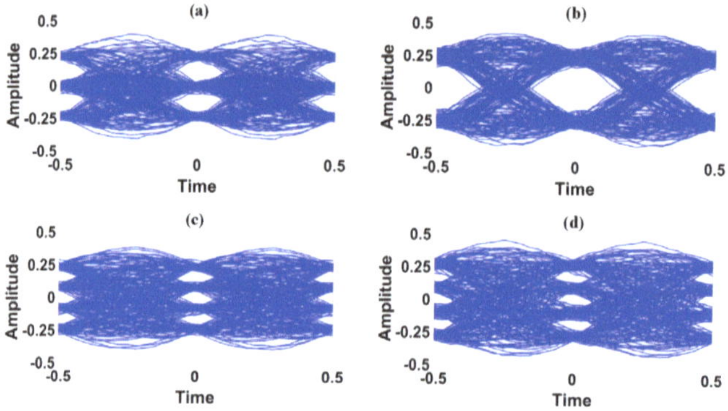

Figure 8. Eye-diagrams of the decoded messages ($S_1^{'}(n)$, $S_2^{'}(n)$, $S_3^{'}(n)$ and $S_4^{'}(n)$). Here, (**a**) the eye-diagram of $S_1^{'}(n)$; (**b**) the eye-diagram of $S_2^{'}(n)$; (**c**) the eye-diagram of $S_3^{'}(n)$; and (**d**) the eye-diagram of $S_4^{'}(n)$.

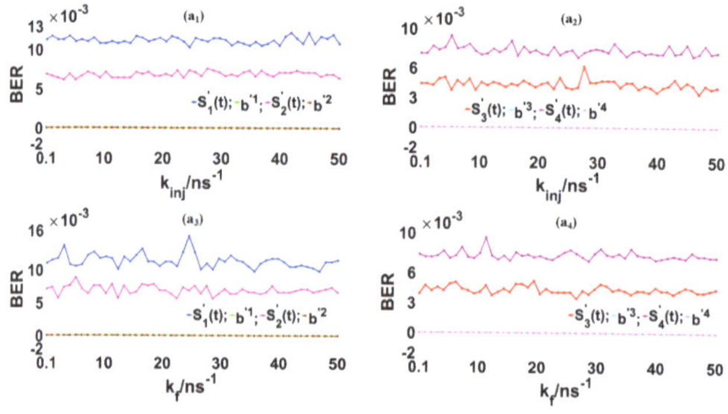

Figure 9. The dependences of the *BER*s for the decoding messages ($S_1^{'}(t)$, $S_2^{'}(t)$, $S_3^{'}(t)$ and $S_4^{'}(t)$) and their corresponding baseband signals ($b_1^{'}$, $b_2^{'}$, $b_3^{'}$ and $b_4^{'}$) on two key parameters k_{inj} and k_f. Here, (**a$_1$**) $S_1^{'}(t)$, $S_2^{'}(t)$, $b_1^{'}$, $b_2^{'}$ via k_{inj}; (**a$_2$**) $S_3^{'}(t)$, $S_4^{'}(t)$, $b_3^{'}$, $b_4^{'}$ via k_{inj}; (**a$_3$**) $S_1^{'}(t)$, $S_2^{'}(t)$, $b_1^{'}$, $b_2^{'}$ via k_f; and (**a$_4$**) $S_3^{'}(t)$, $S_4^{'}(t)$, $b_3^{'}$, $b_4^{'}$ via k_f.

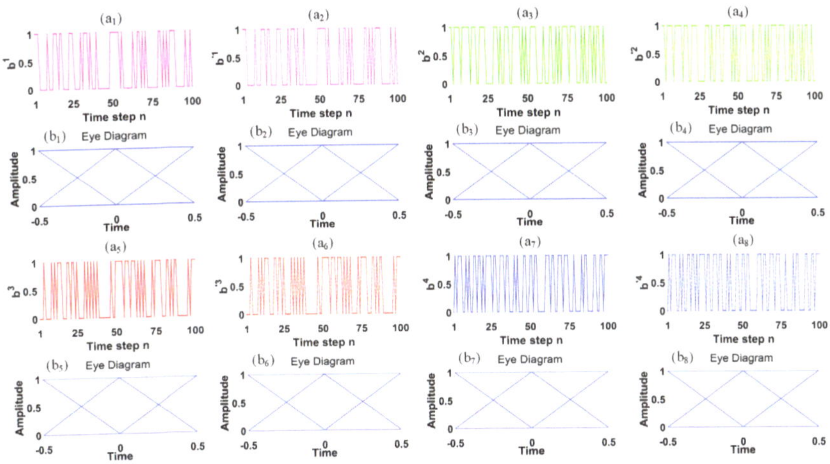

Figure 10. Temporal trajectories and eye-diagrams of the original baseband signals b_1–b_4 and their respectively retrieved baseband signals b_1'–b_4'. Here, (**a$_1$**,**a$_2$**) the temporal trajectories of the b_1 and b_1', respectively, and (**b$_1$**,**b$_2$**) their respectively eye-diagrams. (**a$_3$**,**a$_4$**) the temporal trajectories of the b_2 and b_2', respectively, and (**b$_3$**,**b$_4$**) their corresponding eye-diagrams; (**a$_5$**,**a$_6$**) the temporal trajectories of the b_3 and b_3', respectively, and (**b$_5$**,**b$_6$**) their respectively eye-diagrams; and (**a$_7$**,**a$_8$**) the temporal trajectories of the b_4 and b_4', respectively, and (**b$_7$**,**b$_8$**) their corresponding eye-diagrams.

4. Conclusions

In conclusion, we introduce a novel four-channel coherent optical chaotic secure communication (COCSC) system that integrates four simultaneous photonic reservoir computers with a coherent demodulation apparatus. This system utilizes a QD-spin-VCSEL as the driving laser, while an autonomous QD-spin-VCSEL acts as the reservoir laser. Individually, the four encoded messages, QPSK, 4QAM, 8QAM and 16QAM are modulated onto four distinct polarization components derived from the ground state (GS) and excited state (ES) emissions in the drive QD-spin-VCSEL. Moreover, we construct four concurrent photonic reservoirs using the polarization components originating from the GS and ES of the reservoir QD-spin-VCSEL. Our system achieves a four-channel COCSC system with a capacity of 4 × 100 GHz through chaos synchronization founded on these four parallel photonic reservoirs. Within this arrangement, we ensure robust wideband chaos synchronization between corresponding polarization components of the driving and reservoir lasers. This precise synchronization allows for the accurate decoding of the four distinct messages (QPSK, 4QAM, 8QAM and 16QAM), each masked within different chaotic polarization components. The decoded messages are then demodulated via correlation techniques and further processed using various digital signal processing methodologies, successfully reconstructing the four separate baseband signals encapsulated within the QPSK, 4QAM, 8QAM and 16QAM formats. Through detailed analysis with eye diagrams, bit error rates, and temporal trajectories of the coherently demodulated baseband signals, we observe that each baseband-signal set is impeccably recovered, evidenced by large eye openings in the eye diagrams and a bit error rate that approaches zero for each baseband-signal set. This innovative approach, which harnesses the power of reservoir computing based on a QD spin-VCSEL, paves the way towards advancing multi-channel coherent optical chaotic communications with enhanced security features.

Author Contributions: Conceptualization, D.Z. and T.W.; methodology, D.Z. and T.W.; software, T.W., Q.W. and Y.C.; formal analysis, T.W., Y.C., Q.W., C.Q., Y.W., H.Z. and J.X.; writing—original draft preparation, T.W.; writing—review and editing, D.Z.; and visualization, T.W. All authors have read and agreed to the published version of the manuscript.

Funding: This research was funded by the National Natural Science Foundation of China (62075168); the Special Project in Key Elds of Guangdong Universities: the New Generation of Communication Technology (2020ZDZX3052); the Guang dong Basic and Applied Basic Research Foundation (2023A1515010726); the Innovation Team Project of Colleges and Universities in Guangdong Province (2023KCXTD041, 2021KCXTD051); and the Major Projects of Guangdong Education Department for Foundation Research and Applied Research (2017KZDX086).

Institutional Review Board Statement: No applicable.

Informed Consent Statement: No applicable.

Data Availability Statement: The raw data supporting the conclusions of this article will be made available by the authors on request.

Conflicts of Interest: The authors declare no conflicts of interest.

References

1. Kikuchi, K. Fundamentals of Coherent Optical Fiber Communications. *J. Light. Technol.* **2016**, *34*, 157–179. [CrossRef]
2. Zhang, S.; Kam, P.Y.; Yu, C.; Chen, J. Decision-Aided Carrier Phase Estimation for Coherent Optical Communications. *J. Light. Technol.* **2010**, *28*, 1597–1607. [CrossRef]
3. Tao, Z.; Li, L.; Liu, L.; Yan, W.; Nakashima, H.; Tanimura, T.; Oda, S.; Hoshida, T.; Rasmussen, J.C. Improvements to Digital Carrier Phase Recovery Algorithm for High-Performance Optical Coherent Receivers. *IEEE J. Sel. Top. Quantum Electron.* **2010**, *16*, 1201–1209. [CrossRef]
4. Chen, Y.; Xiao, J.; Dong, Z. Delivering Dual Polarization-Division-Multiplexing Millimeter-Wave Signals at W-Band by One Pair of Antennas. *IEEE Photonics J.* **2019**, *11*, 1–10. [CrossRef]
5. Shao, W.; Fu, Y.; Cheng, M.; Deng, L.; Liu, D. Chaos Synchronization Based on Hybrid Entropy Sources and Applications to Secure Communication. *IEEE Photonics Technol. Lett.* **2021**, *33*, 1038–1041. [CrossRef]
6. Lindemann, M.; Pusch, T.; Michalzik, R.; Gerhardt, N.C.; Hofmann, M.R. Frequency tuning of polarization oscillations: Toward high-speed spin-lasers. *Appl. Phys. Lett.* **2016**, *108*, 042404. [CrossRef]
7. Ledentsov, N.; Agustin, M.; Shchukin, V.; Kropp, J.; Ledentsov, N.; Chorchos, Ł.; Turkiewicz, J.; Khan, Z.; Cheng, C.; Shi, J.; et al. Quantum dot 850 nm VCSELs with extreme high temperature stability operating at bit rates up to 25 Gbit/s at 150 °C. *Solid-State Electron.* **2019**, *155*, 150–158. [CrossRef]
8. Paranthoen, C.; Levallois, C.; Brévalle, G.; Perrin, M.; Le Corre, A.; Chevalier, N.; Turban, P.; Cornet, C.; Folliot, H.; Alouini, M. Low Threshold 1550-nm Emitting QD Optically Pumped VCSEL. *IEEE Photonics Technol. Lett.* **2021**, *33*, 69–72. [CrossRef]
9. Lindemann, M.; Xu, G.; Pusch, T.; Michalzik, R.; Hofmann, M.R.; Žutić, I.; Gerhardt, N.C. Ultrafast spin-lasers. *Nature* **2019**, *568*, 212–215. [CrossRef]
10. Frougier, J.; Baili, G.; Alouini, M.; Sagnes, I.; Jaffrès, H.; Garnache, A.; Deranlot, C.; Dolfi, D.; George, J.M. Control of light polarization using optically spin-injected vertical external cavity surface emitting lasers. *Appl. Phys. Lett.* **2013**, *103*, 252402. [CrossRef]
11. Schires, K.; Seyab, R.A.; Hurtado, A.; Korpijärvi, V.M.; Guina, M.; Henning, I.D.; Adams, M.J. Optically-pumped dilute nitride spin-VCSEL. *Opt. Express* **2012**, *20*, 3550–3555. [CrossRef] [PubMed]
12. Alharthi, S.S.; Orchard, J.; Clarke, E.; Henning, I.D.; Adams, M.J. 1300 nm optically pumped quantum dot spin vertical external-cavity surface-emitting laser. *Appl. Phys. Lett.* **2015**, *107*, 151109. [CrossRef]
13. Rafayelyan, M.; Dong, J.; Tan, Y.; Krzakala, F.; Gigan, S. Large-scale optical reservoir computing for spatiotemporal chaotic systems prediction. *Phys. Rev. X* **2020**, *10*, 041037. [CrossRef]
14. Kong, L.W.; Weng, Y.; Glaz, B.; Haile, M.; Lai, Y.C. Reservoir computing as digital twins for nonlinear dynamical systems. *Chaos Interdiscip. J. Nonlinear Sci.* **2023**, *33*, 033111. [CrossRef]
15. Hou, Y.; Xia, G.; Yang, W.; Wang, D.; Jayaprasath, E.; Jiang, Z.; Hu, C.; Wu, Z. Prediction performance of reservoir computing system based on a semiconductor laser subject to double optical feedback and optical injection. *Opt. Express* **2018**, *26*, 10211–10219. [CrossRef] [PubMed]
16. Amil, P.; Soriano, M.C.; Masoller, C. Machine learning algorithms for predicting the amplitude of chaotic laser pulses. *Chaos Interdiscip. J. Nonlinear Sci.* **2019**, *29*, 113111. [CrossRef] [PubMed]
17. Bao, X.; Zhao, Q.; Yin, H.; Qin, J. Recognition of the optical packet header for two channels utilizing the parallel reservoir computing based on a semiconductor ring laser. *Mod. Phys. Lett. B* **2018**, *32*, 1850150. [CrossRef]
18. Appeltant, L.; Soriano, M.C.; der Sande, G.V.; Danckaert, J.; Massar, S.; Dambre, J.; Schrauwen, B.; Mirasso, C.; Fischer, I. Information processing using a single dynamical node as complex system. *Nat. Commun.* **2011**, *2*, 468. [CrossRef] [PubMed]
19. Adams, M.J.; Alexandropoulos, D. Analysis of Quantum-Dot Spin-VCSELs. *IEEE Photonics J.* **2012**, *4*, 1124–1132. [CrossRef]
20. Shen, Z.; Huang, Y.; Zhu, X.; Zhou, P.; Mu, P.; Li, N. Broad Tunable and High-Purity Photonic Microwave Generation Based on an Optically Pumped QD Spin-VCSEL with Optical Feedback. *Photonics* **2023**, *10*, 326. [CrossRef]

21. Oestreich, M.; Hübner, J.; Hägele, D.; Bender, M.; Gerhardt, N.; Hofmann, M.; Rühle, W.W.; Kalt, H.; Hartmann, T.; Klar, P.; et al. Spintronics: Spin electronics and optoelectronics in semiconductors. In *Advances in Solid State Physics*; Springer: Berlin/Heidelberg, Germany, 2001; pp. 173–186.
22. Zhong, D.; Xu, G.; Luo, W.; Xiao, Z. Real-time multi-target ranging based on chaotic polarization laser radars in the drive-response VCSELs. *Opt. Express* **2017**, *25*, 21684–21704. [CrossRef]
23. VPIphotonics. VPIcomponentMaker Photonic Circuits User's Manual. Available online: https://www.vpiphotonics.com/Tools/PhotonicCircuits (accessed on 10 June 2023).
24. Nakayama, J.; Kanno, K.; Uchida, A. Laser dynamical reservoir computing with consistency: An approach of a chaos mask signal. *Opt. Express* **2016**, *24*, 8679–8692. [CrossRef] [PubMed]
25. San Miguel, M.; Feng, Q.; Moloney, J.V. Light-polarization dynamics in surface-emitting semiconductor lasers. *Phys. Rev. A* **1995**, *52*, 1728–1739. [CrossRef]
26. Georgiou, P.; Tselios, C.; Mourkioti, G.; Skokos, C.; Alexandropoulos, D. Effect of excited state lasing on the chaotic dynamics of spin QD-VCSELs. *Nonlinear Dyn.* **2021**, *106*, 3637–3646. [CrossRef]
27. Guo, X.X.; Xiang, S.Y.; Zhang, Y.H.; Lin, L.; Wen, A.J.; Hao, Y. Four-channels reservoir computing based on polarization dynamics in mutually coupled VCSELs system. *Opt. Express* **2019**, *27*, 23293–23306. [CrossRef] [PubMed]
28. Ke, J.; Yi, L.; Yang, Z.; Yang, Y.; Zhuge, Q.; Chen, Y.; Hu, W. 32 Gb/s chaotic optical communications by deep-learning-based chaos synchronization. *Opt. Lett.* **2019**, *44*, 5776–5779. [CrossRef] [PubMed]
29. Huang, Y.; Zhou, P.; Li, N. High-speed secure key distribution based on chaos synchronization in optically pumped QD spin-polarized VCSELs. *Opt. Express* **2021**, *29*, 19675–19689. [CrossRef]
30. Hou, T.T.; Yi, L.L.; Yang, X.L.; Ke, J.X.; Hu, Y.; Yang, Q.; Zhou, P.; Hu, W.S. Maximizing the security of chaotic optical communications. *Opt. Express* **2016**, *24*, 23439–23449. [CrossRef]

photonics

Article

Enhanced PON and AMCC Joint Transmission with GMM-Based Probability Shaping Techniques

Haipeng Guo [1,2], Chuanchuan Yang [1,2,*], Zhangyuan Chen [1] and Hongbin Li [1]

1 State Key Laboratory of Advanced Optical Communication Systems and Networks, Peking University, Beijing 100871, China
2 Peking University Shenzhen Graduate School, Nanshan District, Shenzhen 518055, China
* Correspondence: yangchuanchuan@pku.edu.cn

Abstract: In ITU-T standards, auxiliary management and control channels (AMCCs), as defined, facilitate the rapid deployment and efficient management of wavelength division multiplexing passive optical network (WDM-PON) systems. The super-imposition of an AMCC introduces additional interference to a PON signal, resulting in the degradation of the performance of the overall transmission. In prior research, we proposed employing a Gaussian mixture model (GMM) to fit a baseband-modulated AMCC signal. Following the analysis of the interference model and the distribution characteristics of received signal errors, we propose a combined optimization method for a transmitter and receiver in this paper. This method, grounded in probabilistic shaping (PS) techniques, optimizes the probability distribution of the transmitted signal based on the AMCC interference model, with the objective of reducing the error rate in PON signal transmission. We have validated this approach within a 50G-PON experimental system by utilizing PAM4 modulation. The experimental results demonstrate the effectiveness of this method for mitigating the impact of baseband-modulated AMCC, thereby reducing the error rate in PON signal transmission. The approach presented in this paper can further minimize the performance degradation introduced by baseband-modulated AMCC in WDM-PON systems, enhancing the efficiency of WDM-PON deployment.

Keywords: AMCC; WDM-PON; GMM; interference model; probabilistic shaping

1. Introduction

According to the latest whitepapers from mobile network operators [1,2], the current demand for internet capacity has experienced explosive growth. In the next 3–5 years, there will still be significant demand for mobile internet, as evidenced by the planned upgrade from the fifth-generation (5G) mobile network architecture to the sixth-generation (6G). This upgrade is expected to potentially increase network capacity by up to 100 times, with end-to-end latency reduced to less than 1 millisecond. Optical communication stands out as an excellent choice for high-speed communication networks, and wavelength division multiplexing passive optical network (WDM-PON) is a particularly attractive optical communication technology that is applicable in various scenarios, including 5G mobile fronthaul (MFH) networks [3,4]. The deployment of WDM-PON in networks requires an auxiliary management and control channel (AMCC) to achieve efficient network deployment. This channel has been defined in the ITU-T G.989 series standards [5–7]. In recent years, extensive research has been conducted on AMCC by various institutions. This research encompasses modulation and transmission techniques, as well as methods to enhance transmission performance [8–15]. Through these advancements, the transmission speed of an AMCC has been elevated to the level of 20 Mbps under specific conditions. Studies in the realm of system applications suggest that an AMCC can play a role in wavelength management and control [16,17], with NTT conducting research on the application of AMCCs in all optical networks [18]. However, as future networks demand higher capacity and lower transmission latency, it is necessary to increase the transmission speed of WDM-PON systems further. Research indicates that PON systems are currently in a new upgrade cycle, and as early as 2020, ITU-T proposed initiating the development of mobile-centric WDM-PON standards [19]. Studies on standardization progress suggest that the introduction of forward error correction (FEC) coding will raise the line rate of PON systems to over 50 Gbps, providing widespread benefits for both regular and latency-sensitive wireless applications [20]. In order to achieve the goal of enhancing PON transmission rates, various methods can be considered, including increasing channel bandwidth, applying simplified coherent techniques, and employing advanced modulation schemes, such as four-level pulse amplitude modulation (PAM4). Consequently, the AMCC responsible for operation administration and maintenance (OAM) data also requires new technologies to support PON upgrades. Research indicates the feasibility of transmitting AMCCs in a 50 Gbps PAM4 network [21,22]. An AMCC can also be employed in simplified coherent systems, as suggested in [23], which introduces a block-based digital signal processing method used to extract AMCC signals in a 25 Gbps QPSK coherent communication system, resulting in a power penalty reduction of 0.2 dB for 128 kbps AMCC signals. Compared to on-off keying (OOK) signals, PAM4 signals carry double information per symbol, reducing the link bandwidth requirements at the same bit rate but making them more susceptible to interference. Thus, the signal attenuation caused by the overlay of AMCC on a PAM4 system is more severe than in an OOK system at the same rate. In order to achieve channel management functions, high-speed AMCC signals need to be correctly demodulated, requiring the maintenance of sufficient AMCC signal amplitude to meet the signal-to-noise ratio (SNR) requirements. However, the larger the amplitude of the AMCC signal, the more interference it introduces to the PON, making it challenging to meet the power penalty requirements specified by the ITU standards [6,7]. In response to these challenges, in [24], we proposed a modeling method for interference signals to assess mixed signals, and we introduce a novel joint demodulation receiver structure capable of simultaneously demodulating PON and AMCC signals while maintaining excellent demodulation performance. In this article, we conduct an analysis of the distribution parameters of the received signal using the AMCC interference model in a multi-level modulation PON system. Based on this analysis, we propose an enhanced transmission system for PON-AMCC signals. This improved method is grounded in probability shaping (PS) techniques and is optimized using a Gaussian mixture model (GMM). By employing a GMM fitting method to acquire signal distribution parameters, we scrutinize interference intensity and error symbol probabilities at different positions in the channel. This adjustment aims to strategically position more signals in areas with lower interference, thereby reducing the transmission error rate and enhancing the overall system's transmission performance. We further apply this method to optimize the joint demodulation receiver proposed in [24]. The effectiveness of this approach is validated in an experimental system with a 50G-PON with an AMCC superimposed, demonstrating improved transmission performance compared to signals transmitted with equal probability.

2. Interference Modeling and Error Symbol Analysis

The transmission of baseband AMCC signals in a multi-level modulation PON signal system can be achieved using a distributed feedback laser (DFB) and a Mach-Zehnder modulator (MZM) [22]; its basic structure is illustrated in Figure 1. The blue segment represents electrical signals, while the green segment represents optical signals.

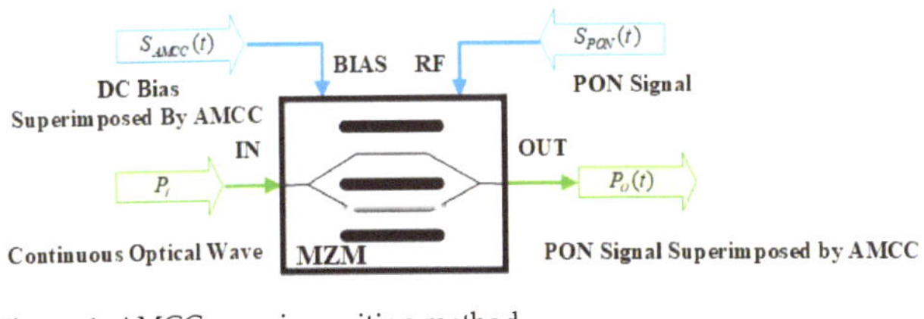

Figure 1. AMCC superimposition method.

The PON signal is input through the RF port of the modulator, while the AMCC signal is connected to the DC bias port. Under appropriate operating conditions, a superimposed baseband AMCC can be achieved, resulting in the output signal being superimposed in the form of Equation (1).

$$P_O(t) = \frac{G * P_I(t)}{2}\left(1 + \cos\left[\frac{\pi}{V_\pi}[A_P * S_{PON}(t) + V_{BAIS} + A_M * S_{AMCC}(t)]\right]\right) + n(t). \quad (1)$$

In the equation, $S_{PON}(t)$ and $S_{AMCC}(t)$ are the values of the PON signal and AMCC signal, respectively, and A_P and A_M represent the amplitude of the PON and AMCC signals, respectively. The modulation index of AMCC can be expressed as A_M/A_P. The parameter G is the insertion loss of MZM, with a value of less than 1, and V_π is the half-wave voltage of MZM. V_{BIAS} indicates the DC bias voltage of the modulator, which determines the operating conditions of the modulator. $P_I(t)$ donates the input optical signal power of the modulator.

Through mathematical derivation, it can be inferred that the distortion caused by AMCC varies for different values of the PON signal. The essential reason for this phenomenon is that the relationship between the input and output signals of the MZM is a cosine mapping rather than a linear mapping. This characteristic results in a significantly lower distortion value for PON signals near the top and bottom of the cosine curve when AMCC is superimposed when compared to signals near the middle of the curve. This feature can be validated through the eye diagram of the signal. For example, Figure 2 illustrates an eye diagram of a PAM4-PON signal overlaid with OOK-AMCC. In the eye diagram, it can be observed that the 4-level PON signal splits into eight levels. However, it is evident that the splitting amplitudes of the signal levels at the top and bottom are much smaller than those at the two middle levels. This observation aligns with the mathematical analysis presented earlier in Equation (1).

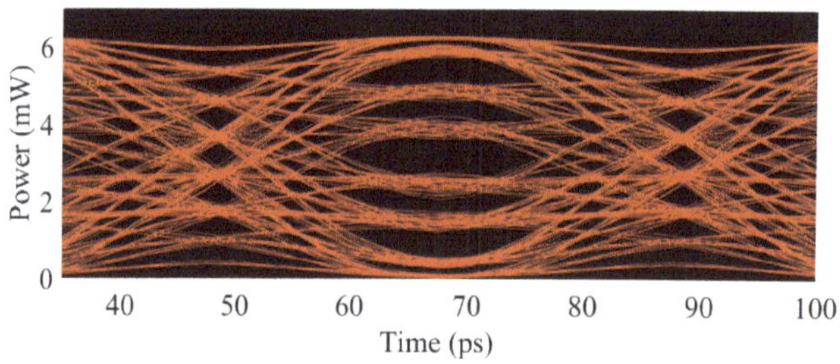

Figure 2. An eye diagram of the transmitted signal after superimposing the AMCC signal.

With the increasing bandwidth of AMCC, traditional interference elimination methods have limited effectiveness. After analyzing the interference characteristics of AMCC, a new interference model based on GMM was proposed in [24]. By analyzing the interference model, we understand that introducing an AMCC results in the splitting of each level of the PAM4 signal into multiple levels. Consequently, the statistical characteristics of the corresponding signals are more complex, impacting the demodulation of the signals given by the receiver. In a joint demodulation receiver, the receiving end utilizes a GMM fitting module to model the received signal and estimate the distribution parameters. For an additive white Gaussian noise (AWGN) channel, the received signal conforms to a Gaussian mixture model, which can be expressed as [25]

$$p(x) = \sum_{k=1}^{K} \rho_k N(x|\mu_k, \sigma_k), \quad (2)$$

where $N(x|\mu_k, \sigma_k)$ is called the k-th component in the mixture model, and ρ_k corresponds to the weight of that component and satisfies

$$\sum_{k=1}^{K} \rho_k = 1 \quad 0 \leq \rho_k \leq 1. \tag{3}$$

The μ_k and σ_k for each level reflect the channel's impact on the signal at that position, representing the SNR of the channel at that location. By evaluating the parameters of the GMM for the received signal, we can assess the SNR for different levels. When the SNR is high at a particular position, signals appearing at that position should exhibit good transmission performance, characterized by a relatively low error rate. Conversely, when the SNR is poor at a particular position, the signals appearing at that position should exhibit a higher error rate. For instance, in a transmission system with a PAM4-PON signal overlaid with OOK-AMCC, according to Equation (1), the received signal should conform to a Gaussian mixture distribution with eight peaks. After fitting a GMM to the received signal using Equation (2), the signal distribution might resemble the one shown in Figure 3. From the figure, it can be observed that when the signal levels are PAM4 level 2 and level 3, the received signal has a higher SNR. Accordingly, a received signal that conforms to this model and has its level at positions 2 to 3 should exhibit better transmission performance, with a lower error rate compared to signals at other positions. Therefore, by transforming the transmitted signal to place more information at positions with a higher SNR, it is possible to effectively reduce the error rate and enhance the overall transmission performance of the system. In order to achieve this purpose, a feasible method is to use probability-shaping techniques to adjust the distribution probability of the transmitted signal.

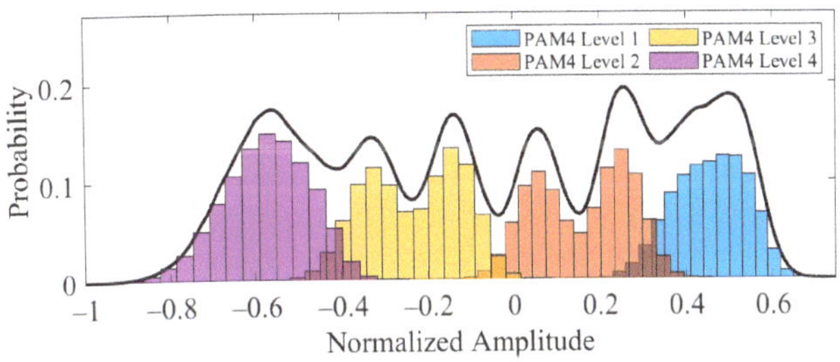

Figure 3. Received signal GMM fitting result.

3. Enhanced Transmission System Utilizing GMM-Based Probability Shaping Techniques

3.1. GMM-Based Probabilistic Shaping Techniques

Probabilistic Shaping (PS) is a technique utilized in digital communication systems to improve information transmission performance over noisy channels. This approach involves meticulous signal design, incorporating probabilistic considerations to mitigate the impact of noise and interference. The core concept is to assign probabilities to different symbol outputs, prioritizing symbols with lower error rates for transmission. The foundational implementation method of the PS techniques has been extensively discussed in previous research, with constant composition distribution matching (CCDM) being a common approach for PS [26,27]. In a PAM modulation system, each symbol represents a specific amplitude level, and the probability distribution of symbol points is uniform in traditional systems, meaning each symbol point has an equal probability of occurrence. However, PS techniques optimize the probability distribution and reduce the probability of occurrence for levels with inferior signal-to-noise ratios (SNRs) while elevating the probability of levels with superior SNRs. Assuming that a PAM-L-modulated signal is generated from the level set $S = \{1, 2, \ldots, L\}$, the probability density function (PDF) $D_S(x)$ tends to align the signal distribution with a Maxwell-Boltzmann (M-B) distribution [28],

$$D(x) = \frac{e^{-v|x|^2}}{\sum_{x' \in S} e^{-v|x'|^2}}, \tag{4}$$

where v is a rate parameter utilized to control the kurtosis of MB distribution. Due to the change in signal distribution, the value of v also undergoes a change, indicating that the non-uniform probability distribution reduces the entropy of the transmitted signal. PS technology has been shown to improve transmission performance in optical communication systems [29–32].

In the PON-AMCC system, we have re-modeled the signal, constructing it as a Gaussian mixture model. In the GMM-based signal model, as described in Equation (2), each PAM level has an associated parameter, σ_l, reflecting the noise level at the l-th PAM amplitude. Through this parameter, we can assess the quality of the channel. Therefore, we can directly use σ_l to adjust the transmitted signal, reducing the probability of higher levels of σ_l occurring. For the l-th level in PAM-L, its PDF should satisfy

$$D(l) = \frac{e^{-\sigma_l^2|l|^2}}{\sum_{l' \in S} e^{-\sigma_l^2|l'|^2}}, \tag{5}$$

The noise level parameter σ_l can be extracted from the results of the GMM fitting. Additionally, $D(l)$ needs to satisfy another condition:

$$\sum_{l=1}^{L} D(l) = 1, \quad 0 \leq D(l) \leq 1, \tag{6}$$

which means the total probability across all levels is equal to 1. With GMM-based PS techniques, the system can be optimized by appropriately selecting the parameter σ_l of the Gaussian distribution according to the channel conditions and transmission requirements. A lower value σ_l increases the occurrence probability of more likely symbol points, thus improving the system capacity. Conversely, a smaller σ_l value makes specific amplitude levels more likely to be used, improving transmission reliability.

3.2. Enhanced Transmission Systems with Joint Transmitter-Receiver Optimization

After proposing a new AMCC interference model in [24], a joint demodulation receiver based on this new interference model is also introduced. The GMM-based PS technique can be used to optimize the transmitter and receiver together in this system, as illustrated in Figure 4. At the transmitter, the module responsible for generating PAM signals is replaced with a PS-PAM signal generation module. $S_{PON}(x)$ and $S_{AMCC}(x)$ are combined using the method shown in Figure 1 through an MZM modulator to produce the transmitted signal $S(x)$. $S(x)$ undergoes the channel treatment to become $R(x)$ and enters the receiver. The basic structure of the receiver is based on a GMM-HMM joint demodulation receiver, consisting of a parameter estimation workflow and joint demodulation workflow. The training sequence from $R(x)$ enters the parameter estimation workflow. The parameter estimation module first fits the signal model using the GMM method to obtain the model parameters for the received signal. The probability distribution parameters of the signal will be entered into the demodulation process as the **P** matrix mentioned in Equation (7) for PON signal demodulation. The σ parameter in the GMM parameters is used to analyze the channel, and according to the method mentioned in Equation (5), the amplitude distribution of the PAM signal is calculated. The parameters obtained, $D(l)$, are then passed to both the HMM transition matrix estimation module in the receiver and the PS-PAM mapping adjustment module in the transmitter. Initially, the transmitter still generated PON signals, $S_{PON}(x)$, with an equiprobable distribution. After obtaining the new parameters, $D(l)$, the PS-PAM mapping adjustment module in the transmitter adjusts the PAM signal using the

CCDM technique proposed in [26] to generate a new PON signal. After receiving the $D(l)$ parameters, the HMM transition matrix estimation module in the receiver calculates the new transition matrix parameters, $\mathbf{H}$, and inputs them into the joint demodulation process. The joint demodulation of PON-AMCC signals using the HMM method can be defined by the following formula [33,34]:

$$\begin{aligned}\lambda &= (\mathbf{\Pi},\ \mathbf{H},\ \mathbf{P}) \\ &= ([\pi_m]_{N*1},\ [h_{mn}]_{N*N},\ [p_m(x)]_{N*1}),\end{aligned} \tag{7}$$

where π is the initial probability distribution, $\mathbf{H}$ is the state transition probability matrix, and $\mathbf{P}$ is the observation probability matrix. In our interference model, the observation probability distribution is obtained by the GMM fitting process; therefore, $p_m(x)$ in the P matrix should be expressed as Equation (2). The transition matrix $\mathbf{H}$ can be described as

$$\mathbf{H} = [h_{mn}]_{N*N}, \quad 1 \leq m, n \leq N. \tag{8}$$

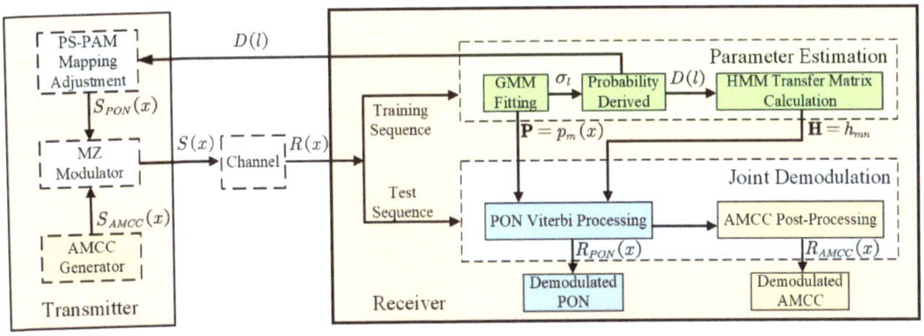

Figure 4. Joint optimization approach based on GMM-based probability shaping techniques.

Considering the randomness of the signals, the transition probabilities are the same when m, n belongs to different groups. Meanwhile, the probability of occurrence for L levels in a single group is equal. Thus, the jump probability for each level needs to be further divided by L. On the other hand, when m and n belong to the same group, the transition probabilities are also equal. Together with the restriction that $\sum_n h_{mn} = 1$, we can obtain

$$h_{mn} = \begin{cases} D(l) \times \frac{1}{K} \times \frac{R_{AMCC}}{R_{PON}}, & m \not\equiv n \ (mod\ K) \\ D(l) \times (1 - \frac{R_{AMCC}}{R_{PON}} \times \frac{K-1}{K}), & m \equiv n \ (mod\ K). \end{cases} \tag{9}$$

When the transmitter uses an equiprobable distribution of L-order PAM signals, the coefficient $D(l)$ can be represented as the constant $1/L$. However, when the transmitted signal undergoes the GMM-based PS method, the probabilities of occurrence for each level are no longer equal. Consequently, the coefficients, $D(l)$, in the transition parameters, h_{mn}, used in the HMM demodulation module of the joint demodulation receiver also need to be adjusted accordingly. As mentioned above, these coefficients are updated to reflect the actual probabilities of level occurrence, which are derived from the GMM fitting step of the receiver and are consistent with the probability parameters used in the transmitter. After updating the transition matrix, $\mathbf{H}$, in the HMM module of the receiver, we proceed with the joint demodulation process for PON and AMCC, where the PON Viterbi processing uses the demodulation method described by Equation (7) to demodulate the received signal, resulting in the output signal $R_{PON}(x)$. The transition parameters during the Viterbi process are input into the AMCC post-processing module for demodulating the AMCC signal $R_{AMCC}(x)$.

In practical transmission systems, the channel conditions may vary over time, and using fixed parameters may not meet the requirements of all scenarios. The parameters for probability shaping at the transmitter can be obtained through the GMM fitting step of the joint demodulation receiver. If we continuously feed back these parameters to the transmitter in real time, the transmitter can adjust the probability shaping parameters based on the current channel conditions. After adjusting the parameters of the transmitted signal, the transmitter can notify the receiver through the AMCC to synchronize and update the demodulation parameters.

4. Experimental Results and Discussion

4.1. Experimental System

In order to validate the signal adjustment method proposed in this article, we constructed an experimental platform, as illustrated in Figure 5. In this platform, we transmitted a 50G-PON signal mixed with an AMCC. The received signals were subjected to offline processing using the joint demodulation method at the receiver.

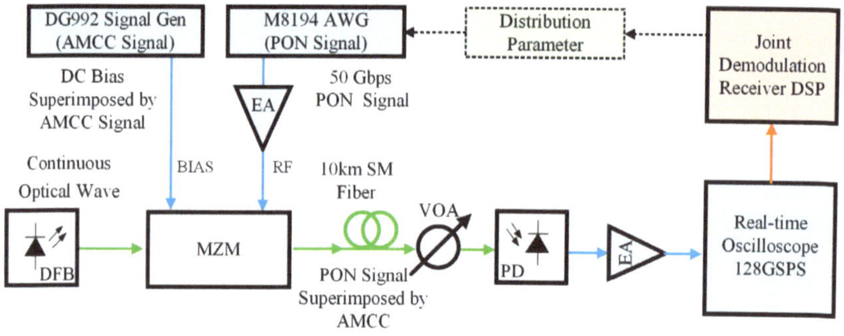

Figure 5. Experimental setup block diagram.

On the transmitter side, the shaped and bandwidth-limited electrical AMCC data are sent to the DC bias port of the modulator via a signal generator (Rigol DG992; 250 MSps), and a pre-generated 50 Gbps PAM4 PON signal is sent from an arbitrary waveform generator (AWG, Keysight M8195A, 65 GSps) to the RF port of the modulator. These two signals are combined in a LiNbO3 intensity modulator (Ixblue MXAN-LN series). An electrical amplifier (EA; Ixblue DR-AN-40-MO) is used in the RF port as a signal driver. The optical power into the fiber is set to 5 dBm. The combined signal is transmitted over a 10 km standard single-mode fiber (SSMF). Simultaneously, a variable optical attenuator (VOA) was employed to control the received optical power (ROP) at the receiver. This was carried out to vary line attenuations to assess the system's performance under different attenuations. Link attenuation will not affect the signal distribution characteristics, and the received signals are consistent with our proposed model under different link losses. At the receiver, the optical signal is detected by the photodetector (Thorlabs DXM30BF), and the electrical signal is connected to a real-time oscilloscope (Keysight UXR0334A, 33 GHz bandwidth). The channel is sampled at 128 GSps to capture the information of the PON signal. The captured signal undergoes processing through the digital signal processing flow, as illustrated in Figure 4, to derive the interference model and distribution parameters. These parameters are essential for the joint demodulation process in the receiver. Simultaneously, they play a pivotal role in tuning and optimizing the transmitted signals. This dual functionality aims to improve the system's overall transmission performance.

4.2. Distribution and Error Symbol

In order to validate the relationship between error symbols and signal distribution, we first analyze the case of an equal probability distribution for the received signals in the experimental system. When the AMCC modulation index in the experimental system was set to 10% and the received signal power was −10 dBm, we performed a statistical analysis of all demodulation errors in the received PON signals. The results of this analysis are shown in Figure 6, where the horizontal axis represents the PAM levels of the received signals, and the vertical axis represents the number of error symbols corresponding to each level.

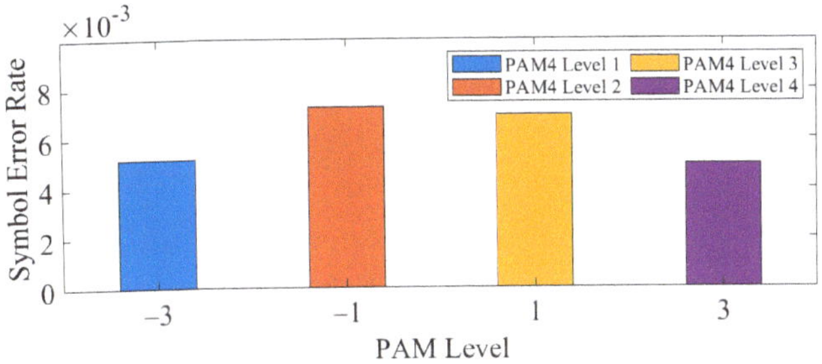

Figure 6. Error symbol statistics of the receiver.

Due to the presence of the AMCC, the original PON signal, which was PAM4-modulated, becomes a Gaussian mixture distribution with eight peaks at the receiver, where every two peaks correspond to a PON signal with the same amplitude value. From the graph, it is evident that the error rate for the PAM4 signal levels of −1 and +1 is significantly lower than for the PAM signal levels of −3 and +3. This observation aligns with the signal-to-noise ratio relationship depicted in Figure 3 based on the Gaussian mixture model. For this channel, we calculated the probability of the PON transmission signal as [0.33, 0.17, 0.17, 0.33] using the proposed method. Because the symbol probability of PAM4 has been modified, resulting in a decrease in the entropy of the non-uniform signal, in order to maintain the same effective bit rate for PON signals, we adjusted the signal rate after using probability shaping. According to the probability distribution, we adjusted the transmission rate of the PON signal to 26 GBaud while maintaining an effective data rate of 50 Gbps.

4.3. System Performance

After optimizing the transmission signal using probability shaping techniques, we performed tests on the transmission performance of PON signals. By using the new method, Figure 7a illustrates the relationship curve between the bit error rate (BER) and the ROP of the PON signal with the different modulation indices of the AMCC superimposed. The points in the figure represent the raw data obtained from the experiments, while the curves are the result of fitting these data points. As the modulation index of AMCC increases, the interference received by the PON also increases. Therefore, the performance of the PON signal is best when there is no interference from AMCC. In the ITU-T standard, it is mentioned that 50G-PON introduces FEC to meet higher transmission requirements, including both Reed–Solomon (RS) and low-density parity-check (LDPC) codes [20]. When RS coding is used, the PHY layer needs to meet a BER of 10^{-3} to achieve error-free transmission by FEC, and this requirement can be reduced to 10^{-2} when LDPC is used. Under this condition, error-free transmission can be realized by using FEC if the proposed method can keep the BER of the PON signal below this threshold. Therefore, we use 3.8×10^{-3} (7% hard-decision forward error correction, HD-FEC), which is the BER threshold required for RS coding, as a comparison criterion in the following discussion. This is shown (black curve) in the figure. Compared to the BER curve of the joint demodulation receiver under equal probability transmitted signals, as presented in [24], the new method proposed in this article results in a reduction in the BER of the PON signal. Additionally, there is a slight increase in the power budget within the 1 dB penalty range. Simultaneously, under these conditions, the BER curve of the AMCC is shown in Figure 7b. The greater modulation depth of the AMCC translates to stronger signal strength and a higher corresponding signal-to-noise ratio (SNR), leading to improved performance. A modulation depth of 5% represents the minimum for the AMCC, thus resulting in the poorest performance. All the modulation indices of the AMCC meet the BER threshold requirements for HD-FEC, indicating that error-free transmission can be achieved.

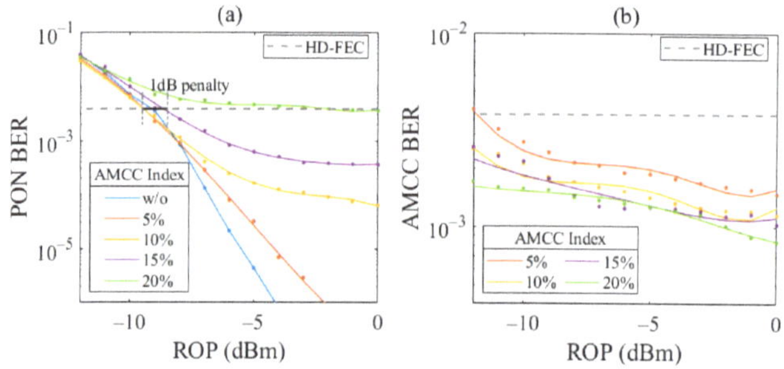

Figure 7. PON and AMCC BER performance. (**a**) PON BER performance with 10 Mbps AMCC interference, and (**b**) 10 Mbps AMCC BER performance.

The curves shown in Figure 8 were obtained under the conditions of an AMCC signal rate of 10 Mbps and a modulation index of 10%. Similar to Figure 7, the points in the figure represent the raw data obtained from the experiments, while the curves are obtained through fitting. The blue curve represents the relationship between the BER of the PON signal, demodulated by the joint demodulation receiver, and the ROP when transmitting signals with an equal probability distribution. The red curve illustrates the BER vs. ROP when the GMM-based PS technique proposed in this article is applied to adjust the transmitted signals, with the transition matrix in the joint demodulation receiver corrected. It can be observed that by employing the GMM-based PS technique for joint transmitter-receiver optimization, the power range of the received PON signal can be increased by nearly 1 dB at an error rate of 10^{-3}.

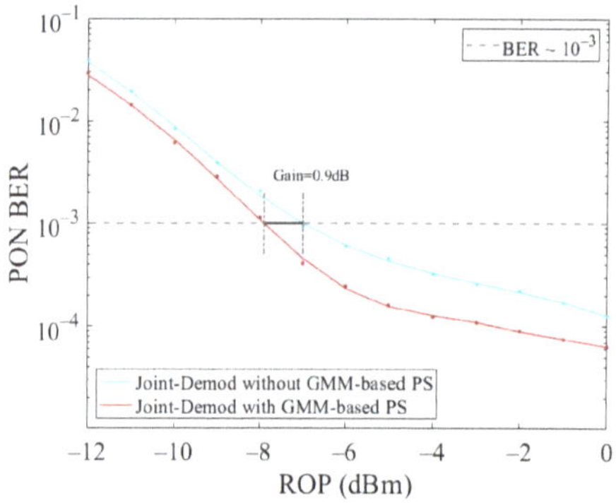

Figure 8. PON BER performance comparison with 10 Mbps 10% AMCC interference.

5. Conclusions

In this paper, we propose a method to enhance PON signal transmission performance through an analysis of the model of received signals in the PON-AMCC transmission system and the distribution characteristics of transmission error symbols. Firstly, we conduct GMM fitting regarding the received signals and analyze their distribution characteristics. Subsequently, we analyze the relationship between the error symbols in the received signals and the GMM distribution, which allows us to establish the objectives for adjusting the probabilities of the transmitted signals. Following these objectives, we employed probability-shaping techniques to modify the signal transmission probability. Due to the changes in the transmitted signals within the joint demodulation receiver, it is necessary to update the HMM transition matrix based on the modified transmission signal distribution characteristics. Finally, we conducted experiments using the proposed method in a 50G-PON experimental system with PAM4 modulation. The experimental results indicate that this method can reduce the error rate of PON signals at an equivalent data rate. This method can further mitigate the impact of AMCCs on PON, thus facilitating the efficient deployment of WDM-PON systems with AMCC. This approach also enables dynamic optimization tailored to the changing channel conditions, facilitating the rapid deployment of PON transmission systems in different channels.

Author Contributions: Conceptualization, H.G. and C.Y.; Data curation, H.G.; Formal analysis, C.Y. and Z.C.; Funding acquisition, C.Y., Z.C. and H.L.; Investigation, H.G. and C.Y.; Methodology, H.G. and C.Y.; Project administration, H.L.; Resources, Z.C. and H.L.; Software, H.G.; Supervision, Z.C. and H.L.; Validation, H.G. and C.Y.; Visualization, H.G.; Writing—original draft, H.G.; Writing—review & editing, C.Y. All authors have read and agreed to the published version of the manuscript.

Funding: This research was funded by Shenzhen Science and Technology R&D Funds (JSGGZD 20220822095601003) and the National Natural Science Foundation of China (U21A20454).

Institutional Review Board Statement: Not applicable.

Informed Consent Statement: Not applicable.

Data Availability Statement: Data are contained within the article.

Conflicts of Interest: The authors declare no conflict of interest.

References

1. NTT DOCOMO. White Paper 5G Evolution and 6G, Version 5.0. 2023. Available online: https://www.docomo.ne.jp/english/corporate/technology/whitepaper_6g/ (accessed on 1 February 2023).
2. China Mobile; China Telecom; China Unicom. 5G-Advanced Technology Evolution from a Network Perspective 2.0. 2022. Available online: https://www-file.huawei.com/-/media/corporate/pdf/news/5g-advanced%20technology%20evolution%20from%20a%20network%20perspective(2022).pdf?la=en (accessed on 1 June 2022).
3. Wang, X.; Ji, Y.; Zhang, J.; Bai, L.; Zhang, M. Low-Latency Oriented Network Planning for MEC-Enabled WDM-PON Based Fiber-Wireless Access Networks. *IEEE Access* **2019**, *7*, 183383–183395. [CrossRef]
4. Larsen, L.M.P.; Checko, A.; Christiansen, H.L. A Survey of the Functional Splits Proposed for 5G Mobile Crosshaul Networks. *IEEE Commun. Surv. Tutor.* **2019**, *21*, 146–172. [CrossRef]
5. ITU-T. Rec. G.989.1 40-Gigabit-Capable Passive Optical Networks (NG-PON2): General Requirements Amendment 1. 2013. Available online: https://www.itu.int/rec/T-REC-G.989.1 (accessed on 1 March 2013).
6. ITU-T. Rec. G.989.2 40-Gigabit-Capable Passive Optical Networks 2 (NG-PON2): Physical Media Dependent (PMD) Layer Specification-Amendment 1. 2019. Available online: https://www.itu.int/rec/T-REC-G.989.2 (accessed on 1 February 2019).
7. ITU-T. Rec. G.989.3 40-Gigabit-Capable Passive Optical Networks (NG-PON2): Transmission Convergence Layer Specification. 2015. Available online: https://www.itu.int/rec/T-REC-G.989.3 (accessed on 1 May 2021).
8. Suzuki, N.; Yoshima, S.; Miura, H.; Motoshima, K. Demonstration of 100-Gb/s/λ-Based Coherent WDM-PON System Using New AGC EDFA Based Upstream Preamplifier and Optically Superimposed AMCC Function. *J. Light. Technol.* **2017**, *35*, 1415–1421. [CrossRef]
9. Honda, K.; Kobayashi, T.; Nishihara, S.; Shimada, T.; Terada, J.; Otaka, A. Experimental Analysis of LTE Signals in WDM-PON Managed by Embedded Pilot Tone. *IEEE Photonics Technol. Lett.* **2017**, *29*, 431–434. [CrossRef]
10. Nakagawa, G.; Sone, K.; Yoshida, S.; Oda, S.; Hirose, Y.; Hoshida, T. Multi-Vendor Interoperation of SFP+ Transceivers for CPRI Signal Transmission with Superimposed AMCC for Mobile Fronthaul. In Proceedings of the 2018 Optical Fiber Communications Conference and Exposition (OFC), San Diego, CA, USA, 11–15 March 2018; p. Tu3L.5.
11. Herrera, L.E.Y.; Calliari, F.; Caballero, D.V.; Amaral, G.C.; Urban, P.J.; von der Weid, J.P. Transmitter-Embedded AMCC, LTE-A and OTDR signal for Direct Modulation Analog Radio over Fiber Systems. In Proceedings of the 2018 Optical Fiber Communications Conference and Exposition (OFC), San Diego, CA, USA, 11–15 March 2018; p. W1F.4.
12. Tan, Z.; Yang, C.; Xu, Z.; Chen, L.; Huang, X.; Guo, H.; Zheng, Z.; Zhang, F.; Wang, Z. Experimental Demonstration for over Mbps Baseband-over-Modulation AMCC Implementation in PtP WDM-PON. In Proceedings of the 2018 Optical Fiber Communications Conference and Exposition (OFC), San Diego, CA, USA, 11–15 March 2018; p. Tu3L.3.
13. Xu, Z.; Yang, C.; Tan, Z.; Guo, H.; Zhang, F. AMCC Superimposition and Extraction With Interference Elimination for 5G Mobile Fronthaul. *IEEE Photonics Technol. Lett.* **2018**, *30*, 1214–1217. [CrossRef]
14. Villafani Caballero, D.; Herrera, L.Y.; Calliari, F.; Urban, P.; von der Weid, J. Embedded time-multiplexed AMCC and OTDR signals for analog radio over fiber links. *Opt. Commun.* **2019**, *452*, 195–199. [CrossRef]
15. Guo, H.; Yang, C.; Gao, Y.; Li, H. AMCC nonlinear baseband superimposition and extraction aided by proposed interference cancellation for WDM-PON used in 5G mobile fronthaul. *Opt. Express* **2022**, *30*, 31602–31613. [CrossRef]

16. Honda, K.; Nakamura, H.; Hara, K.; Sone, K.; Nakagawa, G.; Hirose, Y.; Hoshida, T.; Terada, J. Wavelength control method of upstream signals using AMCC in WDM-PON for 5G mobile fronthaul. *Opt. Express* **2019**, *27*, 26749–26756. [CrossRef] [PubMed]
17. Honda, K.; Hara, K.; Nakamura, H.; Sone, K.; Nakagawa, G.; Hirose, Y.; Hoshida, T.; Terada, J. WDM-PON Management and Control by Auxiliary Management and Control Channel for 5G Mobile Fronthaul. *Opt. Express* **2021**, *29*, 42457–42470. [CrossRef]
18. Tanaka, Y.; Kanai, T.; Hara, K.; Chen, M.; Honda, K.; Shindo, T.; Kaneko, S.; Nakamura, H.; Kani, J.i.; Sano, K.; et al. Extraction of AMCC Signal Superposed by SOA-Integrated EA-DFB Laser for In-Service Monitoring in All-Photonics Network. *J. Light. Technol.* **2022**, *40*, 5783–5792. [CrossRef]
19. Zhang, D.; Liu, D.; Wu, X.; Nesset, D. Progress of ITU-T higher speed passive optical network (50G-PON) standardization. *J. Opt. Commun. Netw.* **2020**, *12*, D99–D108. [CrossRef]
20. Wey, J.S. The Outlook for PON Standardization: A Tutorial. *J. Light. Technol.* **2020**, *38*, 31–42. [CrossRef]
21. Zhang, J.; Wang, K.; Wei, Y.; Zhao, L.; Zhou, W.; Xiao, J.; Liu, B.; Xin, X.; Yu, J. Symmetrical 50-Gb/s/λ PAM-4 TDM-PON at O-Band Supporting 26 dB+ Loss Budget using Low-Bandwidth Optics and Semiconductor Optical Amplifier. In Proceedings of the 2020 Optical Fiber Communications Conference and Exhibition (OFC), San Diego, CA, USA, 8–12 March 2020; p. Th1B.3.
22. Guo, H.; Yang, C.; Qin, X.; Gao, Y.; Zheng, Z.; Li, H. Up to 20 Mb/s Auxiliary Management and Control Channel Signal Transmission in 50 Gb/s PON System. In Proceedings of the 2021 Optical Fiber Communications Conference and Exhibition (OFC), Washington, DC, USA, 6–11 June 2021; p. W1H.4.
23. Igarashi, R.; Koma, R.; Hara, K.; ichi Kani, J.; Yoshida, T. Simultaneous reception of AMCC signals and QPSK signals by a single coherent receiver with DSP. *Opt. Express* **2022**, *30*, 48030–48041. [CrossRef]
24. Guo, H.; Yang, C.; Gao, Y.; Li, H. Interference Modeling and Joint Demodulation Receiver for Multi-level PON with Baseband AMCC Superimposition. *J. Light. Technol.* **2023**, 1–11. [CrossRef]
25. Reynolds, D.A.; Quatieri, T.F.; Dunn, R.B. Speaker Verification Using Adapted Gaussian Mixture Models. *Digit. Signal Process.* **2000**, *10*, 19–41. [CrossRef]
26. Böcherer, G.; Steiner, F.; Schulte, P. Bandwidth Efficient and Rate-Matched Low-Density Parity-Check Coded Modulation. *IEEE Trans. Commun.* **2015**, *63*, 4651–4665. [CrossRef]
27. Schulte, P.; Böcherer, G. Constant Composition Distribution Matching. *IEEE Trans. Inf. Theory* **2016**, *62*, 430–434. [CrossRef]
28. He, Z.; Bo, T.; Kim, H. Probabilistically shaped coded modulation for IM/DD system. *Opt. Express* **2019**, *27*, 12126–12136. [CrossRef] [PubMed]
29. Sun, Z.; Tang, D.; Zuo, W.; Cui, H.; Wu, Z.; Qiao, Y. Triple-Convex Probabilistic Constellation Shaping PAM8 in IM/DD System. *IEEE Photonics Technol. Lett.* **2023**, *35*, 846–849. [CrossRef]
30. Hossain, M.S.B.; Böcherer, G.; Rahman, T.; Wettlin, T.; Stojanović, N.; Calabrò, S.; Pachnicke, S. Probabilistic Shaping for High-Speed Unamplified IM/DD Systems with an O-Band EML. *J. Light. Technol.* **2023**, *41*, 5373–5382. [CrossRef]
31. Che, D.; Cho, J.; Chen, X. Does Probabilistic Constellation Shaping Benefit IM-DD Systems without Optical Amplifiers? *J. Light. Technol.* **2021**, *39*, 4997–5007. [CrossRef]
32. Cho, J.; Winzer, P.J. Probabilistic Constellation Shaping for Optical Fiber Communications. *J. Light. Technol.* **2019**, *37*, 1590–1607. [CrossRef]
33. Rabiner, L.; Juang, B. An introduction to hidden Markov models. *IEEE ASSP Mag.* **1986**, *3*, 4–16. [CrossRef]
34. De Leon, P.L.; Pucher, M.; Yamagishi, J.; Hernaez, I.; Saratxaga, I. Evaluation of Speaker Verification Security and Detection of HMM-Based Synthetic Speech. *IEEE Trans. Audio, Speech, Lang. Process.* **2012**, *20*, 2280–2290. [CrossRef]

photonics

MDPI

Review

Advanced Neural Network-Based Equalization in Intensity-Modulated Direct-Detection Optical Systems: Current Status and Future Trends

Zhaopeng Xu [1], Tonghui Ji [1], Qi Wu [1,2], Weiqi Lu [3], Honglin Ji [1], Yu Yang [1], Gang Qiao [1], Jianwei Tang [1], Chen Cheng [1], Lulu Liu [1], Shangcheng Wang [1], Junpeng Liang [1], Zhongliang Sun [1], Linsheng Fan [1], Jinlong Wei [1] and William Shieh [3,*]

[1] Peng Cheng Laboratory, Shenzhen 518055, China; xuzhp@pcl.ac.cn (Z.X.); jith@pcl.ac.cn (T.J.); sjtuwuqi@sjtu.edu.cn (Q.W.); jihl@pcl.ac.cn (H.J.); yangy07@pcl.ac.cn (Y.Y.); qgstayfoolish@stu.pku.edu.cn (G.Q.); tangjw@pcl.ac.cn (J.T.); 20b952015@stu.hit.edu.cn (C.C.); liull@pcl.ac.cn (L.L.); wangshch@pcl.ac.cn (S.W.); liangjp@pcl.ac.cn (J.L.); sunzhl@pcl.ac.cn (Z.S.); fanlsh@pcl.ac.cn (L.F.); weijl01@pcl.ac.cn (J.W.)

[2] State Key Lab of Advanced Optical Communication Systems and Networks, Shanghai Jiao Tong University, Shanghai 200240, China

[3] School of Engineering, Westlake University, Hangzhou 310024, China; luweiqi@hust.edu.cn

* Correspondence: shiehw@westlake.edu.cn

Abstract: Intensity-modulated direct-detection (IM/DD) optical systems are most widely employed in short-reach optical interconnects due to their simple structure and cost-effectiveness. However, IM/DD systems face mixed linear and nonlinear channel impairments, mainly induced by the combination of square-law detection and chromatic dispersion, as well as the utilization of low-cost non-ideal transceivers. To solve this issue, recent years have witnessed a growing trend of introducing machine learning technologies such as neural networks (NNs) into IM/DD systems for channel equalization. NNs usually present better system performance than traditional approaches, and various types of NNs have been investigated. Despite the excellent system performance, the associated high computational complexity is a major drawback that hinders the practical application of NN-based equalizers. This paper focuses on the performance and complexity trade-off of NNs employed in IM/DD systems, presenting a systematic review of the current status of NN-based equalizers as well as a number of effective complexity reduction approaches. The future trends of leveraging advanced NN in IM/DD links are also discussed.

Keywords: intensity-modulated direct-detection; neural network; training; equalization; computational complexity

1. Introduction

With the exponential growth of Internet Protocol (IP) traffic, there is an ever-increasing demand for capacity in data centers. According to International Data Corporation, global data center traffic will reach 175 Zettabytes (ZB) per year by the end of 2025, up from 33 ZB per year in 2018. Inspired by this dramatic demand for capacity, data centers have become a hot topic for both academia and industry, which drive the research on short-reach optical fiber interconnects within the data centers [1–9]. Compared to coherent detection, intensity-modulated direct-detection (IM/DD) optical links are ideal for such short-reach applications due to their cost-effectiveness and simple structure [10–17]. However, the intensity-only direct-detection, or the simple square-law detection of the optical field, creates a nonlinear channel when combined with channel chromatic dispersion (CD). Additionally, to maintain low costs, bandwidth-limited transceivers and inexpensive lasers, such as directly modulated lasers (DMLs), are preferred, which present non-ideal frequency responses and chirp impairments [18–22]. These mixed linear and nonlinear impairments can significantly degrade bit error rate (BER) performance and limit the system's achievable capacity. Therefore, effective nonlinear equalization techniques are crucial to ensure the desired system BER.

Traditional digital signal processing (DSP) approaches such as decision feedback equalization (DFE) and Volterra series-based equalization, are decades old. They have been widely applied in IM/DD systems to deal with the nonlinear impairments [23–29]. In recent years, advances in machine learning (ML) [30–39] have led to the introduction and growing popularity of numerous ML algorithms in the field of optical fiber communication. These algorithms have found applications across various aspects of optical communications, such as optical performance monitoring [40–60] and channel equalization for different types of optical systems [61–87]. For IM/DD channel equalization, ML algorithms, especially neural networks (NNs), have been found superior to traditional approaches in terms of system performance. Due to the introduction of different nonlinear activation functions and the layered DSP design, NNs are extremely suitable to solve nonlinear problems. Among the broad topic of applying ML in optical communications, this paper specifically focuses on leveraging NN for nonlinear equalization in short-reach IM/DD systems. Different types of NNs and their variants are presented targeting at improved system performance.

While introducing the cutting-edge NNs trying to explore better system performance, it is also important to pay special attention to the computational complexity (CC) [88–90]. Complicated NN equalizers with increased CC can lead to higher latency and greater power consumption in the receiver, which hinders their practical implementation. CC is particularly relevant for NN-based equalization, where it impacts both the training and equalization (inference) processes. The training process for NNs typically requires a substantial number of training symbols and epochs. When the link scenario changes, the performance of the previously trained NNs may degrade, necessitating retraining to adapt to the new conditions, which is computationally inefficient. During the equalization process, the computational load is significant as well, with the number of multiplications per equalized symbol needing to be limited to a few tens to enable real-time DSP implementation [91–93]. Given these considerations, it is highly desirable to reduce CC in both NN training and the equalization processes. We can make a trade-off between the performance and CC of NN-based equalizers according to different link requirements.

In this paper, we provide a systematic review of the application of NN for equalization in short-reach IM/DD optical links, taking both system performance and CC into account. The remainder of this paper is organized as follows. Section 2 provides the introduction and the mathematical model of typical IM/DD systems, discussing the benefits and bottlenecks. Section 3 presents different performance-oriented advanced NN-based equalization structures, providing a comprehensive summary of existing works. Section 4 gives a detailed overview of a number of techniques effectively addressing both training and equalization CC of NN-based equalizers. Finally, Section 5 concludes this paper and discusses future perspectives.

2. Short-Reach IM/DD Systems

2.1. IM/DD System Structure

This paper discusses the traditional double-sideband (DSB) IM/DD systems which possess the simplest structure among various designs of optical transmission systems. A general illustration of a typical IM/DD communication system as well as the associated DSP processes are depicted in Figure 1. In this system, a laser serves as the light source, and the transmitted electrical signal is directly modulated onto the optical intensity. Various types of laser/modulator modules can be employed at the transmitter, including DML; a vertical-cavity surface-emitting laser (VCSEL); an electro-absorption modulated laser (EML), which consists of a laser combined with a separate electro-absorption modulator (EAM); a laser combined with a Mach-Zehnder Modulator (MZM); and other advanced silicon photonic modulators. Pulse amplitude modulation (PAM) formats with different levels are usually adopted for intensity-only optical transmission, such as PAM-2, PAM-4, and PAM-8. At the transmitter DSP, the PAM signals are generated and passed through a root raised cosine (RRC) filter for pulse shaping. The signals are sampled at a proper sampling rate before being sent out for transmission. The choice of fiber can vary based on the transmitter type. A single-mode fiber (SMF) is commonly used for most of the transmitters, while a multi-mode fiber (MMF) is selected for systems utilizing a VCSEL-based transmitter.

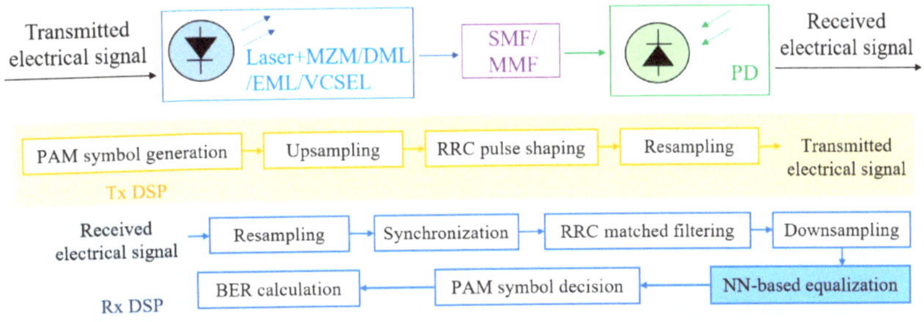

Figure 1. A typical structure of IM/DD systems with transceiver DSP procedures.

At the receiver, as shown in Figure 1, only a single-ended photo-detector (PD) is employed to convert the optical signal into electrical power. Unlike coherent detection, which could preserve both amplitude and phase of the signal, the simple square-law direct-detection can only preserve the amplitude information, and that is why PAM is usually employed for IM/DD systems. The received electrical signal is further processed by a series of commonly employed DSP procedures such as resampling, synchronization, and matched filtering. The signals are then fed into the NN-based equalization module for nonlinearity mitigation. Finally, hard-decision is performed and the system BER is calculated.

2.2. IM/DD System Model

Most IM/DD systems face intrinsic nonlinearity problem when performing square-law detection over signals affected by a dispersive channel [94]. When the IM/DD system is not operated at zero-dispersion wavelength, the CD effects is not negligible. The frequency response of CD can be expressed by

$$H(\omega) = e^{j\frac{1}{2}\beta_2\omega^2 L}, \tag{1}$$

where β_2 is the group velocity dispersion coefficient, L denotes the fiber length, and ω denotes the signal angular frequency. Assuming an ideal transmitter is employed, after intensity-modulation, the output optical power of the transmitter laser, denoted by $P_{Tx}(t)$, is given by

$$P_{Tx}(t) = \eta(S_0 + S_{Tx}(t)), \tag{2}$$

where $S_{Tx}(t)$ represents the transmitted electrical signal, S_0 denotes the bias current, and η denotes the modulation coefficient. If we omit the phase impact, the optical field of the transmitter laser, denoted by $E_{Tx}(t)$, can be written as

$$E_{Tx}(t) = \sqrt{P_{Tx}(t)} = \sqrt{\eta(S_0 + S_{Tx}(t))} = \sqrt{\eta S_0}\sqrt{1 + \frac{S_{Tx}(t)}{S_0}}. \tag{3}$$

Note that the bias current S_0 normally needs to be large enough to make the signal located at the linear modulation range of lasers. As such, we can perform Taylor series expansion over $E_{Tx}(t)$, and $E_{Tx}(t)$ can be rewritten as

$$E_{Tx}(t) = \sqrt{\eta S_0}\left(1 + \sum_{n=1}^{\infty} c_n \left(\frac{S_{Tx}(t)}{S_0}\right)^n\right), \tag{4}$$

where the Taylor expansion coefficients are calculated by

$$c_n = \frac{(-1)^{n-1}(2n)!}{2^{2n}(n!)^2(2n-1)}. \tag{5}$$

After transmission through the optical fiber channel, the received optical field, denoted by $E_{Rx}(t)$, is modeled by the convolution of the transmitted optical field $E_{Tx}(t)$ and the CD response in time domain denoted by $h(t)$, which is shown as

$$\begin{aligned} E_{Rx}(t) &= E_{Tx}(t) \otimes h(t) \\ &= \sqrt{\eta S_0}\left(1+\sum_{n=1}^{\infty} c_n\left(\frac{S_{Tx}(t)}{S_0}\right)^n\right) \otimes h_R(t) + j\sqrt{\eta S_0}\left(1+\sum_{n=1}^{\infty} c_n\left(\frac{S_{Tx}(t)}{S_0}\right)^n\right) \otimes h_I(t)' \end{aligned} \tag{6}$$

where $h_R(t)$ and $h_I(t)$ represents the real and imaginary part of $h(t)$, and $\otimes$ represents the convolution operation. Equation (6) can be simplified by calculating $1 \otimes h_R(t)$ and $1 \otimes h_I(t)$ based on Equation (1), where the simplified version is written by

$$E_{Rx}(t) = \sqrt{\eta S_0}\left(1+\sum_{n=1}^{\infty} c_n\left(\frac{S_{Tx}(t)}{P_0}\right)^n \otimes h_R(t)\right) + j\sqrt{\eta S_0}\sum_{n=1}^{\infty} c_n\left(\frac{S_{Tx}(t)}{P_0}\right)^n \otimes h_I(t). \tag{7}$$

The received square-law detected electrical signal, denoted by $S_{Rx}(t)$, is shown as

$$S_{Rx}(t) = R|E_{Rx}(t)|^2, \tag{8}$$

where R represents the responsivity of the PD. With simple mathematical derivation, $S_{Rx}(t)$ can be expanded and written as (note that $c_1 = \frac{1}{2}$)

$$\begin{aligned} S_{Rx}(t) = \; & R\eta S_0 + R\eta S_{Tx}(t) \otimes h_R(t) + 2R\eta S_0 \sum_{n=2}^{\infty} c_n\left(\frac{S_{Tx}(t)}{S_0}\right)^n \otimes h_R(t) + \\ & R\eta S_0\left[\left(\sum_{n=1}^{\infty} c_n\left(\frac{S_{Tx}(t)}{S_0}\right)^n \otimes h_R(t)\right)^2 + \left(\sum_{n=1}^{\infty} c_n\left(\frac{S_{Tx}(t)}{S_0}\right)^n \otimes h_I(t)\right)^2\right] \end{aligned}. \tag{9}$$

As shown in Equation (9), the received signal $S_{Rx}(t)$ is separated into four parts. The first term denotes the direct current, which is constant and can be easily removed. The second term is a linear convolution of the transmitted signal $S_{Tx}(t)$ and the real part of time-domain CD response $h_R(t)$. This is known as the power fading effect, where the IM/DD signals suffer from destructive frequencies especially when the data rate and fiber length increase. The third term is the convolution of the high order signal term with the real part of time-domain CD response $h_R(t)$, while the fourth term shows the signal-to-signal beating interference (SSBI). The first two terms are linear, while the last two terms show nonlinear impacts. Even in the ideal case, we find that IM/DD channel is intrinsically nonlinear. In practical applications, the laser, modulator, and PD can introduce more severe nonlinear impairments. The mixed linear and nonlinear impairments significantly degrade system performance, which necessitate advanced equalization methods such as powerful NNs.

3. Performance-Oriented NN-Based Equalizers

3.1. FNN-Based Equalizer

The NN-based equalizers for IM/DD links are first investigated targeting at improved system BER performance. As the simplest form of NN, feedforward NNs (FNNs) are widely employed for equalization in IM/DD systems [95–106]. A typical two-layer FNN equalization structure is depicted in Figure 2. Assuming $n^{[0]}$ inputs and $n^{[1]}$ hidden neurons are employed for the FNN, the DSP process is operated by

$$y = f^{[2]}\left(\mathbf{W}^{[2]} f^{[1]}(\mathbf{W}^{[1]}\mathbf{x} + \mathbf{b}^{[1]}) + b^{[2]}\right), \tag{10}$$

where $\mathbf{x} \in \mathbb{R}^{n^{[0]}}$ and y represents the inputs and output of FNN, $\mathbf{W}^{[1]} \in \mathbb{R}^{n^{[1]} \times n^{[0]}}/\mathbf{b}^{[1]} \in \mathbb{R}^{n^{[1]}}/f^{[1]}$ and $\mathbf{W}^{[2]} \in \mathbb{R}^{n^{[2]} \times n^{[1]}}/b^{[2]}/f^{[2]}$ denotes the weights/biases/activation functions of the hidden and the output layer. The NN is operated in a sliding-window manner, which predicts the received symbol sequentially. Different NN parameters can be selected and optimized to yield different system performance.

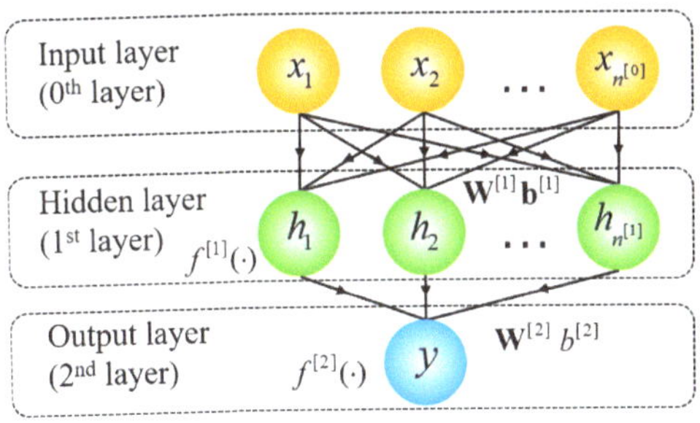

Figure 2. Schematic of a two-layer FNN-based equalizer.

The first application introducing FNN into IM/DD systems is observed in [95], where a simple two-layer FNN is deployed to infer simultaneously the linear and non-linear channel response. The NN has four outputs, each corresponding to one level of the PAM-4 signal and the entire NN function as a classifier. With the help of the FNN, a 168-Gb/s PAM-4 signal is successfully transmitted over 1.5-km SMF, achieving up to 10 times BER reduction over conventional FFE. In [96], FNN is implemented to attain a 64-Gb/s PAM4 4-km MMF link employing 850-nm VCSEL. FNN outperforms 3rd order Volterra series in their VCSEL-based IM/DD setup. Recorded high 256-Gb/s·km data-rate distance product is achieved supported by FNN-based equalization. In [97], FNN is used for nonlinear equalization in IM/DD passive optical network (PON) scenarios. With the aid of FNN, 50-Gb/s PAM4 IM/DD PON transmission via 20 km SMF is realized using 10-GHz class optical devices, where the end-to-end 3-dB bandwidth is only 3.6 GHz. FNN shows its superiority over conventional approaches, and shows its effectiveness in resolving bandwidth problems. In [98], a DML-based IM/DD link is shown using FNN at the receiver. A 20-Gb/s 18-km O-band PAM4 transmission is realized, where the FNN nonlinear equalizer is found to increase the channel capacity and drastically reduce the impact of nonlinear penalties. In [99], the authors extend their [98] and increase the data rate to 54 Gb/s. Different modulation formats are used, where the FNN-based equalizers work well for all the cases. FNN is adopted in wavelength division multiplexing (WDM) IM/DD links in [100], where 4×50-Gb/s PAM4 signal is transmitted over 80-km SMF. In this work, a dispersion compensation fiber (DCF) is used to pre-compensate the CD impacts. FNN shows about 2 dB power sensitivity improvement over conventional nonlinear DSP methods.

The demonstration of FNN in IM/DD systems has not gone away in years. More recently, FNN is applied in 137-Gb/s PAM4 link using 25-GHz class 850-nm optical devices [101]. The signal is transmitted over an in-house fabricated 40 cm optical backplane. The 112-Gbps 100-m VCSEL-MMF optical interconnects are demonstrated in [102], and the FNN achieve more than one order of magnitude BER improvement compared with Volterra series in such system. Similar as [97], a 50-Gb/s 20-km link is shown in [103] using bandwidth-limited transceivers. Under bandwidth constraints, the FNN-based equalizer again presents superior performance. IM/DD link using the simple OOK modulation format is shown in [104], where a 50-Gb/s OOK signal is transmitted over 30-km SMF. The FNN is also successfully demonstrated in real-time field-programmable gate arrays (FPGAs) in this work. The IM/DD link data rate is increased to as high as 160 Gb/s in [105], employing a GeSi EAM. The highest single-wavelength PAM4 data rate is achieved based on a single EAM, supported by FNN-based nonlinear mitigation. A more generalizable FNN-based equalizer is shown in [106], where a 56-Gb/s PAM4 signal is transmitted over

20/30/40-km SMFs using the proposed FNN. All the above works prove that FNN is effective in mitigating the channel impairments of short-reach IM/DD systems.

3.2. CNN-Based Equalizer

Following the introduction of FNN, more powerful NNs are employed for equalization in IM/DD systems. Convolutional NNs (CNNs) are employed to explore deeper into the system performance in [107–111]. CNN is a regularized type of FNN that learns feature engineering by filter optimization with the help of convolutional and pooling layers, which are widely used for image classification tasks. The schematic of a CNN-based equalizer is illustrated in Figure 3. Considering only one-dimension data (which is the case of signal processing for channel equalization), assuming the input, filter, and output of the convolutional layer are represented by **x**, **f**, and **y**, the convolution operation is expressed as

$$y_i = \sum_{k=1}^{L} x_{i+k-1} f_k, \tag{11}$$

where L denotes the filter length. The pooling operation reduces the number of data by combining the outputs of neuron clusters at one layer into a single neuron in the next layer. Max pooling and average pooling are the most commonly used, which take either the maximum or the average value of each local cluster of neurons in the feature map. A typical CNN consists of many stacks of convolutional and pooling layers, where each stack represents one feature of the input data. The features are collected and fed into fully connected layers same as the FNN to give the final outputs.

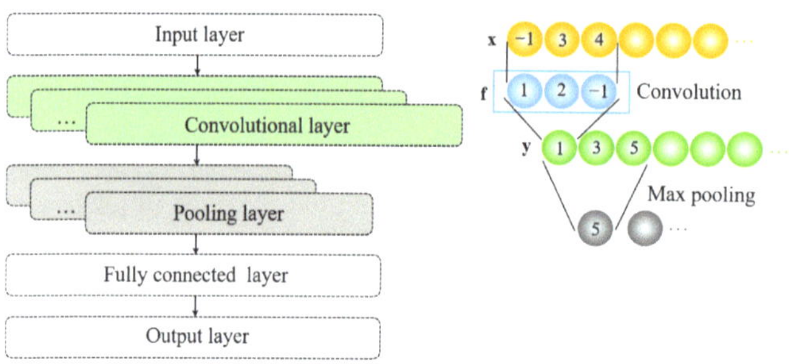

Figure 3. Schematic of a CNN-based equalizer.

CNN is applied for equalization in a 112-Gb/s 40-km PAM4 optical link using EML in [107]. The CNN has one input layer, three convolutional layers, two fully connected layers, and one output layer for classification of PAM signals. It has been shown that the performance of the proposed CNN model outperforms Volterra series and FNN equalizers. In [108–110], the same group use different types of CNN for equalization in different IM/DD systems. The system is varied in modulation formats (PAM-4, PAM-8, PAM-16), transmission bands (C-band, O-band), data rates (56 Gb/s, 100 Gb/s) and system bandwidth (10-G class, 20-G class). The CNN is changed with different number of convolutional layers and the number of neurons in each layer. All the different demonstrations show strong equalization ability of CNN. A temporal CNN (TCNN) is proposed in [111], which introduce dilated convolutions and residual connections onto the traditional CNN. Better system performance is observed compared with traditional CNN equalization architecture, and the proposed scheme enables as far as 100-km SMF IM/DD transmission of a 56-Gb/s PAM4 signal.

3.3. RNN-Based Equalizer

Although CNN presents better performance, it also requires much larger network structures, which is deemed too complex for real-time application. The investigation on

CNN-based equalization seems to have disappeared in recent years, and researchers focus more on recurrent NN (RNN)-based equalization [112–124]. Four types of RNNs are found in IM/DD applications, which are auto-regressive RNN (AR-RNN), layer-recurrent NN (L-RNN), long short-term memory (LSTM) and gate recurrent unit (GRU) networks. These RNNs are built on top of the traditional FNN.

The schematic of a two-layer AR-RNN-based equalizer is shown in Figure 4. On top of the FNN, a few output delays are sent back to the hidden layer and serve as new inputs. Assuming the number of output delays used in the feedback loop is denoted by k, and the output delays and the associated weights are represented by $\mathbf{y}_d$ and $\mathbf{W}^d \in \mathbb{R}^{n^{[1]} \times k}$, the DSP process is given by

$$y = f^{[2]}\left(\mathbf{W}^{[2]} f^{[1]}\left(\left[\mathbf{W}^{[1]}, \mathbf{W}^{d}\right]\left[\mathbf{x}^T, \mathbf{y}_d^T\right]^T + \mathbf{b}^{[1]}\right) + b^{[2]}\right). \quad (12)$$

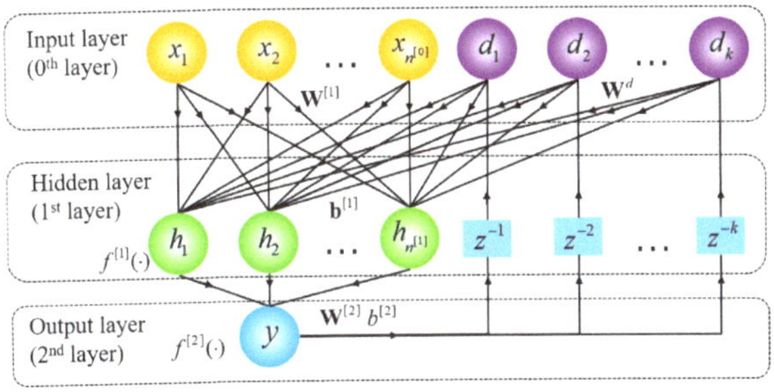

Figure 4. Schematic of a two-layer AR-RNN-based equalizer.

In terms of the equalization process, the operation of using past predicted output symbols as additional inputs provides more information when predicting the current output symbol. As such, better performance can normally be achieved with the help of this information. AR-RNN is first used for equalization in IM/DD systems in [112]. The 50-Gb/s PAM-2 and 100-Gb/s PAM-4 signals are transmitted over 20-km SMF, where the receiver adopt AR-RNN with seven feedbacks to read the historical decision results. In [113,114], the AR-RNN is implemented using FPGAs in a parallel manner, and a 100-Gb/s IM/DD PON system is demonstrated. It is shown that the AR-RNN can beat FNN equalizers with the same input/output size and the number of training parameters, achieving better receiver sensitivity performance.

The structure of L-RNN is depicted in Figure 5. Different from AR-RNN, which uses output feedbacks, L-RNN collects the delays from the outputs of hidden neurons and sends them back to the hidden layer again for data processing. Assuming the number of rounds of hidden layer delays used in the feedback loop is denoted by k, and the hidden layer delays and the associated weights are represented by $\mathbf{h}_d$ and $\mathbf{W}^h \in \mathbb{R}^{n^{[1]} \times kn^{[1]}}$, the DSP process of L-RNN is given by

$$y = f^{[2]}\left(\mathbf{W}^{[2]} f^{[1]}\left(\left[\mathbf{W}^{[1]}, \mathbf{W}^{h}\right]\left[\mathbf{x}^T, \mathbf{h}_d^T\right]^T + \mathbf{b}^{[1]}\right) + b^{[2]}\right). \quad (13)$$

Similar to AR-RNN, additional useful information about former predictions is also provided in L-RNN when predicting the current symbol. An L-RNN-based equalizer is proposed for equalization in a VCSEL-MMF optical interconnect in [115]. It has been shown that L-RNN is more powerful than ANN in dealing with sequential signals, and has the potential of reaching much lower BER with similar complexity. In [116], the authors extend their work in [115] by employing hidden feature extraction before sequence training. The input features are first extracted using principal component analysis or other dimensionality reduction approaches before sending into the L-RNN equalizer. Aided by the feature-enhanced L-RNN, single-lane 288-Gb/s PAM-8 signal transmission over 100-m MMF is realized with BER well below the 20% SD-FEC threshold.

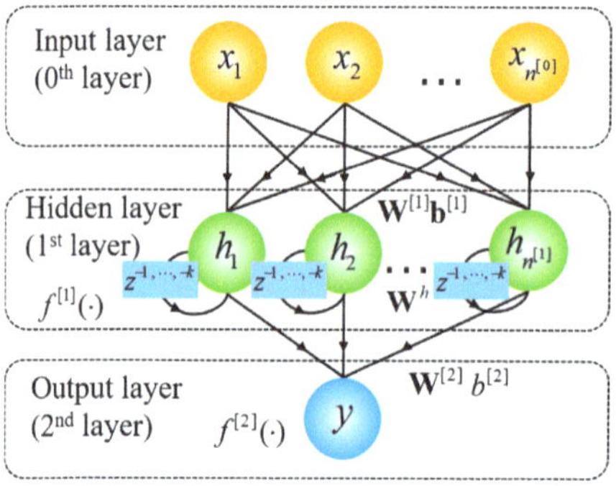

Figure 5. Schematic of a two-layer L-RNN-based equalizer.

The architecture of LSTM and GRU networks are given in Figure 6. Compared with traditional FNN, an LSTM/GRU layer is added, where inside contains a number of LSTM/GRU cells. Both LSTM and GRU address the vanishing gradient problem in traditional RNNs by introducing gating mechanisms that allow them to capture long-term dependencies more effectively. The detailed complicated LSTM/GRU cell structure will not be discussed in this paper. Interested readers can refer to [117–124] for more information.

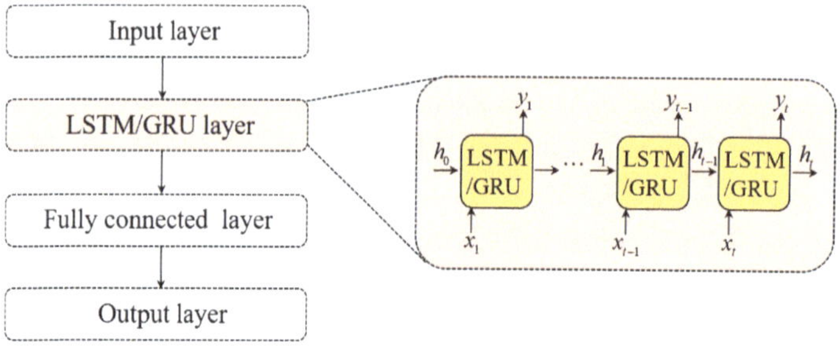

Figure 6. Schematic of LSTM- and GRU-based RNN equalizers.

LSTM is used for equalization in both classification and regression manners in [117]. Both cases work well for a 50-Gb/s 100-km PAM4 optical system. In [118,119], a 160-Gb/s 1-km PAM4 link is conducted using a silicon-microring-modulator (Si-MRM) and an LSTM-based equalizer. Two LSTM layers and two fully connected layers are employed. The nonlinearity induced by the modulator is effectively mitigated by the proposed powerful equalizer. In [120], the authors extend their work in [118,119] with updated experimental configuration. The LSTM now supports 270-Gb/s PAM-8 signal to transmit 1-km SMF using the Si-MRM, which greatly increases the achievable data rate. LSTMs are also employed in [121,122] to achieve 200+ Gb/s per single lane. Note that non-zero dispersion-shifted fiber (NZDSF) is used to reduce the impact of CD. In addition to LSTM, the performance of GRU is also tested in [122], where it achieves slightly higher BER than LSTM. More works on GRU-based equalization can be found in [123,124], where the GRU is proposed to resolve the patterning effect of the semiconductor optical amplifiers (SOAs) applied in IM/DD systems. The input power dynamic range of SOA can be greatly extended with the help of the GRU-based equalization.

3.4. Cascade NN-Based Equalizer

The introduction of different recurrent structures of RNN significantly improves the system performance. However, the training and equalization complexity also increase, especially for the LSTM and GRU ones. Another variant of FNN is the cascade NN, which is computationally friendly. The structure of cascade FNN is shown in Figure 7. On top of the traditional FNN, cascade connections are included, which connect the input and every previous layer to the following layers. For a two-layer cascade FNN, the input layer is simply connected to the output layer. Assuming the cascaded weights and are represented by $\mathbf{W}^c \in \mathbb{R}^{n^{[0]}+k}$, the equalization process of cascade FNN is given by

$$y = f^{[2]}\left(\left[\mathbf{W}^{[2]}, \mathbf{W}^c\right]\left[\left(f^{[1]}\left(\mathbf{W}^{[1]}\mathbf{x} + \mathbf{b}^{[1]}\right)\right)^T, \mathbf{x}^T\right]^T + b^{[2]}\right). \quad (14)$$

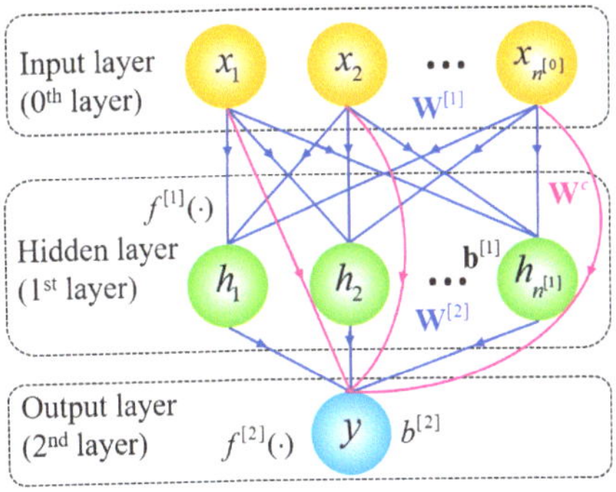

Figure 7. Schematic of a two-layer cascade FNN-based equalizer.

The cascade connections produce a pure linear path for direct mapping of the inputs to the output. This enables an efficient joint linear and nonlinear effect estimation and results in better system performance when used for equalization. Both cascade FNN and cascade RNN are proposed in [125,126]. A 100-Gb/s 15-km PAM4 link is built using a band-limited DML, where the cascade structure help improves the receiver sensitivity by 1 dB compared with NNs without cascade connections. It is also demonstrated that cascade NN-based equalizers have a much faster training speed. A more recent work is found in [127], where the cascade structure is shown as "skip connections". The experimental setup is similar as used in [123,124], where the NN performs well with skip connections. The effect of simplified training is also verified in this work.

3.5. Other Types of NN-Based Equalizers

In addition to the above-mentioned NNs, there are also many different type of NN-based equalizers demonstrated in IM/DD systems. Radial basis function NN (RBF-NN) is shown in [128] in a 4 × 50-Gb 80-km PAM-4 IM/DD link. The RBF-NN employs Gaussian activation function in the hidden layer, and achieves better network stability and fitting ability compared with traditional Volterra series or FNN. There are many discussions on the application of spiking NN (SNN) in IM/DD systems recently, as shown in [129–133]. In addition to neuronal and synaptic state used in traditional NNs, SNNs incorporate the concept of time into their operating model. The SNN-based equalizers have been implemented in application-specific integrated circuits (ASICs), and have been verified in both simulation and experiments. Interested readers can refer to [129–133] for more details about SNN-based equalization in IM/DD systems. The different types of NN-based equalizers and IM/DD links are summarized and shown in Table 1.

Table 1. Different types of NN-based equalization for various short-reach IM/DD links.

NN Type	Ref.	Modulation	Data Rate	Fiber Length	Tx Type	Wavelength
FNN	[95]	PAM4	168 Gb/s	1.5 km SMF	MZM (35 GHz)	~1550 nm
	[96]	PAM4	64 Gb/s	4 km MMF	VCSEL (25 GHz)	~850 nm
	[97]	PAM4	50 Gb/s	20 km SMF	MZM (10 GHz)	~1550 nm
	[98]	PAM4	20 Gb/s	18 km SMF	DML (10 GHz)	~1310 nm
	[99]	PAM2/PAM4/PAM8	54 Gb/s	25 km SMF	DML (10 GHz)	~1550 nm
	[100]	PAM4	4 × 50 Gb/s	80 km SMF	DML (20 GHz)	~1550 nm
	[101]	PAM4	137 Gb/s	40 cm MMF	MZM (25 GHz)	~850 nm
	[102]	PAM4	112 Gb/s	100 m MMF	VCSEL (NA)	~850 nm
	[103]	PAM4	50 Gb/s	20 km SMF	DML (10 GHz)	~1310 nm
	[104]	PAM2	50 Gb/s	30 km SMF	MZM (35 GHz)	~1310 nm
	[105]	PAM4	160 Gb/s	2 km SMF	GeSi EAM (30 GHz)	~1550 nm
	[106]	PAM4	56 Gb/s	20/30/40 km SMF	MZM (40 GHz)	~1550 nm
CNN	[107]	PAM4	112 Gb/s	40 km SMF	EML (25 GHz)	~1310 nm
	[108]	PAM4	56 Gb/s	25 km SMF	DML (10 GHz)	~1310 nm
	[109,110]	PAM8/PAM16	100 Gb/s	25 km SMF	DML (20 GHz)	~1310/1550 nm
	[111]	PAM4	56 Gb/s	100 km SMF	MZM (40 GHz)	~1550 nm
RNN	[112]	PAM2/PAM4	60/100 Gb/s	20 km SMF	MZM (40 GHz)	~1550 nm
	[113,114]	PAM4	100 Gb/s	20 km SMF	MZM (NA)	~1310 nm
	[115]	PAM4	56 Gb/s	100 m MMF	VCSEL (18 GHz)	~850 nm
	[116]	PAM8	288 Gb/s	100 m MMF	VCSEL (23 GHz)	~850 nm
	[117]	PAM4	50 Gb/s	100 km SMF	DML (18 GHz)	~1550 nm
	[118,119]	PAM4	160 Gb/s	1 km SMF	Si MRM (47 GHz)	~1550 nm
	[120]	PAM8	270 Gb/s	1 km SMF	Si MRM (55 GHz)	~1550 nm
	[121,122]	PAM4	212 Gb/s	1 km NZDSF	EML (40 GHz)	~1550 nm
	[123,124]	PAM4	100 Gb/s	5.4 km SMF	MZM (NA)	~1550 nm
Cascade NN	[125,126]	PAM4	50/100 Gb/s	25/15 km SMF	DML (16 GHz)	~1550 nm
	[127]	PAM4	100 Gb/s	4.8 km SMF	MZM (33 GHz)	~1550 nm
RBF-NN	[128]	PAM4	4 × 50 Gb/s	80 km SMF	DML (18 GHz)	~1550 nm
SNN	[129,130]	PAM4	224 Gb/s	4 km SMF	NA	~1270 nm
	[131,132]	PAM4	100 Gb/s	2 km SMF	NA	~1310 nm
	[133]	PAM4	200 Gb/s	5 km SMF	NA	~1270 nm

Besides the direct utilization of NN for equalization, NNs are often combined with sequence decoders to achieve better system performance. Maximum likelihood sequence estimation (MLSE) based on NN is proposed in [134], where the NN is used to estimate the nonlinear channel responses and to calculate the metrics for the Viterbi algorithm. Similarly, an NN-BCJR equalization scheme is proposed in [135], where an NN-based nonlinear channel emulator is adopted to calculate the transition metric in the BCJR algorithm. In [136,137], duobinary training strategy is proposed. The NN equalizer is first trained targeting at the duobinary form of the signal, and MLSE is followed to recover the enforced ISI. This approach is particularly effective in addressing the bandwidth limitations problems in IM/DD systems. In [138], the NN equalizer is trained targeting at adaptive duobinary form of the signal. The optimal partial-response parameter is learned through NN training, where the system performance can be further improved compared with equalization with conventional partial-response target.

3.6. Performance and Complexity Comparison of FNN-, L-RNN-, Cascade FNN-, and AR-RNN-Based Equalizers

An IM/DD experiment is conducted to verify the performance and complexity of above-mentioned NN-based equalizers. Here we only show the results of four types, i.e., FNN-, L-RNN-, cascade FNN-, and AR-RNN-based equalizers, since other types such as CNN- or LSTM/GRU-based ones are considered much more complex, which makes them difficult to be applied in real-time applications. Interested readers can refer to [95–133] for more details on the performance of different types of NN-based equalizers. The IM/DD experiment is based on a DML with a 3-dB bandwidth around 16 GHz, where a 50-Gb/s PAM4 signal is generated and transmitted over 25-km SMF [125,126]. A variable optical attenuator (VOA) is applied at the receiver to tune the received optical power (ROP). The NNs only have two layers, where tanh activation function is selected for the hidden layer and linear activation function is used for the output layer. The NNs are used in a regression manner which means that only one output is adopted. A total of 20,000 random PAM4 symbols are used for training the NNs, while an additional 1.2 million PAM4 symbols are collected for NN-based nonlinear equalization and BER calculation.

We first show the best system performance of each NN-based equalizers, where the complexity constraint is omitted. As many as 15 inputs and nine hidden neurons are selected, which can guarantee that the NNs achieve their best performance. The BER-ROP curves of different NN-based equalizers are shown in Figure 8a. We use the form (the number of inputs, the number of hidden neurons) to represent the size of the different NNs, as shown in the figure. It can be observed that the performance of NNs follows the order of AR-RNN, cascade FNN (denoted by C-FNN in the figure), L-RNN, and FNN. Compared with traditional FNN, the three FNN variants all improve the system performance. The receiver sensitivity is improved by approximately 2/1/0.5 dB by AR-RNN/cascade FNN/L-RNN. Figure 8b illustrates the number of multiplications (denoted by N_{mul}) of all the NNs adopted in Figure 8a. Considering the same number of inputs and hidden neurons, L-RNN is obviously more complicated than the other employed NNs. Cascade FNN and AR-RNN, however, show limited additional complexity compared with traditional FNN.

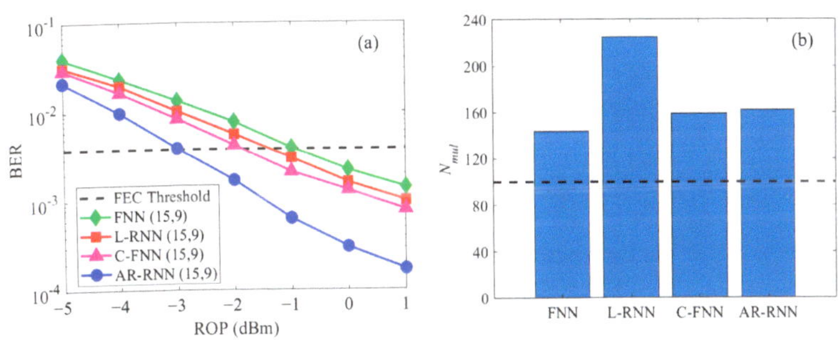

Figure 8. (**a**) BER versus ROP of different NN-based equalizers with 15 inputs and 9 hidden neurons; (**b**) The number of multiplications for the NNs used in (**a**) to recover one symbol.

Figure 9a depicts the system BER performance of the NN-based equalizers under the complexity constraint, where the N_{mul} of all the NNs are all kept below 100, shown in Figure 9b. The number of inputs and hidden neurons of the different types of NNs are carefully chosen to achieve the best system performance with only a few tens of multiplications involved to recover one symbol, showing the potential for real-time implementation. When the N_{mul} of NNs are lower than 100, L-RNN becomes the worst equalizer since its size is affected most by the complexity constraint. Cascade FNN and AR-RNN, however, still present superior BER performance over FNN, increasing the receiver sensitivity by about 1.5 and 2 dB, respectively.

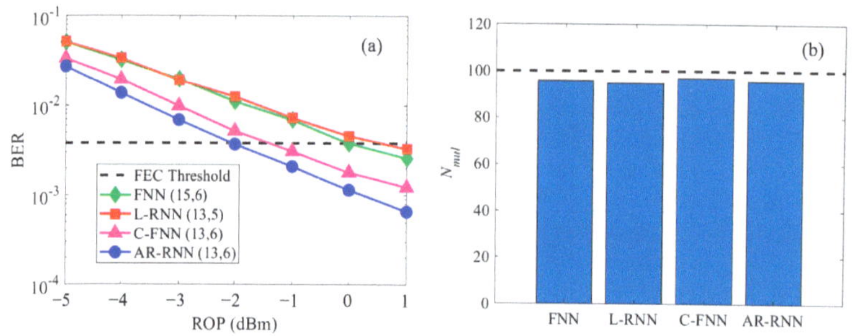

Figure 9. (**a**) BER versus ROP of different NN-based equalizers under the complexity constraint; (**b**) The number of multiplications for the NNs used in (**a**) to recover one symbol.

3.7. Possible Pitfalls When Applying NN-Based Equalizers

One thing we need to pay special attention to when using NNs is the so-called possible pitfalls or overestimation traps [139–143]. It has been observed that NNs are capable of learning the operational logic of pseudo-random bit sequences (PRBSs). This ability may lead to an overestimation of the NNs' performance, as the performance improvements might stem from predicting the sequence patterns rather than from mitigating channel impairments. Such overestimation is not limited to PRBS but can also occur with other data types that follow specific patterns, including short repeated sequences.

To mitigate this overestimation problem, several approaches can be adopted. First is employing pure random data, i.e., true random numbers generated through unpredictable physical processes, to ensure that each transmitted symbol is independent. This method prevents NNs from learning any underlying patterns. Second is using different mixtures of PRBSs with varying orders to train the NNs. This can also prevent the recognition of consistent patterns across the combined sequences. Lastly, ensuring that the number of NN inputs does not exceed the PRBS order can naturally address the issue. For PRBS transmission, the NNs require at least as many inputs as the PRBS order to fully grasp the PRBS operational logic. In the context of equalization in small-scale optical transmission systems such as IM/DD links, where only a limited number of NN inputs are needed for equalization, one viable strategy is to use sufficiently long PRBSs.

4. Computationally Efficient NN-Based Equalizers

The NNs indeed greatly exploit the performance of IM/DD system. However, it is also obvious that the CC is largely increased, which makes NN receivers less practicable for real-time implementation. Much progress has been made on resolving the complexity issue when applying NN for IM/DD equalization. This section will review all the techniques, focusing on both NN training and equalization.

4.1. Transfer Learning

The training process of NN-based equalizers is usually time-consuming, which involves many iterations of forward- and backward-propagation calculations. When there are many optical links needed for equalization, the training of different NN-based equalizers becomes a big problem. Transfer learning is proposed to speed up the NN training process [144]. Transfer learning is a machine learning strategy that involves repurposing a model designed for one task to serve as the foundation for a different, but related, task. This method capitalizes on the insights gained from the initial task and applies them to a new challenge. It is especially advantageous when the new task has a limited amount of labeled data, as it enables the model to utilize the extensive data and computational resources already invested in training the original model. Transfer learning has been introduced into optical communications for optical performance monitoring [145–147], and for equalization of coherent or single-sideband (SSB) signals [148–151]. It is first introduced for equalization in IM/DD systems in [152,153], where the flow diagram is given in Figure 10. For NN training of the target IM/DD system, we can leverage the NN trained from different source IM/DD systems and use transfer learning. Since the source NN-based equalizers preserve channel information that are related to the target system, they can serve as a better starting point for NN training in the target system, instead of training purely from scratch. Transfer learning-aided fast equalization is demonstrated for both FNN and RNN-based equalizers in a 50-Gb/s 20-km PAM-4 target IM/DD system [152,153]. The target system equalization is accelerated by adopting NNs from a number of source systems with different data rates and fiber lengths. The 60-Gb/s 15-km source system is found closest to the target one, where significant reduction of 90%/87.5% in training epochs and 62.5%/53.8% in training symbols are achieved. The study also reveals that FNNs can be smoothly transferred to RNNs for equalization in the target system, whereas the reverse adaptation is not practical.

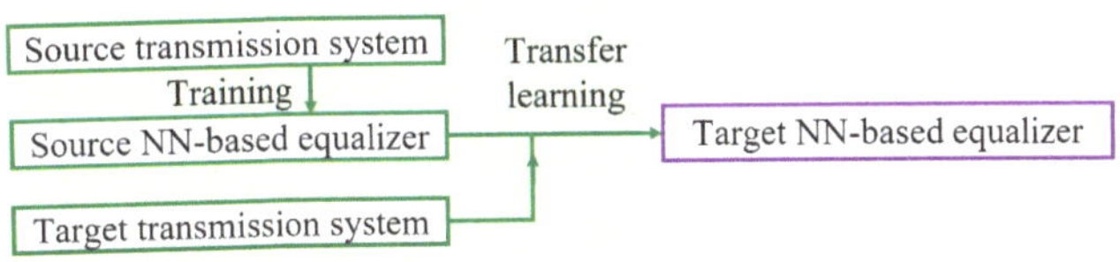

Figure 10. Flow diagram of transfer learning-aided equalization.

In addition to FNN- and RNN-based equalizers, transfer learning can be smoothly applied for CNN-based equalization in IM/DD systems, as shown in [154]. Similar to [152,153], source systems with varying data rates and fiber lengths are employed, and transfer learning again shows its effectiveness in reducing the number of training epochs and the size of the training dataset. The iterative pruning technique is introduced into the transfer learning-aided equalization for IM/DD links in [155,156], where the convergence speed can be further enhanced during TL between the source and target links. By fine-tuning the pruning parameters, an optimal balance between performance stability and complexity can be attained. Transfer learning is set to be pivotal in the advancement of optical-switched data center networks, where the dynamic reconfiguration of optical link parameters is crucial. Utilizing transferred NN receivers, new optical interconnects can be rapidly deployed.

4.2. Pruning

When the optical links are fixed and do not change dynamically, the training complexity of NNs can be omitted since the well-trained equalizer can be stably used without the need for retraining. The equalization complexity becomes the primary concerns. Pruning is one possible technique to reduce the size and equalization complexity of an NN by removing less important weights. As shown in Figure 11, after pruning, a sparse NN structure is presented compared with its fully connected counterpart. This process helps in making the model more efficient, often leading to faster inference times and reduced computational resources, without significantly compromising the model's performance. Pruning can also enhance generalization by preventing overfitting and is particularly useful for deploying models on resource-constrained devices. The pruning techniques have been applied in Volterra-series-based equalization [157–160] for IM/DD systems, as well as in NN-based equalization for coherent [161] and SSB signals [162]. Pruning of NN-based equalizers in IM/DD systems is found in [98,163–167].

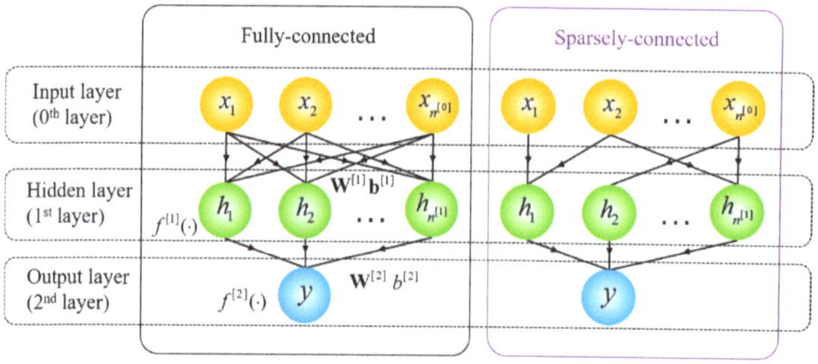

Figure 11. Schematic of fully and sparsely connected FNN-based equalizers.

The pruning process include the importance assessment of all the weights in the NN-based equalizer. This is commonly performed by setting a threshold, where the weights with absolute values lower than the threshold are considered insignificant and can be pruned. In [98], the traditional pruning method is applied in a DML-based IM/DD link. It has been demonstrated that the BER curves of the pruned NN are close to that of the unpruned NN, showing the ability of pruning in reducing receiver complexity without degrading much of the system performance. In [163], an iterative pruning algorithm is proposed for NN-based equalization in VCSEL-based IM/DD links. Compared with traditional one-shot pruning, which prunes the NN only one time, the iterative pruning method prunes the NN many times. The NN can be fine-tuned accordingly, which leads to a better complexity reduction efficiency. Ref. [164] presents the real-time pruned NN in FPGAs for VCSEL-based optical interconnects. The included hardware resources are minimized by pruning, showing the potential of applying NN receivers in practical applications. In [165], pruning is applied in a different cascade RNN-based equalization structure in a DML-based IM/DD link. It is shown that the receiver complexity is largely decreased, despite the utilization of NN structures. The importance of cascade and recurrent connections are also verified in the pruning process. In [166,167], adaptive L2-regularization is introduced to facilitate pruning in EML-based optical interconnects. A two-step training scheme is proposed, where the first step involves using L2-regularization during training to encourage sparsity in weight representations, and the second step applies the traditional pruning mechanism to remove the insignificant weights. The proposed L2-regularization-aided pruning approach shows better performance compared with conventional direct pruning.

4.3. Multi-Task Learning

Multi-task learning is also an efficient technique to address the equalization complexity issue. Multi-task learning is a machine learning approach where a model is trained simultaneously on multiple related tasks, leveraging shared representations to improve performance on each task. By learning commonalities and differences among tasks, the model can lead to improved accuracy and efficiency. Considering the equalization tasks in optical transmission systems, multi-task learning is referred to as multi-symbol prediction, shown in Figure 12, where multiple symbols are recovered simultaneously rather than processed sequentially, one at a time. By dealing with multiple symbol using only one NN, a better utilization of weights and biases can be realized. The information provided by the weights and biases in the traditional single-output NN to recover the current symbol can still be useful for predicting the following symbols. Part of the NN parameters can be shared to enable a more efficient equalization structure.

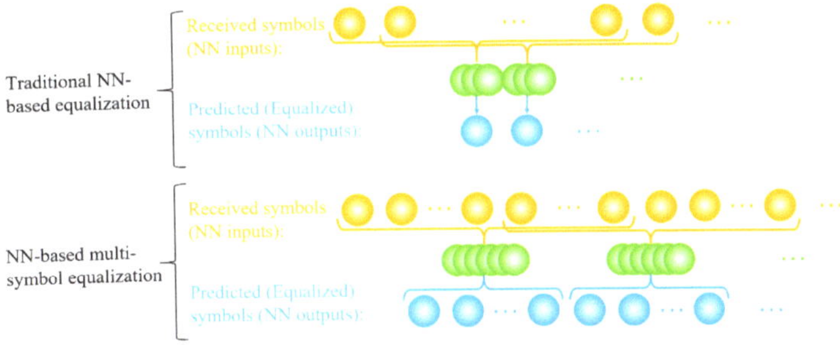

Figure 12. Schematic of traditional and multi-symbol NN-based equalization.

NNs with multi-outputs are adopted in IM/DD links in [104,113,114,124,168], where the main purpose is to enable high throughputs. By increasing the number of NN outputs, parallel computing can be realized. At the same FPGA clock frequency, higher throughputs can be achieved while the number of employed FPGAs remains the same. For the complexity reduction purpose, multi-symbol IM/DD equalization is first proposed in [169,170], where FNN-, cascade NN-, and RNN-based multi-symbol equalizations are demonstrated. All the cases reduce the number of multiplications for one symbol recovery to about a few tens, which indicates the potential real-time implementation of NNs with multi-output selections. The work also finds that there exist an optimal number of NN outputs that reduce the computational complexity most. The multi-symbol equalization idea is then introduced into LSTM and GRU-based IM/DD equalization in [121,122], as well as reservoir computing-based equalization [171,172], where similar conclusions about complexity reduction are given. The multi-symbol equalization scheme can even be combined with pruning techniques to jointly reduce the receiver complexity [173]. Recent hardware demonstrations of multi-output NN-based equalizers further indicates their effectiveness in reducing the equalization complexity [174,175], where the chip areas are considerably saved.

4.4. Quantization

Another approach to relax the equalization complexity requirement of NN is quantization. NN quantization is a method that reduces the computational and memory demands of an NN by changing its parameters and activations from high-precision (e.g., 32-bit floating-point) to lower-precision formats (e.g., 8-bit integers). A quantized NN-based equalizer is shown in Figure 13, where bit shifters and quantizers are adopted between each layer. The quantization approach shrinks the model size and speeds up inference, enhancing efficiency for hardware deployment. Although quantization can lead to some loss of accuracy, careful tuning helps preserve the performance while significantly lowering computational costs and power usage. A few works [176–179] have already demonstrated the computationally efficient quantized NN-based equalization in coherent optical transmission systems.

For equalization in short-reach IM/DD scenarios, fixed-precision quantization of NNs is employed in [114,168], targeting FPGA implementation NN receivers. Both works quantize the floating-point (32-bit)-based NNs to integer-based ones, where only a few bits are used to represent each weight and bias. Negligible BER penalty is observed when reducing the quantized bits, which suggests the floating-point-based NNs are actually redundant in precisions. By reducing the number of bits, the floating-point calculations all change to integers, and the memory needed for NN parameter storage is drastically decreased. In [180,181], a mixed-precision quantization method is proposed for IM/DD equalization to further decrease the number of quantized bits compared with the fixed-precision counterpart. A straightforward input neuron partitioning approach is applied to determine the high- and low-precision weights. The proposed mixed-precision quantization is verified in both traditional FNN and advanced cascade RNN scenarios. In [182], joint mixed-precision quantization and pruning is proposed to squeeze out more bits in the NN-based equalizer.

The connections of NN are either directly eliminated or represented by a suitable number of quantization bits through weight clustering, creating a hybrid compressed sparse network structure that computes much faster and consumes less hardware resources. The system performance can still be upheld using the pruned mixed-precision-quantized NN receivers.

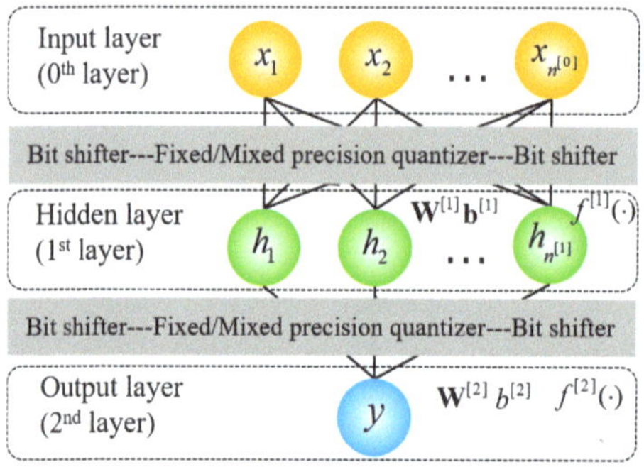

Figure 13. Schematic of a quantized FNN-based equalizer.

5. Conclusions and Future Perspectives

This paper presents a comprehensive overview of the current status of applying NNs for equalization in short-reach IM/DD optical links, considering both system performance and complexity. Traditional FNN and a series of advanced NNs are adopted to effectively mitigate the linear and nonlinear impairments in IM/DD channels. Transfer learning, pruning, multi-task learning, and quantization approaches are introduced to make the NN-based equalizer more computationally efficient, considering both the training and equalization phases.

Future directions of NN-based equalization in short-reach IM/DD systems still focus on improving the performance and reducing the complexity. One key area of interest is the continuous exploration of more advanced and powerful NN-based equalizers to enhance system performance. One thing we need to mention is that the works present in this paper mainly consider only post-equalization for simplicity. Joint optimizations of both pre- and post-NN-based equalization, or the so-called end-to-end learning structures, may be viable solutions to further improve BER. Considering different short-reach links, it is also important to develop approaches to improve the ability of generalization for NN-based equalizers. Another area of interest lies in the development of more efficient and intelligent approaches for complexity reduction. Algorithms that enable faster NN training and equalization are vital for realizing real-time receiver implementations. Moreover, the NN-based equalizers shown in this paper are all used as black boxes, relying purely on data-driven methods. It is also important to incorporate physical interpretations into the equalization model and develop physics-informed NN receivers for short-reach IM/DD applications.

Author Contributions: Conceptualization, Z.X., T.J., Q.W., W.L., and W.S.; methodology, Z.X., T.J., Q.W., W.L., and W.S.; investigation, Z.X., H.J., Y.Y., G.Q., J.T., C.C., L.L., S.W., J.L., Z.S., L.F., and J.W.; writing—original draft preparation, Z.X.; writing—review and editing, Z.X., T.J., Q.W., W.L., and W.S.; visualization, Z.X.; supervision, W.S. All authors have read and agreed to the published version of the manuscript.

Funding: This work was supported by National Key Research and Development Program of China (2020YFB1806400).

Institutional Review Board Statement: Not applicable.

Informed Consent Statement: Not applicable.

Data Availability Statement: Data underlying the results presented in this paper are not publicly available at this time but may be obtained from the authors upon reasonable request.

Conflicts of Interest: The authors declare no conflicts of interest.

References

1. Cartledge, J.; Karar, A. 100 Gb/s Intensity Modulation and Direct Detection. *J. Light. Technol.* **2014**, *32*, 2809–2814. [CrossRef]
2. Chagnon, M. Optical Communications for Short Reach. *J. Light. Technol.* **2019**, *37*, 1779–1797. [CrossRef]
3. Zhong, K.; Zhou, X.; Huo, J.; Yu, C.; Lu, C.; Lau, A.P.T. Digital Signal Processing for Short-Reach Optical Communications: A Review of Current Technologies and Future Trends. *J. Light. Technol.* **2018**, *36*, 377–400. [CrossRef]
4. Wu, Q.; Xu, Z.; Zhu, Y.; Ji, H.; Yang, Y.; Qiao, G.; Liu, L.; Wang, S.; Liang, J.; Wei, J.; et al. Gradient-Descent Noise Whitening Techniques for Short-Reach IM-DD Optical Interconnects with Severe Bandwidth Limitation. *Opt. Express* **2024**, *32*, 1715–1727. [CrossRef]
5. Xu, Z.; Sun, C.; Manton, J.; Shieh, W. Joint Equalization of Linear and Nonlinear Impairments for PAM4 Short-Reach Direct Detection Systems. *IEEE Photon. Technol. Lett.* **2021**, *33*, 425–428. [CrossRef]
6. Tan, Z.; Yang, C.; Zhu, Y.; Xu, Z.; Zou, K.; Zhang, F.; Wang, Z. A 70 Gbps NRZ Optical Link Based on 850 nm Band-Limited VCSEL for Data-Center Intra-Connects. *Sci. China. Info. Sci.* **2018**, *61*, 080406. [CrossRef]
7. Tan, Z.; Yang, C.; Zhu, Y.; Xu, Z.; Zou, K.; Zhang, F.; Wang, Z. High speed band-limited 850-nm VCSEL link based on time-domain interference elimination. *IEEE Photon. Technol. Lett.* **2017**, *29*, 751–754. [CrossRef]
8. Lu, W.; Liu, Z.; Xu, Z.; Liu, L.; Zou, Y.; Dai, X.; Yang, Q.; Shieh, W. Impact of Non-Gaussian Noise on GMI and LDPC Performance in Neural Network Equalized Systems. *Opt. Lett.* **2024**, *49*, 923–926. [CrossRef]
9. Zhou, J.; Qiao, Y.; Huang, X.; Yu, C.; Cheng, Q.; Tang, X.; Guo, M.; Liu, W.; Li, Z. Joint FDE and MLSD algorithm for 56-Gbit/s optical FTN-PAM4 system using 10G-class optics. *J. Light. Technol.* **2019**, *37*, 3343–3350. [CrossRef]
10. Zhong, K.; Zhou, X.; Wang, Y.; Gui, T.; Yang, Y.; Yuan, J.; Wang, L.; Chen, W.; Zhang, H.; Man, J.; et al. Recent Advances in Short Reach Systems. In Proceedings of the Optical Fiber Communication Conference (OFC), Los Angeles, CA, USA, 19–23 March 2017.
11. Wu, Q.; Xu, Z.; Zhu, Y.; Ji, H.; Yang, Y.; Liang, J.; Qiao, G.; Wang, S.; Liu, L.; Wei, J.; et al. 405-GBd OOK and 201-GBd PAM-4 IM/DD Optics at Record Faster-Than-Nyquist Ratios of 226.6% and 62.1% Enabled by Advanced Noise Whitening. In Proceedings of the Asia Communications and Photonics Conference (ACP), Wuhan, China, 4–7 November 2023; pp. 1–4.
12. Xu, X.; Zhou, E.; Liu, G.N.; Zuo, T.; Zhong, Q.; Zhang, L.; Bao, Y.; Zhang, X.; Li, J.; Li, Z. Advanced Modulation Formats for 400-Gbps Short-Reach Optical Inter-Connection. *Opt. Express* **2015**, *1*, 492–500. [CrossRef]
13. Shi, J.; Zhang, J.; Li, X.; Chi, N.; Chang, G.; Yu, J. 112 Gb/s/λ CAP Signals Transmission over 480 km in IM-DD System. In Proceedings of the Optical Fiber Communications Conference (OFC), San Diego, CA, USA, 11–15 March 2018.
14. Yekani, A.; Rusch, L.A. Interplay of Bit Rate, Linewidth, Bandwidth, and Reach on Optical DMT and PAM with IMDD. *IEEE Trans. Commun.* **2019**, *4*, 2908–2913. [CrossRef]
15. Wu, Q.; Xu, Z.; Zhu, Y.; Ji, T.; Ji, H.; Yang, Y.; Liang, J.; Cheng, C.; Qiao, G.; He, Z.; et al. Beyond 200-Gb/s O-band Intensity Modulation and Direct Detection Optics with Joint Look-Up-Table-Based Predistortion and Digital Resolution Enhancement for Low-Cost Data Center Interconnects. *Adv. Photon. Nexus* **2024**, *3*, 036007. [CrossRef]
16. Liu, N.; Xiao, J.; Liu, C.; Shen, G. IM/DD transmission for Data Center Optical Interconnects. In Proceedings of the Asia Communications and Photonics Conference (ACP), Shanghai, China, 24–27 October 2021.
17. Jiang, L.; Yan, L.; Yi, A.; Pan, Y.; Zhang, B.; Hu, Q.; Pan, W.; Luo, B. Integrated Components and Solutions for High-Speed Short-Reach Data Transmission. *Photonics* **2021**, *8*, 77. [CrossRef]
18. Wu, Q.; Xu, Z.; Zhu, Y.; Ji, H.; Yang, Y.; Liang, J.; Ji, T.; Qiao, G.; Wang, S.; Liu, L.; et al. Dispersion-Unmanaged Transmission of 144-Gb/s PAM-8 and 110-Gb/s PAM-6 Signals Over 10/20-km SSMF for DML-Based IM-DD Optics at C Band. In Proceedings of the Optical Fiber Communication Conference (OFC), San Diego, CA, USA, 24–28 March 2024.
19. Wu, Q.; Xu, Z.; Zhu, Y.; Ji, T.; Ji, H.; Yang, Y.; Qiao, G.; Cheng, C.; Tang, J.; Zhao, T.; et al. High-Speed Dispersion-Unmanaged DML-Based IM-DD Optics at C-band with Advanced Nonlinear Equalization and Noise Whitening. *J. Light. Technol.* **2024**, *42*, 5591–5598. [CrossRef]
20. Gao, Y.; Cartledge, J.C.; Yam, S.S.H.; Rezania, A.; Matsui, Y. 112 Gb/s PAM-4 Using a Directly Modulated Laser with Linear Pre-Compensation and Nonlinear Post-Compensation. In Proceedings of the European Conference and Exhibition on Optical Communication (ECOC), Dusseldorf, Germany, 18–22 September 2016.
21. Zhang, K.; Zhuge, Q.; Xin, H.; Morsy, M.; El-Fiky, E.; Yi, L.; Hu, W.; Plant, D. Intensity Directed Equalizer for the Mitigation of DML Chirp Induced Distortion in Dispersion-Unmanaged C-band PAM Transmission. *Opt. Express* **2017**, *23*, 28123–28135. [CrossRef]
22. Zhang, K.; Zhuge, Q.; Xin, H.; Hu, W.; Plant, D.V. Performance Comparison of DML, EML and MZM in Dispersion-Unmanaged Short Reach Transmissions with Digital Signal Processing. *Opt. Express* **2018**, *26*, 34288–34304. [CrossRef] [PubMed]
23. Stojanovic, N.; Karinou, F.; Qiang, Z.; Prodaniuc, C. Volterra and Wiener Equalizers for Short-Reach 100G PAM-4 Applications. *J. Light. Technol.* **2017**, *21*, 4583–4594. [CrossRef]

24. Watts, P.M.; Mikhailov, V.; Savory, S.; Bayvel, P.; Glick, M.; Lobel, M.; Christensen, B.; Kirkpatrick, P.; Shang, S.; Killey, R.I. Performance of single-mode fiber links using electronic feed-forward and decision feedback equalizers. *IEEE Photon. Technol. Lett.* **2005**, *17*, 2206–2208. [CrossRef]
25. Wang, H.; Zhou, J.; Li, F.; Liu, L.; Yu, C.; Yi, X.; Huang, X.; Liu, W.; Li, Z. Variable-step DD-FTN algorithm for PAM8-based short-reach optical interconnects. In Proceedings of the CLEO: Science and Innovations, San Jose, CA, USA, 5–10 May 2019.
26. Miao, X.; Bi, M.; Yu, J.; Li, L.; Hu, W. SVM-Modified-FFE Enabled Chirp Management for 10G DML-based 50Gb/s/λ PAM4 IM-DD PON. In Proceedings of the Optical Fiber Communications Conference (OFC), San Diego, CA, USA, 3–7 March 2019.
27. Tang, X.; Liu, S.; Xu, X.; Qi, J.; Guo, M.; Zhou, J.; Qiao, Y. 50-Gb/s PAM4 over 50-km Single Mode Fiber Transmission Using Efficient Equalization Technique. In Proceedings of the Optical Fiber Communications Conference (OFC), San Diego, CA, USA, 3–7 March 2019.
28. Zhou, H.; Li, Y.; Liu, Y.; Yue, L.; Gao, C.; Li, W.; Qiu, J.; Guo, H.; Hong, X.; Zuo, Y.; et al. Recent Advances in Equalization Technologies for Short-Reach Optical Links Based on PAM4 Modulation: A Review. *Appl. Sci.* **2019**, *9*, 2342. [CrossRef]
29. Li, D.; Song, H.; Cheng, W.; Deng, L.; Cheng, M.; Fu, S.; Tang, M.; Liu, D. 180 Gb/s PAM8 Signal Transmission in Bandwidth-Limited IMDD System Enabled by Tap Coefficient Decision Directed Volterra Equalizer. *IEEE Access* **2020**, *8*, 19890–19899. [CrossRef]
30. LeCun, Y.; Bengio, Y.; Hinton, G. Deep learning. *Nature* **2015**, *521*, 436–444. [CrossRef] [PubMed]
31. Goodfellow, I.; Bengio, Y.; Courville, A. *Deep Learning*; MIT Press: London, UK, 2016.
32. Bishop, C.M. *Pattern Recognition and Machine Learning*; Springer: New York, NY, USA, 2006.
33. Schmidhuber, J. Deep learning in neural networks: An overview. *Neural Netw.* **2015**, *61*, 85–117. [CrossRef]
34. Hecht-Nielsen, R. Theory of the backpropagation neural network. In *Neural Networks for Perception*; Academic Press: Cambridge, MA, USA, 1992; pp. 65–93.
35. Duchi, J.; Hazan, E.; Singer, Y. Adaptive subgradient methods for online learning and stochastic optimization. *J. Mach. Learn. Res.* **2011**, *12*, 2121–2159.
36. Kingma, D.P.; Ba, J. Adam: A method for stochastic optimization. *arXiv* **2014**, arXiv:1412.6980.
37. Raich, R.; Zhou, G.T. Orthogonal polynomials for complex Gaussian processes. *IEEE Trans. Signal Process.* **2004**, *52*, 2788–2797. [CrossRef]
38. Domingos, P.M. A few useful things to know about machine learning. *Commun. ACM* **2012**, *55*, 78–87. [CrossRef]
39. Han, S.; Mao, H.; Dally, W.J. Deep compression: Compressing deep neural networks with pruning, trained quantization and huffman coding. *arXiv* **2015**, arXiv:1510.00149.
40. Musumeci, F.; Rottondi, C.; Nag, A.; Macaluso, I.; Zibar, D.; Ruffini, M.; Tornatore, M. An overview on application of machine learning techniques in optical networks. *IEEE Commun. Surv. Tutor.* **2018**, *21*, 1383–1408. [CrossRef]
41. Khan, F.N.; Zhou, Y.; Lau, A.P.T.; Lu, C. Modulation Format Identification in Heterogeneous Fiber-Optic Networks Using Artificial Neural Networks. *Opt. Express* **2012**, *20*, 12422–12431. [CrossRef]
42. Zibar, D.; Piels, M.; Jones, R.; Schaeffer, C.G. Machine Learning Techniques in Optical Communication. *J. Light. Technol.* **2016**, *34*, 1442–1452. [CrossRef]
43. Wu, X.; Jargon, J.A.; Skoog, R.A.; Paraschis, L.; Willner, A.E. Applications of Artificial Neural Networks in Optical Performance Monitoring. *J. Light. Technol.* **2009**, *27*, 3580–3589.
44. Tanimura, T.; Hoshida, T.; Kato, T.; Watanabe, S.; Morikawa, H. Data-Analytics-Based Optical Performance Monitoring Technique for Optical Transport Networks. In Proceedings of the Optical Fiber Communication Conference (OFC), San Diego, CA, USA, 11–15 March 2018.
45. You, Y.; Jiang, Z.; Janz, C. Machine Learning-Based EDFA Gain Model. In Proceedings of the 2018 European Conference on Optical Communication (ECOC), Rome, Italy, 23–27 September 2018.
46. Morais, R.M.; Pedro, J. Machine Learning Models for Estimating Quality of Transmission in DWDM Networks. *J. Opt. Commun. Netw.* **2018**, *10*, D84–D99. [CrossRef]
47. Yankov, M.; Ros, F.; Moura, U.; Carena, A.; Zibar, D. Flexible Raman Amplifier Optimization Based on Machine Learning-Aided Physical Stimulated Raman Scattering Model. *J. Light. Technol.* **2017**, *35*, 1887–1893. [CrossRef]
48. Panayiotou, T.; Savva, G.; Shariati, B.; Tomkos, I.; Ellinas, G. Machine Learning for QoT Estimation of Unseen Optical Network States. In Proceedings of the Optical Fiber Communication Conference (OFC), San Diego, CA, USA, 3–7 March 2019.
49. Wang, D.; Zhang, M.; Li, J.; Li, Z.; Li, J.; Song, C.; Chen, X. Intelligent Constellation Diagram Analyzer Using Convolutional Neural Network-Based Deep Learning. *Opt. Express* **2017**, *25*, 17150–17166. [CrossRef]
50. Musumeci, F.; Rottondi, C.E.M.; Corani, G.; Shahkarami, S.; Cugini, F.; Tornatore, M. A Tutorial on Machine Learning for Failure Management in Optical Networks. *J. Light. Technol.* **2019**, *37*, 4125–4139. [CrossRef]
51. Shahkarami, S.; Musumeci, F.; Cugini, F.; Tornatore, M. Machine-Learning-Based Soft-Failure Detection and Identification in Optical Networks. In Proceedings of the Optical Fiber Communications Conference and Exposition (OFC), San Diego, CA, USA, 11–15 March 2018; pp. 1–3.
52. Khan, F.N.; Fan, Q.; Lu, C.; Lau, A.P.T. An Optical Communication's Perspective on Machine Learning and Its Applications. *J. Light. Technol.* **2019**, *37*, 493–516. [CrossRef]
53. Zhang, K.; Fan, Y.; Ye, T.; Tao, Z.; Oda, S.; Tanimura, T.; Akiyama, Y.; Hoshida, T. Fiber Nonlinear Noise-to-Signal Ratio Estimation by Machine Learning. In Proceedings of the Optical Fiber Communication Conference (OFC), San Diego, CA, USA, 3–7 March 2019.

54. Schaedler, M.; Kuschnerov, M.; Pachnicke, S.; Bluemm, C.; Pittala, F.; Changsong, X. Subcarrier Power Loading for Coherent Optical OFDM Optimized by Machine Learning. In Proceedings of the Optical Fiber Communication Conference (OFC), San Diego, CA, USA, 3–7 March 2019.
55. Wang, C.; Fu, S.; Xiao, Z.; Tang, M.; Liu, D. Long Short-Term Memory Neural Network (LSTM-NN) Enabled Accurate Optical Signal-to-Noise Ratio (OSNR) Monitoring. *J. Light. Technol.* **2019**, *37*, 4140–4146. [CrossRef]
56. Jones, R.; Eriksson, T.; Yankov, M.; Zibar, D. Deep Learning of Geometric Constellation Shaping Including Fiber Nonlinearities. In Proceedings of the 2018 European Conference on Optical Communication (ECOC), Rome, Italy, 23–27 September 2018; pp. 1–3.
57. Jones, R.; Yankov, M.; Zibar, D. End-to-end learning for GMI optimized geometric constellation shape. In Proceedings of the 2018 European Conference on Optical Communication (ECOC), Dublin, Ireland, 22–26 September 2019; pp. 1–3.
58. Khan, F.N.; Zhong, K.; Zhou, X.; Al-Arashi, W.H.; Yu, C.; Lu, C.; Lau, A.P.T. Joint OSNR Monitoring and Modulation Format Identification in Digital Coherent Receivers Using Deep Neural Networks. *Opt. Express* **2017**, *25*, 17767–17776. [CrossRef] [PubMed]
59. Tanimura, T.; Hoshida, T.; Kato, T.; Watanabe, S.; Morikawa, H. Convolutional Neural Network-Based Optical Performance Monitoring for Optical Transport Networks. *J. Opt. Commun. Netw.* **2019**, *11*, A52–A59. [CrossRef]
60. Wong, E.; Mondal, S.; Ruan, L. Machine Learning Enhanced Next-Generation Optical Access Networks—Challenges and Emerging Solutions [Invited Tutorial]. *J. Opt. Commun. Netw.* **2023**, *14*, A49–A62. [CrossRef]
61. Rajbhandari, S.; Ghassemlooy, Z.; Angelova, M. Effective denoising and adaptive equalization of indoor optical wireless channel with artificial light using the discrete wavelet transform and artificial neural network. *J. Light. Technol.* **2009**, *27*, 4493–4500. [CrossRef]
62. Haigh, P.A.; Ghassemlooy, Z.; Rajbhandari, S.; Papakonstantinou, I.; Popoola, W. Visible light communications: 170 Mb/s using an artificial neural network equalizer in a low bandwidth white light configuration. *J. Light. Technol.* **2014**, *32*, 1807–1813. [CrossRef]
63. Sorokina, M.; Sergeyev, S.; Turitsyn, S. Fiber echo state network analogue for high-bandwidth dual-quadrature signal processing. *Opt. Express* **2019**, *27*, 2387–2395. [CrossRef] [PubMed]
64. Wu, Q.; Zhu, Y.; Li, X.; Jiang, H.; Cheng, C.; Fu, M.; Zhang, Y.; Zhuge, Q.; Li, Z.; Hu, W. Four-dimensional direct detection receiver enabling Jones-space field recovery with phase and polarization diversity. *Photon. Res.* **2024**, *12*, 399–410. [CrossRef]
65. Chen, G.; Du, J.; Sun, L.; Zhang, W.; Xu, K.; Chen, X.; Reed, G.T.; He, Z. Nonlinear Distortion Mitigation by Machine Learning of SVM Classification for PAM-4 andPAM-8 Modulated Optical Interconnection. *J. Light. Technol.* **2018**, *3*, 650–657. [CrossRef]
66. Chen, G.; Du, J.; Sun, L.; Zheng, L.; Xu, K.; Tsang, H.K.; Chen, X.; Reed, G.T.; He, Z. Machine Learning Adaptive Receiver for PAM-4 Modulated Optical Interconnection Based on Silicon Microring Modulator. *J. Light. Technol.* **2018**, *18*, 4106–4113. [CrossRef]
67. Xie, Y.; Wang, Y.; Kandeepan, S.; Wang, K. Machine Learning Applications for Short Reach Optical Communication. *Photonics* **2022**, *9*, 30. [CrossRef]
68. Ranzini, S.M.; Da Ros, F.; Bülow, H.; Zibar, D. Tunable Optoelectronic Chromatic Dispersion Compensation Based on Machine Learning for Short-Reach Transmission. *Appl. Sci.* **2019**, *9*, 4332. [CrossRef]
69. Chagnon, M.; Karanov, B.; Schmalen, L. Experimental Demonstration of a Dispersion Tolerant End-to-End Deep Learning-Based IM-DD Transmission System. In Proceedings of the European Conference on Optical Communication (ECOC), Roma, Italy, 23–27 September 2018; pp. 1–3.
70. Younus, O.I.; Hassan, N.B.; Ghassemlooy, Z.; Zvanovec, S.; Alves, L.N.; Le-Minh, H. The Utilization of Artificial Neural Network Equalizer in Optical Camera Communications. *Sensors* **2021**, *21*, 2826. [CrossRef]
71. Jin, R.; Wei, Y.; Zhang, J.; Shi, J.; Chi, N. Neural-Network-Based End-To-End Learning for Adaptive Optimization of Two-Dimensional Signal Generation in UVLC Systems. *Opt. Express* **2024**, *32*, 6309–6328. [CrossRef]
72. Jarajreh, M.A.; Giacoumidis, E.; Aldaya, I.; Le, S.T.; Tsokanos, A.; Ghassemlooy, Z.; Doran, N.J. Artificial Neural Network Nonlinear Equalizer for Coherent Optical OFDM. *IEEE Photon. Technol. Lett.* **2015**, *27*, 387–390. [CrossRef]
73. Wu, Q.; Xu, Z.; Zhu, Y.; Zhang, Y.; Ji, H.; Yang, Y.; Qiao, G.; Liu, L.; Wang, S.; Liang, J.; et al. Machine Learning for Self-Coherent Detection Short-Reach Optical Communications. *Photonics* **2023**, *10*, 1001. [CrossRef]
74. Fang, X.; Chen, X.; Yang, F.; Zhang, L.; Zhang, F. 6.4Tb/s SSB WDM Transmission over 320km SSMF with Linear Network Assisted LSTM. *IEEE Photon. Technol. Lett.* **2021**, *33*, 1407–1410. [CrossRef]
75. Orsuti, D.; Antonelli, C.; Chiuso, A.; Santagiustina, M.; Mecozzi, A.; Galtarossa, A.; Palmieri, L. Deep Learning-Based Phase Retrieval Scheme for Minimum-Phase Signal Recovery. *J. Light. Technol.* **2023**, *41*, 578–592. [CrossRef]
76. An, S.; Zhu, Q.; Li, J.; Ling, Y.; Su, Y. 112-Gb/s SSB 16-QAM signal transmission over 120-km SMF with direct detection using a MIMO-ANN nonlinear equalizer. *Opt. Express* **2019**, *27*, 12794–12805. [CrossRef] [PubMed]
77. Sozos, K.; Deligiannidis, S.; Mesaritakis, C.; Bogris, A. Self-Coherent Receiver Based on a Recurrent Optical Spectrum Slicing Neuromorphic Accelerator. *J. Light. Technol.* **2023**, *41*, 2666–2674. [CrossRef]
78. Chen, Z.; Wang, W.; Zou, D.; Ni, W.; Luo, D.; Li, F. Real-Valued Neural Network Nonlinear Equalization for Long-Reach PONs Based on SSB Modulation. *IEEE Photon. Technol. Lett.* **2022**, *35*, 167–170. [CrossRef]
79. Karanov, B.; Chagnon, M.; Thouin, F.; Eriksson, T.; Bülow, H.; Lavery, D.; Bayvel, P.; Schmalen, L. End-to-End Deep Learning of Optical Fiber Communications. *J. Light. Technol.* **2018**, *36*, 4843–4855. [CrossRef]
80. Karanov, B.; Lavery, D.; Bayvel, P.; Schmalen, L. End-To-End Optimized Transmission over Dispersive Intensity-Modulated Channels Using Bidirectional Recurrent Neural Networks. *Opt. Express* **2019**, *27*, 19650–19663. [CrossRef]

81. Srinivasan, M.; Song, J.; Grabowski, A.; Szczerba, K.; Iversen, H.; Schmidt, M.; Zibar, D.; Schroder, J.; Larsson, A.; Hager, C.; et al. End-to-End Learning for VCSEL-based Optical Interconnects: State-of-the-Art, Challenges, and Opportunities. *J. Light. Technol.* **2023**, *41*, 3261–3277. [CrossRef]
82. Nevin, J.; Nallaperuma, S.; Shevchenko, N.; Li, X.; Faruk, M.; Savory, S. Machine Learning for Optical Fiber Communication Systems: An Introduction and Overview. *APL Photon.* **2021**, *6*, 121101. [CrossRef]
83. Ros, F.; Ranzini, S.; Dischler, R.; Cem, A.; Aref, V.; Bulow, H.; Zibar, D. Machine-Learning-Based Equalization for Short-Reach Transmission: Neural Networks and Reservoir Computing. In Proceedings of the SPIE, Metro and Data Center Optical Networks and Short-Reach Links IV, San Francisco, CA, USA, 6–12 March 2021; p. 1171205.
84. Kalla, S.; Gagne, C.; Zeng, M.; Rusch, L. Recurrent Neural Networks Achieving MLSE Performance for Optical Channel Equalization. *Opt. Express* **2021**, *29*, 13033–13047. [CrossRef] [PubMed]
85. Schädler, M.; Böcherer, G.; Pachnicke, S. Soft-Demapping for Short Reach Optical Communication: A Comparison of Deep Neural Networks and Volterra Series. *J. Light. Technol.* **2021**, *39*, 3095–3105. [CrossRef]
86. Huang, L.; Jiang, W.; Xu, Y.; Hu, W.; Yi, L. Low-Complexity Volterra-Inspired Neural Network Equalizer in 100-G Band-Limited IMDD PON System. *Opt. Lett.* **2022**, *47*, 5692–5695. [CrossRef]
87. Ahmad, S.T.; Kumar, K.P. Radial basis function neural network nonlinear equalizer for 16-QAM coherent optical OFDM. *IEEE Photonics Technol. Lett.* **2016**, *28*, 2507–2510. [CrossRef]
88. Xu, Z.; Shieh, W. On the Computational Complexity of Artificial Neural Networks for Short-Reach Optical Communication. In Proceedings of the Opto-Electronics and Communications Conference (OECC), Shanghai, China, 2–6 July 2023; pp. 1–4.
89. Xu, Z.; Sun, C.; Manton, J.; Shieh, W. Computational Complexity Analysis of Neural Network-Based Nonlinear Equalization for Short Reach Direct Detection Systems. In Proceedings of the Asia Communications and Photonics Conference (ACP), Chengdu, China, 2–5 November 2019.
90. Xu, Z.; Sun, C.; Manton, J.; Shieh, W. Performance and Computational Complexity Evaluation for Neural Network-Based Short-Reach Optical Links. In Proceedings of the Conference on Lasers and Electro-Optics Pacific Rim (CLEO-PR), Sydney, Australia, 2–6 August 2020.
91. Perin, J.K.; Shastri, A.; Kahn, J.M. Design of Low-Power DSP-Free Coherent Receivers for Data Center Links. *J. Light. Technol.* **2017**, *35*, 4650–4662. [CrossRef]
92. Xu, Z.; Sun, C.; Ji, T.; Manton, J.; Shieh, W. Computational Complexity Comparison of Feedforward/Radial Basis Function/Recurrent Neural Network-Based Equalizer for a 50-Gb/s PAM4 Direct-Detection Optical Link. *Opt. Express* **2019**, *27*, 36953–36964. [CrossRef]
93. Xu, Z.; Sun, C.; Ji, T.; Dong, S.; Zhou, X.; Shieh, W. Investigation on the Impact of Additional Connections to Feedforward Neural Networks for Equalization in PAM4 Short-Reach Direct Detection Links. In Proceedings of the Asia Communications and Photonics Conference (ACP), Beijing, China, 24–27 October 2020.
94. Zhang, L.; Zhu, Y.; Zhang, F.; Chen, Z. Trends and Enabling Technologies on High-Speed Metro and Short Reach Optical Transmission Systems. *Telecommun. Sci.* **2019**, *35*, 62–73.
95. Estarán, J.; Müller, R.; Mestre, M.; Jorge, F.; Mardoyan, H.; Konczykowska, A.; Dupuy, J.; Bigo, S. Artificial Neural Networks for Linear and Non-linear Impairment Mitigation in High-Baudrate IMDD Systems. In Proceedings of the 2016 European Conference on Optical Communication (ECOC), Düsseldorf, Germany, 18–22 September 2016.
96. Chuang, C.; Wei, C.; Lin, T.; Chi, K.; Liu, L.; Shi, J.; Chen, Y.; Chen, J. Employing Deep Neural Network for High Speed 4-PAM Optical Interconnect. In Proceedings of the 2017 European Conference on Optical Communication (ECOC), Gothenburg, Sweden, 17–21 September 2017.
97. Ye, C.; Zhang, D.; Huang, X.; Feng, H.; Zhang, K. Demonstration of 50Gbps IM/DD PAM4 PON over 10GHz Class Optics Using Neural Network Based Nonlinear Equalization. In Proceedings of the 2017 European Conference on Optical Communication (ECOC), Gothenburg, Sweden, 17–21 September 2017.
98. Reza, A.; Rhee, J. Nonlinear Equalizer Based on Neural Networks for PAM-4 Signal Transmission Using DML. *IEEE Photon. Technol. Lett.* **2018**, *30*, 1416–1419. [CrossRef]
99. Reza, A.; Costas, M.; Browning, C.; Otero, F.; Barry, L. Single-Lane 54-Gbit/s PAM-4/8 Signal Transmissions Using 10G-Class Directly Modulated Lasers Enabled by Low-Complexity Nonlinear Digital Equalization. *IEEE Phonton. J.* **2022**, *14*, 7226909. [CrossRef]
100. Luo, M.; Gao, F.; Li, X.; He, Z.; Fu, S. Transmission of 4×50-Gb/s PAM-4 Signal over 80-km Single Mode Fiber Using Neural Network. In Proceedings of the Optical Fiber Communication Conference (OFC), San Diego, CA, USA, 11–15 March 2018.
101. Zhang, Q.; Jiang, Y.; Zhou, H.; Deng, C.; Duan, S.; Wang, Z.; Li, Y.; Song, Y.; Chen, J.; Zhang, J.; et al. 137 Gb/s PAM-4 Transmissions at 850 nm over 40 cm Optical Backplane with 25 G Devices with Improved Neural Network-Based Equalization. *Appl. Sci.* **2019**, *9*, 5095. [CrossRef]
102. Ge, L.; Zhang, W.; Zhang, Y.; Liang, C.; Du, J.; He, Z. Compressed Nonlinear Equalizers for Optical Interconnects: Efficiency and Stability. In Proceedings of the Optical Fiber Communication Conference (OFC), San Diego, CA, USA, 8–12 March 2020.
103. Yi, L.; Liao, T.; Xue, L.; Hu, W. Neural Network-Based Equalization in High-Speed PONs. In Proceedings of the Optical Fiber Communication Conference (OFC), San Diego, CA, USA, 8–12 March 2020.

104. Kaneda, N.; Zhu, Z.; Chuang, C.; Mahadevan, A.; Farah, B.; Bergman, K.; Veen, D.; Houtsma, V. FPGA Implementation of Deep Neural Network Based Equalizers for High-Speed PON. In Proceedings of the Optical Fiber Communication Conference (OFC), San Diego, CA, USA, 8–12 March 2020.
105. Zhang, L.; Yang, F.; Ming, H.; Zhu, Y.; Ruan, X.; Li, Y.; Zhang, F. 160-Gb/s Nyquist PAM-4 Transmission with GeSi-EAM Using Artificial Neural Network Based Nonlinear Equalization. In Proceedings of the Optical Fiber Communication Conference (OFC), San Diego, CA, USA, 8–12 March 2020.
106. Chen, H.; Liu, X.; Fan, Z.; Hu, C.; Zhao, J. Dual-Learning Based Neural Networks for Short-reach Optical Communications. In Proceedings of the Conference on Lasers and Electro-Optics (CLEO), San Jose, CA, USA, 9–14 May 2021.
107. Chuang, C.; Liu, L.; Wei, C.; Liu, J.; Henrickson, L.; Huang, W.; Wang, C.; Chen, Y.; Chen, J. Convolutional Neural Network based Nonlinear Classifier for 112-Gbps High Speed Optical Link. In Proceedings of the Optical Fiber Communications Conference (OFC), San Diego, CA, USA, 11–15 March 2018.
108. Li, P.; Yi, L.; Xue, L.; Hu, W. 56 Gbps IM/DD PON based on 10G-Class Optical Devices with 29 dB Loss Budget Enabled by Machine Learning. In Proceedings of the Optical Fiber Communications Conference (OFC), San Diego, CA, USA, 11–15 March 2018.
109. Yi, L.; Li, P.; Liao, T.; Hu, W. 100Gb/s/λ IM-DD PON Using 20G-class Optical Devices by Machine Learning Based Equalization. In Proceedings of the European Conference and Exhibition on Optical Communication (ECOC), Rome, Italy, 23–27 September 2018.
110. Li, P.; Yi, L.; Xue, L.; Hu, W. 100Gbps IM/DD Transmission over 25km SSMF using 20Gclass DML and PIN Enabled by Machine Learning. In Proceedings of the Optical Fiber Communications Conference (OFC), San Diego, CA, USA, 11–15 March 2018.
111. Zhang, J.; Yan, L.; Jiang, L.; Yi, A.; Pan, Y.; Pan, W.; Luo, B. 56-Gbit/s PAM-4 Optical Signal Transmission over 100-km SMF Enabled by TCNN Regression Model. *IEEE Photon. J.* **2021**, *13*, 7200606. [CrossRef]
112. Ye, C.; Zhang, D.; Hu, X.; Huang, X.; Feng, H.; Zhang, K. Recurrent Neural Network (RNN) Based End-to-End Nonlinear Management for Symmetrical 50Gbps NRZ PON with 29dB+ Loss Budget. In Proceedings of the European Conference and Exhibition on Optical Communication (ECOC), Rome, Italy, 23–27 September 2018.
113. Huang, X.; Zhang, D.; Hu, X.; Ye, C.; Zhang, K. Recurrent Neural Network Based Equalizer with Embedded Parallelization for 100Gbps/λ PON. In Proceedings of the Optical Fiber Communication Conference (OFC), San Francisco, CA, USA, 6–10 June 2021.
114. Huang, X.; Zhang, D.; Hu, X.; Ye, C.; Zhang, K. Low-Complexity Recurrent Neural Network Based Equalizer with Embedded Parallelization for 100-Gbit/s/λ PON. *J. Light. Technol.* **2022**, *40*, 1353–1359. [CrossRef]
115. Zhou, Q.; Yang, C.; Liang, A.; Zheng, X.; Chen, Z. Low Computationally Complex Recurrent Neural Network for High Speed Optical Fiber Transmission. *Opt. Commun.* **2019**, *441*, 121–126. [CrossRef]
116. Yang, C.; Gao, Y.; Wang, J.; Li, H.; Hasnain, C. Enhanced Recurrent Neural Network Equalization Based on Hidden Feature Extraction Learning for Optical Interconnect. In Proceedings of the Optical Fiber Communication Conference (OFC), San Diego, CA, USA, 24–28 March 2024.
117. Dai, X.; Li, X.; Luo, M.; You, Q.; Yu, S. LSTM Networks Enabled Nonlinear Equalization in 50-Gb/s PAM4 Transmission Links. *Appl. Opt.* **2019**, *58*, 6079–6084. [CrossRef] [PubMed]
118. Peng, C.; Chan, D.; Tong, Y.; Chow, C.; Liu, Y.; Yeh, C.; Tsang, H. Long Short-Term Memory Neural Network for Mitigating Transmission Impairments of 160 Gbit/s PAM4 Microring Modulation. In Proceedings of the Optical Fiber Communication Conference (OFC), San Francisco, CA, USA, 6–10 June 2021.
119. Peng, C.; Chan, D.; Chow, C.; Hung, T.; Jian, Y.; Tong, Y.; Kuo, P.; Chen, G.; Liu, Y.; Yeh, C.; et al. Long Short Term Memory Neural Network (LSTMNN) and Inter-Symbol Feature Extraction for 160 Gbit/s PAM4 from Silicon Micro-Ring Transmitter. *Opt. Commun.* **2023**, *529*, 129067. [CrossRef]
120. Hung, T.; Chan, D.; Peng, C.; Chow, C.; Yeh, C.; Tsang, H. 300-Gbit/s/λ PAM8 Modulation with a Silicon Microring Modulator using Long Short Term Memory Regression and Deep Neural Network Classification. In Proceedings of the Optical Fiber Communication Conference (OFC), San Diego, CA, USA, 24–28 March 2024.
121. Sang, B.; Zhang, J.; Wang, C.; Kong, M.; Tan, Y.; Zhao, L.; Zhou, W.; Shang, D.; Zhu, Y.; Yi, H.; et al. Multi-Symbol Output Long Short-Term Memory Neural Network Equalizer for 200+ Gbps IM/DD System. In Proceedings of the European Conference and Exhibition on Optical Communication (ECOC), Bordeaux, France, 13–16 September 2021.
122. Sang, B.; Zhou, W.; Tan, Y.; Kong, M.; Wang, C.; Wang, M.; Zhao, L.; Zhang, J.; Yu, J. Low Complexity Neural Network Equalization Based on Multi-Symbol Output Technique for 200+ Gbps IM/DD Short Reach Optical System. *J. Light. Technol.* **2022**, *40*, 2890–2900. [CrossRef]
123. Murphy, S.; Townsend, P.; Antony, C. Recurrent Neural Network Equalizer to Extend Input Power Dynamic Range of SOA in 100Gb/s/λ PON. In Proceedings of the Conference on Lasers and Electro-Optics (CLEO), San Jose, CA, USA, 15–20 May 2022.
124. Murphy, S.; Jamali, F.; Townsend, P.; Antony, C. High Dynamic Range 100 Gbit/s PAM4 PON with SOA Preamplifier using Gated Recurrent Neural Network Equaliser. In Proceedings of the European Conference and Exhibition on Optical Communication (ECOC), Basel, Switzerland, 18–22 September 2022.
125. Xu, Z.; Sun, C.; Ji, T.; Ji, H.; Manton, J.; Shieh, W. Cascade Recurrent Neural Network Enabled 100-Gb/s PAM4 Short-Reach Optical Link Based on DML. In Proceedings of the Optical Fiber Communication Conference (OFC), San Diego, CA, USA, 8–12 March 2020.
126. Xu, Z.; Sun, C.; Ji, T.; Manton, J.; Shieh, W. Cascade Recurrent Neural Network Assisted Nonlinear Equalization for a 100 Gb/s PAM4 Short-Reach Direct Detection System. *Opt. Lett.* **2020**, *45*, 4216–4219. [CrossRef] [PubMed]

127. Murphy, S.; Jamali, F.; Townsend, P.; Antony, C. Adaptive Neural Network Equalisation Using Skip Connections for Future 100 Gbit/s/λ Passive Optical Networks. In Proceedings of the 2023 European Conference on Optical Communication (ECOC), Glasgow, Scotland, 1–5 October 2023.
128. Yang, Z.; Gao, F.; Fu, S.; Li, X.; Deng, L.; He, Z.; Tang, M.; Liu, D. Radial Basis Function Neural Network Enabled C-band 4 × 50 Gb/s PAM-4 Transmission over 80 km SSMF. *Opt. Lett.* **2018**, *15*, 3542–3545. [CrossRef]
129. Arnold, E.; Bocherer, G.; Muller, E.; Spilger, P.; Schemmel, J.; Calabro, S.; Kuschnerov, M. Spiking Neural Network Equalization on Neuromorphic Hardware for IM/DD Optical Communication. In Proceedings of the European Conference and Exhibition on Optical Communication (ECOC), Basel, Switzerland, 18–22 September 2022.
130. Arnold, E.; Bocherer, G.; Strasser, F.; Muller, E.; Spilger, P.; Billaudelle, S.; Weis, J.; Schemmel, J.; Calabro, S.; Kuschnerov, M. Spiking Neural Network Nonlinear Demapping on Neuromorphic Hardware for IM/DD Optical Communication. *J. Light. Technol.* **2023**, *41*, 3424–3431. [CrossRef]
131. Bocherer, G.; Strasser, F.; Arnold, E.; Lin, Y.; Schemmel, J.; Calabro, S.; Kuschnerov, M. Spiking Neural Network Linear Equalization: Experimental Demonstration of 2km 100Gb/s IM/DD PAM4 Optical Transmission. In Proceedings of the Optical Fiber Communication Conference (OFC), San Diego, CA, USA, 5–9 March 2023.
132. Bocherer, G. Spiking Neural Network Equalization: Towards Evaluation of Throughput and Power Consumption. In Proceedings of the 2023 European Conference on Optical Communication (ECOC), Glasgow, Scotland, 1–5 October 2023.
133. Bank, A.; Edelmann, E.; Schmalen, S. Energy-Efficient Spiking Neural Network Equalization for IM/DD Systems with Optimized Neural Encoding. In Proceedings of the Optical Fiber Communication Conference (OFC), San Diego, CA, USA, 24–28 March 2024.
134. Taniguchi, H.; Yamamoto, S.; Masuda, A.; Kisaka, Y.; Kanazawa, S. 800-Gbps PAM-4 2-km Transmission using 4-λ LAN-WDM TOSA with MLSE based on Deep Neural Network. In Proceedings of the Optical Fiber Communication Conference (OFC), San Diego, CA, USA, 6–10 March 2022.
135. Yan, A.; Xing, S.; Li, G.; Li, Z.; Luo, P.; Sun, A.; Shi, J.; Zhang, H.; Xiao, X.; He, Z.; et al. 140-Gbaud PAM-8 IM/DD Transmission and FTN Signal Processing based on Low-Complexity Nonlinear M-BCJR Equalization with Deep Neural-Network Channel Model. In Proceedings of the 2023 European Conference on Optical Communication (ECOC), Glasgow, Scotland, 1–5 October 2023.
136. Bluemm, C.; Liu, B.; Li, B.; Rahman, T.; Hossain, M.; Schaedler, M.; Schlichtmann, U.; Kuschnerov, M.; Calabrò, S. 800Gb/s PAM4 Transmission over 10km SSMF Enabled by Low-complex Duobinary Neural Network Equalization. In Proceedings of the European Conference and Exhibition on Optical Communication (ECOC), Basel, Switzerland, 18–22 September 2022.
137. Bluemm, C.; Liu, B.; Li, B.; Rahman, T.; Hossain, M.; Schaedler, M.; Schlichtmann, U.; Kuschnerov, M.; Calabrò, S. Hardware-Efficient Duobinary Neural Network Equalizers for 800 Gb/s IM/DD PAM4 Transmission Over 10 km SSMF. *J. Light. Technol.* **2023**, *41*, 3783–3790. [CrossRef]
138. Yan, A.; Xing, S.; Li, Z.; Li, G.; Sun, A.; Luo, P.; Shi, J.; Zhang, H.; Wang, D.; Xiao, X.; et al. Adaptive Partial-Response Neural Network Equalization for Bandwidth-limited PAM Transmission in Intra-Datacenter Interconnect. *J. Light. Technol.* **2024**, *42*, 2762–2773. [CrossRef]
139. Eriksson, T.; Bülow, H.; Leven, A. Applying Neural Networks in Optical Communication Systems: Possible Pitfalls. *IEEE Photon. Technol. Lett.* **2017**, *29*, 2091–2094. [CrossRef]
140. Shu, L.; Li, J.; Wan, Z.; Zhang, W.; Fu, S.; Xu, K. Overestimation Trap of Artificial Neural Network: Learning the Rule of PRBS. In Proceedings of the European Conference and Exhibition on Optical Communication (ECOC), Rome, Italy, 23–27 September 2018.
141. Chuang, C.; Liu, L.; Wei, C.; Liu, J.; Henrickson, L.; Wang, C.; Chen, Y.; Chen, J. Study of Training Patterns for Employing Deep Neural Networks in Optical Communication Systems. In Proceedings of the European Conference and Exhibition on Optical Communication (ECOC), Rome, Italy, 23–27 September 2018.
142. Liao, T.; Xue, L.; Huang, L.; Hu, W.; Yi, L. Training Data Generation and Validation for a Neural Network-Based Equalizer. *Opt. Lett.* **2020**, *45*, 5113–5116. [CrossRef] [PubMed]
143. Kim, J.; Kim, H. PRBS Orders Required to Train ANN Equalizer for PAM Signal without Overfitting. *Opt. Express* **2022**, *30*, 25486–25497. [CrossRef]
144. Pan, S.; Yang, Q. A Survey on Transfer Learning. *IEEE Trans. Knowl. Data Eng.* **2010**, *22*, 1345–1359. [CrossRef]
145. Xia, L.; Zhang, J.; Hu, S.; Zhu, M.; Song, Y.; Qiu, K. Transfer Learning Assisted Deep Neural Network for OSNR Estimation. *Opt. Express* **2019**, *27*, 19398–19406. [CrossRef]
146. Azzimonti, D.; Rottondi, C.; Giusti, A.; Tornatore, M.; Bianco, A. Active vs Transfer Learning Approaches for QoT Estimation with Small Training Datasets. In Proceedings of the Optical Fiber Communication Conference (OFC), San Diego, CA, USA, 8–12 March 2020.
147. Wang, D.; Xu, Y.; Li, J.; Zhang, M.; Li, J.; Qin, J.; Ju, C.; Zhang, Z.; Chen, X. Comprehensive Eye Diagram Analysis: A Transfer Learning Approach. *IEEE Phonton. J.* **2019**, *11*, 7205019. [CrossRef]
148. Zhang, J.; Xia, L.; Zhu, M.; Hu, S.; Xu, B.; Qiu, K. Fast Remodeling for Nonlinear Distortion Mitigation Based on Transfer Learning. *Opt. Lett.* **2019**, *44*, 4243–4246. [CrossRef]
149. Zhang, J.; Xu, T.; Jin, T.; Jiang, W.; Hu, S.; Huang, X.; Xu, B.; Yu, Z.; Yi, X.; Qiu, K. Meta-Learning Assisted Source Domain Optimization for Transfer Learning Based Optical Fiber Nonlinear Equalization. *J. Light. Technol.* **2023**, *41*, 1269–1277. [CrossRef]
150. Freire, P.; Abode, D.; Prilepsky, J.; Costa, N.; Spinnler, B.; Napoli, A.; Turitsyn, S. Transfer Learning for Neural Networks-Based Equalizers in Coherent Optical Systems. *J. Light. Technol.* **2021**, *39*, 6733–6745. [CrossRef]

151. Freire, P.; Spinnler, B.; Abode, D.; Prilepsky, J.; Ali, A.; Costa, N.; Schairer, W.; Napoli, A.; Ellis, A.; Turitsyn, S. Domain Adaptation: The Key Enabler of Neural Network Equalizers in Coherent Optical Systems. In Proceedings of the Optical Fiber Communication Conference (OFC), San Diego, CA, USA, 6–10 March 2022.
152. Xu, Z.; Sun, C.; Ji, T.; Ji, H.; Manton, J.; Shieh, W. Transfer Learning Aided Neural Networks for Nonlinear Equalization in Short-Reach Direct Detection Systems. In Proceedings of the Optical Fiber Communication Conference (OFC), San Diego, CA, USA, 8–12 March 2020.
153. Xu, Z.; Sun, C.; Ji, T.; Manton, J.; Shieh, W. Feedforward and Recurrent Neural Network-Based Transfer Learning for Nonlinear Equalization in Short-Reach Optical Links. *J. Light. Technol.* **2021**, *39*, 475–480. [CrossRef]
154. Liu, X.; Li, J.; Fan, Z.; Zhao, J. Transfer-Learning Based Convolutional Neural Network for Short-Reach Optical Interconnects. In Proceedings of the Asia Communications and Photonics Conference (ACP), Shanghai, China, 24–27 October 2021.
155. Xiao, J.; Sun, L.; Liu, C.; Mao, B.; Liu, G. Fastly Converged Transfer Learning Using Neuron-Pruning Nonlinear Equalizer for Intra Data Center Networking. In Proceedings of the Opto-Electronics and Communications Conference (OECC), Toyama, Japan, 3–6 July 2022.
156. Xiao, J.; Sun, L.; Liu, C.; Liu, G. Optimizations and Investigations for Transfer Learning of Iteratively Pruned Neural Network Equalizers for Data Center Networking. *Opt. Express* **2022**, *30*, 36358–36367. [CrossRef]
157. Huang, W.; Chang, W.; Wei, C.; Liu, J.; Chen, Y.; Chi, K.; Wang, C.; Shi, J.; Chen, J. 93% Complexity Reduction of Volterra Nonlinear Equalizer by L1-Regularization for 112-Gbps PAM-4 850-nm VCSEL Optical Interconnect. In Proceedings of the Optical Fiber Communications Conference (OFC), San Diego, CA, USA, 11–15 March 2018.
158. Ge, L.; Zhang, W.; Liang, C.; He, Z. Threshold based Pruned Retraining Volterra Equalization for PAM-4 100 Gbps VCSEL and MMF Based Optical Interconnects. In Proceedings of the Asia Communications and Photonics Conference (ACP), Hangzhou, China, 26–29 October 2018.
159. Chuang, C.; Chang, W.; Wei, C.; Ho, C.; Huang, C.; Shi, J.; Henrickson, L.; Chen, Y.; Chen, J. Sparse Volterra Nonlinear Equalizer by Employing Pruning Algorithm for High-Speed PAM-4 850-nm VCSEL Optical Interconnect. In Proceedings of the Optical Fiber Communications Conference (OFC), San Diego, CA, USA, 3–7 March 2019.
160. Yadav, G.; Chuang, C.; Feng, K.; Yan, J.; Chen, J.; Chen, Y. Reducing Computation Complexity by Using Elastic Net Regularization Based Pruned Volterra Equalization in a 80 Gbps PAM-4 Signal for Inter-Data Center Interconnects. *Opt. Express* **2020**, *28*, 38539–38552. [CrossRef] [PubMed]
161. Fujisawa, S.; Yaman, F.; Batshon, H.; Tanio, M.; Ishii, N.; Huang, C.; Lima, T.; Inada, Y.; Prucnal, P.; Kamiya, N.; et al. Weight Pruning Techniques Towards Photonic Implementation of Nonlinear Impairment Compensation Using Neural Networks. *J. Light. Technol.* **2022**, *40*, 1273–1282. [CrossRef]
162. Wan, Z.; Li, J.; Shu, L.; Luo, M.; Li, X.; Fu, S.; Xu, K. Nonlinear Equalization Based on Pruned Artificial Neural Networks for 112-Gb/s SSB-PAM4 Transmission over 80-km SSMF. *Opt. Express* **2018**, *8*, 10631–10642. [CrossRef] [PubMed]
163. Ge, L.; Zhang, W.; Liang, C.; He, Z. Compressed Neural Network Equalization Based on Iterative Pruning Algorithm for 112-Gbps VCSEL-Enabled Optical Interconnects. *J. Light. Technol.* **2020**, *38*, 1323–1329. [CrossRef]
164. Li, M.; Zhang, W.; Chen, Q.; He, Z. High-Throughput Hardware Deployment of Pruned Neural Network Based Nonlinear Equalization for 100-Gbps Short-Reach Optical Interconnect. *Opt. Lett.* **2021**, *46*, 4980–4983. [CrossRef]
165. Xu, Z.; Dong, S.; Jiang, C.; Manton, J.; Shieh, W. Sparsely-Connected Cascade Recurrent Neural Network-Based Nonlinear Equalizer for a 100-Gb/s PAM4 Optical Interconnect. In Proceedings of the Asia Communications and Photonics Conference (ACP), Shanghai, China, 24–27 October 2021.
166. Yadav, G.; Chuang, C.; Feng, K.; Chen, J.; Chen, Y. Computation Efficient Sparse DNN Nonlinear Equalization for IM/DD 112 Gbps PAM4 Inter-Data Center Optical Interconnects. *Opt. Lett.* **2021**, *46*, 1999–2002. [CrossRef]
167. Yadav, G.; Chuang, C.; Feng, K.; Chen, J.; Chen, Y. Sparsity Learning Deep Neural Network Nonlinear Equalization Method for 112Gbps PAM4 Short-Reach Transmission Links. *J. Light. Technol.* **2023**, *41*, 2333–2342. [CrossRef]
168. Kaneda, N.; Chuang, C.; Zhu, Z.; Mahadevan, A.; Farah, B.; Bergman, K.; Veen, D.; Houtsma, V. Fixed-Point Analysis and FPGA Implementation of Deep Neural Network Based Equalizers for High-Speed PON. *J. Light. Technol.* **2022**, *40*, 1972–1980. [CrossRef]
169. Xu, Z.; Sun, C.; Dong, S.; Manton, J.H.; Shieh, W. Towards Low Computational Complexity for Neural Network-Based Equalization in PAM4 Short-Reach Direct Detection Systems by Multi-Symbol Prediction. In Proceedings of the Optical Fiber Communication Conference (OFC), San Francisco, CA, USA, 6–10 June 2021.
170. Xu, Z.; Dong, S.; Manton, J.; Shieh, W. Low-Complexity Multi-Task Learning Aided Neural Networks for Equalization in Short-Reach Optical Interconnects. *J. Light. Technol.* **2022**, *40*, 45–54. [CrossRef]
171. Osadchuk, Y.; Jovanovic, O.; Zibar, D.; Ros, F. Reservoir Computing-based Multi-Symbol Equalization for PAM 4 Short-reach Transmission. In Proceedings of the Conference on Lasers and Electro-Optics (CLEO), San Jose, CA, USA, 7–12 May 2023.
172. Osadchuk, Y.; Jovanovic, O.; Zibar, D.; Ros, F. Multi-Symbol Reservoir Computing-Based Equalization for PAM-4 IM/DD Transmission. *IEEE Photon. Technol. Lett.* **2024**, *36*, 853–856. [CrossRef]
173. Xu, Z.; Dong, S.; Ji, H.; Manton, J.H.; Shieh, W. Computationally-Efficient Sparsely-Connected Multi-Output Neural Networks for IM/DD System Equalization. In Proceedings of the Optical Fiber Communication Conference (OFC), San Diego, CA, USA, 6–10 March 2022.
174. Liu, B.; Bluemm, C.; Calabro, S.; Li, B.; Schlichtmann, U. Area-Efficient Neural Network CD Equalizer for 4 × 200Gb/s PAM4 CWDM4 Systems. In Proceedings of the Optical Fiber Communication Conference (OFC), San Diego, CA, USA, 5–9 March 2023.

175. Liu, B.; Bluemm, C.; Calabro, S.; Li, B.; Schlichtmann, U. Area-Efficient Hardware Parallelization of Neural Network CD Equalizers for 4×200 Gb/s PAM4 CWDM4 Systems. In Proceedings of the 2023 European Conference on Optical Communication (ECOC), Glasgow, Scotland, 1–5 October 2023.
176. Ron, D.; Freire, P.; Prilepsky, J.; Kopae, M.; Napoli, A.; Turitsyn, K. Experimental Implementation of a Neural Network Optical Channel Equalizer in Restricted Hardware Using Pruning and Quantization. *Sci. Rep.* **2022**, *12*, 8713. [CrossRef] [PubMed]
177. Darweesh, J.; Costa, N.; Napoli, A.; Spinnler, B.; Jaouen, Y.; Yousefi, M. Few-bit Quantization of Neural Networks for Nonlinearity Mitigation in a Fiber Transmission Experiment. In Proceedings of the European Conference and Exhibition on Optical Communication (ECOC), Basel, Switzerland, 18–22 September 2022.
178. Freire, P.; Napoli, A.; Spinnler, B.; Anderson, M.; Ron, D.; Schairer, W.; Bex, T.; Costa, N.; Turitsyn, S.; Prilepsky, J. Reducing Computational Complexity of Neural Networks in Optical Channel Equalization: From Concepts to Implementation. *J. Light. Technol.* **2023**, *41*, 4557–4581. [CrossRef]
179. Akino, T.; Wang, Y.; Kojima, K.; Parsons, K.; Yoshida, T. Zero-Multiplier Sparse DNN Equalization for Fiber-Optic QAM Systems with Probabilistic Amplitude Shaping. In Proceedings of the European Conference and Exhibition on Optical Communication (ECOC), Bordeaux, France, 13–16 September 2021.
180. Xu, Z.; Ji, H.; Yang, Y.; Qiao, G.; Wu, Q.; Li, J.; Lu, W.; Liu, L.; Wang, S.; Wei, J.; et al. Mixed-Precision Integer-Arithmetic-Only Neural Network-Based Equalizers for DML-Based Short-Reach IM/DD Systems. In Proceedings of the 2023 European Conference on Optical Communication (ECOC), Glasgow, Scotland, 1–5 October 2023.
181. Xu, Z.; Ji, H.; Yang, Y.; Qiao, G.; Wu, Q.; Lu, W.; Liu, L.; Wang, S.; Liang, J.; Li, J.; et al. Quantization of Recurrent Neural Network for Low-Complexity High-Speed IM/DD System Equalization Based on Neuron Clustering. In Proceedings of the Asia Communications and Photonics Conference (ACP), Wuhan, China, 4–7 November 2023; pp. 1–4.
182. Xu, Z.; Wu, Q.; Lu, W.; Ji, H.; Chen, H.; Ji, T.; Yang, Y.; Qiao, G.; Tang, J.; Cheng, C.; et al. Weight-Adaptive Joint Mixed-Precision Quantization and Pruning for Neural Network-Based Equalization in Short-Reach Direct Detection Links. *Opt. Lett.* **2024**, *49*, 3500–3503. [CrossRef]

photonics MDPI

Review

Machine Learning in Short-Reach Optical Systems: A Comprehensive Survey

Chen Shao [1,*], Elias Giacoumidis [2], Syed Moktacim Billah [1], Shi Li [2], Jialei Li [2], Prashasti Sahu [3], André Richter [2], Michael Faerber [1] and Tobias Kaefer [1]

1 Department of Economics and Management, Karlsruhe Institute of Technology (KIT), 76131 Karlsruhe, Germany; ge59wah@mytum.de (S.M.B.); michael.faerber@kit.edu (M.F.); tobias.kaefer@kit.edu (T.K.)
2 VPIphotonics GmbH, Hallerstraße 6, 10587 Berlin, Germany; elias.giacoumidis@vpiphotonics.com (E.G.); shi.li@vpiphotonics.com (S.L.); jia.li9@kit.edu (J.L.); andre.richter@vpiphotonics.com (A.R.)
3 Electronic and Information Engineering, Technical University of Chemnitz, Str. der Nationen 62, 09111 Chemnitz, Germany; prashastisahu13@gmail.com
* Correspondence: chen.shao2@kit.edu

Abstract: Recently, extensive research has been conducted to explore the utilization of machine learning (ML) algorithms in various direct-detected and (self)-coherent short-reach communication applications. These applications encompass a wide range of tasks, including bandwidth request prediction, signal quality monitoring, fault detection, traffic prediction, and digital signal processing (DSP)-based equalization. As a versatile approach, ML demonstrates the ability to address stochastic phenomena in optical systems networks where deterministic methods may fall short. However, when it comes to DSP equalization algorithms such as feed-forward/decision-feedback equalizers (FFEs/DFEs) and Volterra-based nonlinear equalizers, their performance improvements are often marginal, and their complexity is prohibitively high, especially in cost-sensitive short-reach communications scenarios such as passive optical networks (PONs). Time-series ML models offer distinct advantages over frequency-domain models in specific contexts. They excel in capturing temporal dependencies, handling irregular or nonlinear patterns effectively, and accommodating variable time intervals. Within this survey, we outline the application of ML techniques in short-reach communications, specifically emphasizing their utilization in high-bandwidth demanding PONs. We introduce a novel taxonomy for time-series methods employed in ML signal processing, providing a structured classification framework. Our taxonomy categorizes current time-series methods into four distinct groups: traditional methods, Fourier convolution-based methods, transformer-based models, and time-series convolutional networks. Finally, we highlight prospective research directions within this rapidly evolving field and outline specific solutions to mitigate the complexity associated with hardware implementations. We aim to pave the way for more practical and efficient deployment of ML approaches in short-reach optical communication systems by addressing complexity concerns.

Keywords: machine learning; optical communications; passive optical network; equalization; optical performance monitoring; modulation format identification; bit-error ratio; optical signal-to-noise ratio; nonlinearities

1. Introduction

Short-reach optical transmission systems have gained substantial attraction owing to their remarkable attributes of high bandwidth and low latency [1]. In the evolving landscape of communication technologies, short-reach optical communication has emerged as an essential domain, driven by the increasing demand for high-speed data transfer in applications such as inter-data centers [2], access/local area networks, and industrial automation [3]. This increasing demand requires efficient, low-latency communication systems tailored to short-reach scenarios, typically up to 100 km. While long-haul, optical communication has been immersive in data transmission, its applicability encounters challenges when adapting to the constraints of shorter distances. This is mainly due to physical and technical limitations that prevent its seamless integration into existing networking environments characterized by the need for energy-efficient and cost-effective data transmission over limited distances. Passive optical networks (PONs) utilize passive optical splitters and combiners, which are less expensive than the active components required in traditional point-to-point fiber networks. This makes PONs a cost-effective fiber-optic solution.

Since PONs rely on passive optical splitters, they inherently introduce power losses, limiting the overall power budget and the number of users that can be supported on a single PON. In addition, effects caused by the fiber, such as chromatic dispersion (CD) and nonlinearity can limit the PON-reach [4], especially when intensity-modulated and direct-detected (IMDD) high baud-rate signals are considered [5].

Ongoing research endeavors are dedicated to advancing optical detection schemes to overcome these limitations and increase the signal bit rate in both short-reach and long-haul optical communication networks [6]. For instance, the regeneration of coherent optical systems in the last decade has been a major breakthrough, as they have gone beyond just using intensity-only modulation [7]. Coherent systems employ external modulators to employ complex baseband signals to the optical field. The optical coherent receiver, equipped with phase diversity, linearly recovers signals and compensates for fiber impairments through digital signal processing (DSP) [8]. Coherent technology enables the transmission of advanced modulation formats and polarization multiplexing to increase the signal bit rate significantly. Additionally, coherent optical systems enable dense wavelength division multiplexing (DWDM) and super-channels, which push long-distance optical networks into the multi-terabit per second capacity range [9].

Except for traditional homodyne-coherent technology, coherent communication strategies include diverse techniques, such as phase detection through heterodyne detection. While this approach has its merits [10], a notably favored incoherent approach such as IMDD is practically preferred due to its inherent simplicity and cost-effectiveness in short-reach communications [11,12].

In contrast to coherent transmission, IMDD operates by encoding information into the intensity of the optical signal, with the modulation signal being real-valued and positive [12]. The implementation of IMDD eliminates the need for complex optical components and local oscillators, reducing hardware complexity. Additionally, IMDD systems are less susceptible to phase noise and polarization-related issues, making them robust and practical for scenarios where cost efficiency and simplicity are paramount [12]. Furthermore, practical considerations like operation and safety can limit the highest and average values of the modulated signal in IMDD systems. These restrictions give IMDD systems specific characteristics in how they function [13]. Various models, such as the Poisson channel, square-root Gaussian channel, and Gaussian channel with input-dependent noise, among others, exist to rapidly assess and characterize IMDD systems [14–16]. In contrast to conventional methodologies that depend on analog components and processing [17], IMDD can potentially integrate machine learning (ML) algorithms at the receiver DSP if required [18], providing a flexible and adaptable solution for enhancing the transmission performance. According to [19], the combination of ML and DSP techniques allows IMDD systems to dynamically adapt and optimize signal parameters. This addresses impairments and variations in real time without needing complex hardware adjustments. This approach represents a significant benefit, as it not only reduces the costs associated with complex hardware setups in short-reach systems, but also highlights the effectiveness of intelligent signal processing [18,20,21].

In this survey, we examine the significant progress made in short-distance optical communications research over the past decade. First, we summarize several key research areas (Section 2). Afterwards, we focus on the equalization problem, introducing benchmark DSP methods (Section 3) and ML algorithms (Section 4). Then, we categorize recent sequence models in the ML field (Section 5), dividing them into convolution-based, transformer-based, and Fourier-based neural networks. We explore the advantages, disadvantages, and complexities of each method in addressing the equalization problem. In the final section, we provide an overview of the model compression field, outlining two approaches to compress models. We see these approaches as potential solutions for addressing hardware complexity concerns.

The primary contribution of this survey is to summarize the existing research on ML implementations for short-reach optical communications across a range of applications. Specifically, our contributions are the following:

1. We review existing deep learning (DL) models, providing a comprehensive understanding of their principles, characteristics, and hypothesis classes. This facilitates an in-depth exploration for researchers seeking supervised neural-network-based ML models suitable for their specific applications.
2. We highlight the features and complexities of these models, elucidating recent developments in the field of DL. This information is valuable for researchers interested in delving deeper into research and staying abreast of current advancements.
3. We discuss the current limitations and research gaps in the ongoing development of DL, addressing the challenges posed by these factors in real-world applications. Furthermore, we provide constructive insights regarding the selection of models and potential future directions.
4. Given the challenge of high hardware complexity, we introduce model compression as a potential solution from the DL field. We present existing works that employ this approach within the optical communication field, aiming to inspire more researchers to pursue research in this domain.

2. Applications in Short-Reach Systems

After systematically organizing recent literature in the past few years, we have categorized ML-based research for short-reach optical systems into four classes based on application tasks: Bandwidth Request and Prediction, Subcarrier Allocation, Equalization, and Fault Detection. We clarify the physical and mathematical aspects of their respective tasks, enumerate several recent works, and provide a summary of current advancements.

Bandwidth Request and Prediction: It aims to leverage network information to predict future bandwidth availability and enable its utilization by related applications. In mathematical terms, the real-time bandwidth forecast at a specific time (t) involves estimating the available bandwidth that will be accessible in the immediate future ($t + \tau$) [22]. One proposed method, known as predictive-dynamic bandwidth allocation (P-DBA), utilizes this concept to predict high-priority traffic during waiting periods, resulting in reduced latency and packet loss rates within a Gigabit PON (GPON) [22]. Another approach demonstrated in [23] leverages the k-nearest neighbor algorithm to predict additional bandwidth requirements for each optical network unit (ONU) in a PON. This adaptive learning-based approach dynamically adjusts the k value based on real-time traffic conditions, showcasing the adaptability of ML in optimizing bandwidth allocation [23]. Artificial neural networks (ANNs) have also shown promise in achieving flexible bandwidth allocations across various application scenarios, particularly emphasizing low-latency objectives [24,25]. For example, feed-forward-based ANNs, explored in [26], are utilized to predict packet arrivals in time-division multiple access (TDMA) ONUs, effectively reducing additional DBA processing delays [26]. Furthermore, Xgboost [27] is employed to predict bandwidth requests for ONUs in Ethernet PON (EPON), optimizing bandwidth utilization across polling periods. This study introduced a dynamic wavelength and bandwidth assignment scheme for time and WDM (TWDM) PONs, incorporating regression techniques for efficient resource allocation [28]. Recent studies show that ML approaches are versatile in addressing challenges related to predicting and managing bandwidth needs. This paves the way for developing more adaptive and efficient short-reach optical communication systems in the near future [22–26,28].

Subcarrier Allocation: The optimization of bandwidth allocation for enhanced spectral efficiency has led to increased interest in subcarrier allocation for PONs. This approach involves mathematically formulating the allocation problem as an integer linear programming (ILP) task, which includes tasks such as optimizing wavelength configurations, assigning subcarriers to transmitters, and minimizing lost traffic and energy costs. To address this challenge, deep reinforcement learning has emerged as a promising technique that enables dynamic subcarrier sharing among ONUs, facilitating efficient DBA. At the medium access control (MAC) layer, the dynamic subcarrier allocation (DSA) algorithm schedules ONU upstream transmissions by considering instantaneous bandwidth requirements and existing traffic conditions [29]. This showcases the adaptability of ML in resource scheduling. Several studies focus on algorithm-level cost reduction and two-dimensional resource scheduling for orthogonal frequency-division multiplexing (OFDM)-PONs including [29–31]. These DSA algorithms address challenges related to latency, throughput, and energy efficiency, highlighting the versatility of ML in enhancing subcarrier allocation strategies [32]. Moreover, the integration of traffic prediction technology and fair-aware DSA algorithms, as proposed in [32,33], further enhances the performance of subcarrier allocation in short-reach optical communication systems. These advancements improve the efficiency and adaptability of subcarrier allocation by applying ML methodologies [34].

Power Budget Limitations: The electric power budgeting issue is about predicting future energy consumption using historical data on power usage and related environmental factors like weather, user behavior, and equipment efficiency. The goal is to forecast power consumption for upcoming time periods. However, the development of large-scale, systematic ML models for this task is limited by the lack of publicly available datasets. Recent research has provided a basic process for constructing the necessary data and has also presented baseline ML models as a starting point. Specifically, the data construction process involves compiling and organizing relevant datasets, including time-series power consumption data, weather information, occupancy patterns, and equipment performance metrics. This standardized data can then be used to develop and test ML models for power consumption forecasting. For instance, the recent work in [35] has introduced baseline ML models that demonstrate the feasibility of using these techniques to predict future power consumption, despite the constraints posed by the scarcity of publicly accessible datasets.

Equalization: The objective of this task is to minimize fiber-induced distortions by employing post-processing techniques that compensate for linear effects, such as CD. Mathematically, the equalizer optimizes the function $f(x)$ to ensure that the equalized output sequence y closely approximates the input signal. Performance evaluation primarily relies on the bit-error ratio (BER). In PON systems, using shallow-based DL models for post-equalizers has shown potential in addressing nonlinear distortions for both IMDD and coherent signals. This is especially useful in scenarios with modulator nonlinearities or high-launched optical power to meet tight power budgets [7]. As the fiber-induced nonlinear effects are increasing in the latter case, in single-channel coherent PONs, this results in self-phase modulation (SPM). In multi-channel PONs, the increased nonlinear effects result in cross-phase modulation (XPM) and four-wave mixing (FWM). In IMDD PONs, low-complexity artificial neural network (ANN)-based equalizers have demonstrated performance comparable to Volterra-based equalizers in pulse amplitude modulation with four levels (PAM4) systems [36]. While post-equalization techniques have proven effective, the computational complexity at the ONU receiver is a challenge. To address this, strategies for centralized pre-equalization at the transmitter side have been proposed. Examples include memory polynomial-based pre-equalizers [36] and trained neural-network-based pre-equalizers [37]. These methods enhance equalization effectiveness while keeping the ONU receiver simple.

Fault Detection: Short-reach optical communication systems, including PONs, are susceptible to failures such as fiber cuts, equipment failures, power outages, natural disasters, and ONU transceiver malfunctions [38]. Service disruptions can result in significant financial losses for service providers. Identifying faulty ONUs presents challenges, especially when nearly equidistant branch terminations lead to overlapping reflections, making it difficult to pinpoint the exact defective branch [38]. Conventional monitoring approaches become less reliable as PON systems grow in complexity. Recent advancements in ML-enabled proactive fault monitoring offer promising solutions to ensure stable network operation. ML-based fault prediction algorithms utilize past network fault data to discover underlying patterns and similarities. By doing so, these algorithms enhance the detection of optical network problems and facilitate proactive repairs, thereby preventing potential issues from occurring. Several research papers propose using ML algorithms for monitoring management in optical networks. Notably, technologies like random forest and ANN algorithms have been employed to continuously monitor the BER, predict network component failures, and assess fault severity [39]. Wang et al. [40] introduced a hybrid approach combining double exponential smoothing and support vector machines for equipment failure prediction in software-defined metropolitan area networks. Bayesian-network-based models have also been developed for diagnosing PON faults [40].

3. DSP for Signal Equalization in Communication Systems

In this section, we provide an overview of conventional signal equalization techniques, ranging from basic zero-forcing equalization to more advanced approaches such as feed-forward equalizers (FFEs), decision-feedback equalizers (DFEs), Viterbi and Volterra equalizers, and adaptive equalizers. We discuss the advantages and limitations of these techniques, comparing the performance of ML models. Table 1 provides the complexity analysis for each method.

Zero Forcing: It is a linear equalizer (LE) derived by minimizing inter-symbol interference (ISI). A study in [41] has established the analytical foundation for optimal zero-forcing and minimum mean-squared error (MSE) equalization in channels with additive white noise and specified frequency response. The study demonstrates that an optimal LE can be implemented as a cascade of filters, with taps spaced at symbol intervals. However, when the channel effect exhibits deep frequency response "valleys", equalization will yield poor performance due to noise enhancement.

Feed-Forward Equalizer: The FFE [42] mitigates ISI in communication channels by processing the received signal forwardly without feedback. Its simplicity makes it suitable for systems where feedback is unstable or challenging for implementation.

Decision-Feedback Equalizer: Due to the noise enhancement, the DFE is designed to reduce ISI by subtracting already-known symbols. In this way, ISI from already detected symbols is eliminated. Adaptation of the forward and feedback filters of DFEs follow the same pattern as for LEs [43]. The disadvantage is that it could potentially lead to accumulated errors from feeding back incorrect detection decisions

Viterbi Equalizer: The Viterbi equalizer seeks to estimate the most likely sequence of transmitted symbols, given the received sequence. By constructing a trellis diagram where nodes represent possible transmitted symbols and transitions denote potential channel transitions, the Viterbi algorithm dynamically optimizes path metrics to identify the most probable sequence. This process involves state transition probabilities and precise calculations to mitigate the impact of channel impairments. Mathematically, the Viterbi equalizer applies the Viterbi algorithm, which belongs to the dynamic programming algorithm for finding the most likely sequence of hidden states in a hidden Markov model. The time complexity of the Viterbi equalizer is determined by the Viterbi algorithm, which depends on the length of the input sequence and the number of states, making it $O(T \cdot N^2)$, where T denotes the length of the sequence and N refers to the number of hidden states [44].

Volterra Equalizer: This is a nonlinear equalizer used in optical communication systems to compensate for nonlinear distortions introduced by the fiber channel [45]. In PAM4 systems, severe ISI can be introduced due to the imperfect bandwidth of optical and electrical components. The main bandwidth bottleneck in IMDD systems comes from the transmitter side, as the achievable bandwidth of receiver-side devices is typically twice as high as the bandwidth of transmitter-side devices. In such scenarios, the Volterra equalizer can be effectively employed to address both a potential nonlinearity from the transmitter and the bandwidth limitations of the optical components. The higher-order Volterra kernels can model the frequency-dependent distortion and nonlinear effects caused by the limited transmitter bandwidth and nonlinear devices, such as Mach-Zehnder modulators. The Volterra equalizer is based on the Volterra series expansion, which allows for the modeling of nonlinear systems. The key idea is to use a set of nonlinear filters, known as Volterra kernels, to capture the nonlinear characteristics of the channel. The structure of a Volterra equalizer consists of multiple stages, each representing a different order of nonlinearity. The first stage corresponds to the linear equalizer, which performs initial equalization to address linear distortions. Subsequent stages of the Volterra equalizer capture and compensate for higher-order nonlinear distortions. These stages involve nonlinear filters that take multiple past symbols as inputs and produce outputs based on their interaction. The number of stages and the complexity of the Volterra equalizer depend on the specific system requirements and the level of nonlinear distortions present. The coefficients of the Volterra kernels are typically adapted or optimized using algorithms such as the least mean squares (LMS) or recursive least squares (RLS) algorithms. These algorithms iteratively adjust the coefficients based on the error between the equalized signal and the desired signal, aiming to minimize the distortion and improve the overall system performance.

Adaptive Filtering: Adaptive filtering [46] is used in communication systems where channel characteristics vary over time. The mathematical interpretation involves using an adaptive algorithm that iteratively modifies the filter parameters to minimize the error signal between the desired output and the actual output, enabling the filter to adapt to changing input conditions. The actual convergence time and the total time complexity over multiple iterations depend on the convergence behavior of the specific algorithm and its sensitivity to the input data. Assuming t taps, the total time complexity for updating all coefficients is $O(t)$. FFEs and DFEs are regarded as adaptive filtering versions designed explicitly for short-reach communications.

Table 1. Complexity analysis for DFE, FFE, LE, Adaptive Filtering, and Viterbi algorithms. t refers to the number of the taps. N in Viterbi denotes the number of the hidden states.

Models	DFE	FFE	LE	Adaptive Filtering	Viterbi
Train	$O(t)$	$O(t)$	$O(t)$	$O(t)$	$O(t \cdot N^2)$
Inference	$O(t)$	$O(t)$	$O(t)$	$O(t)$	$O(t \cdot N^2)$

4. Traditional Sequential ML Methods

With the increasing demand for higher data transmission rates and the limitations of traditional prediction methods reaching their practical limits in terms of accuracy, the need for algorithms with high precision, reliability, and low complexity has become urgent. In this section, we introduce new DL-based models to address this challenge. We overview relevant research studies, providing a chronological exploration of key sequential models, namely, recurrent neural networks (RNN), long short-term memory (LSTM), gated recurrent unit (GRU), and convolutional neural networks (CNN). The key architectural parts of DL models are explained, with clear examples showing how they work, how they are used, and how complex they are.

In 2018, Karanov et al. [47] introduced an end-to-end deep neural network system for optical communications, encompassing the entire chain of a transmitter, receiver, and channel model. This research showed that transceiver optimization can be achieved in a complete, end-to-end way. Owing to the sequential structure of communication systems, sequential models, including LSTM networks [48], RNNs [49], and GRUs [50] have been extensively employed. They are considered as baseline algorithms in order to generate more advanced and efficient algorithms.

RNN: Originally designed for machine translation in natural language processing, this model is based on the Markov assumption about the hidden state and output sequence:

the output sequence depends only on the current potential state h_t. The potential state depends on the previous moment's latent h_{t-1} and input variables x_{t-1} rather than on the historical data $x_{(t-1,\ldots,0)}, h_{(t-1,\ldots,0)}$. Renowned for their adaptability in handling variable-length sequences and preserving state information across elements, these models find valuable applications in diverse communication fields [3]. In recent work, they have shown promising results in equalization compared to benchmark methods based on Volterra and Viterbi equalizers in two-dimensional eight-level PAM (2D-PAM8) links [51].

Despite its great equalization performance, this model suffers from exploding gradient issues caused by the direct gradient flow of multiple layers [52]. In such networks, the backpropagation of the gradient is performed by accumulating the gradient matrix. This can cause the gradient to grow exponentially if the eigenvalues of the gradient matrix are greater than 1, making the training process very difficult to converge. Conversely, when the eigenvalues of the gradient matrix are less than 1, the gradient will decrease over time until it vanishes completely, causing the parameters to stop updating [53].

LSTM: The LSTM architecture can assist in overcoming this issue by extending the hidden state to a cell state, which is built using a gating mechanism. This mechanism has input, forget, and output gates that help control the flow of information [54]. LSTM models have additional internal states beyond just the hidden state. This allows them to learn a weight matrix that can better preserve useful information in the hidden state. The input gate decides what new information from the current input to be stored in the cell state. The forget gate decides what memories from the previous cell state to keep or discard. The output gate controls what information gets passed to the next cell state. This gating mechanism provides the ability to effectively hold onto relevant details from long sequences while filtering out irrelevant information. This makes it easier to learn dependencies between distant parts of the input. As a result, LSTMs have been widely used in short-range communication tasks that require capturing complex long-term relationships in the data [55].

GRU: This architecture simplifies the gating mechanism used in LSTM models. It has an update gate and a reset gate, instead of the three gates in LSTM [56]. The update gate determines what relevant information to retain from the previous state and the current input. The reset gate controls what data to discard. It is useful in scenarios where the temporal dependencies and relationships between adjacent symbols in a sequence are important. For example, in short-range communication systems, GRUs can help mitigate signal distortions caused by CD and nonlinearities [57]. Recent research in 120 Gb/s coherent 64-quadrature amplitude-modulated optical systems for transmission at 375 km has shown that using a bi-directional GRU as a nonlinear equalizer can help improve the quality factor (Q-factor) beyond the 8.52 dB limit (8.52 dB estimated from $Q = 20\log_{10}(\sqrt{2}\text{erfc}^{-1}(2\text{BER}))$ [58], typically required for hard-decision forward error correction (HD-FEC).

CNN: CNNs are not technically considered sequential models. However, they are widely used across many different domains. This is because of its important advantages, such as high parameter efficiency, weight-sharing mechanism, and plug-and-play characteristics [59]. CNNs use a convolutional kernel to scan the input signal in a specific dimension, capturing temporal features that are important for the task at hand. This convolutional layer is typically followed by a pooling layer and a nonlinear activation function. The pooling layer reduces redundancy, while the activation function introduces nonlinearity. The convolutional kernel is designed to extract features that closely match the input signal. Afterwards, backpropagation is used to optimize the weights of the network. This allows the CNN to learn and enhance the features that are most relevant for the target task. The weights in the network's weight matrix are updated through backpropagation to amplify the important features needed for effective performance on the given ML problem. Furthermore, it has been observed that using multiple layers of small convolutional kernels is often more efficient than using large kernels. This approach, known as the inception architecture, was first introduced in the GoogleNet model [60]. Two commonly used blocks in CNN are the inception module and the inception reduction module, which extract temporal dependencies of different scales by employing a concatenation of 1×1, 3×3, and 5×5 convolutional kernels. In addition, It also uses a special type of convolutional kernel with a size of 1×1. This 1×1 convolution serves a unique purpose: it helps to reduce the number of feature map channels or dimensions. It is commonly used between two regular convolution layers or at the output layer of the network.

Summary: In this section, we have introduced the most common building blocks used in ML models for short-reach optical communication systems. These fundamental components are still widely used in current approaches. To summarize the complexity of the models discussed earlier, we have provided a table (Table 2) that outlines the complexity analysis for each of the models. In this complexity analysis, we have focused solely on the computation required per batched sample, without considering the choice of hyperparameters like the number of epochs or batch size. This provides a compact overview of the computational demands of each model on a per-sample basis.

Table 2. Complexity of DNN, GRU, LSTM, RNN, and CNN. t refers to the number of taps. n_s, n_o, n_h, and d denote input, output, hidden neuron, and depth of the DNN, respectively.

Models	DNN	GRU	LSTM	RNN	CNN
Train	$O(d-1)n^2)$	$O((3n_h^2+6n_h)n_s)$	$O(4n_h^2+7n_h)n_s$	$O(n_h^2 n_s)$	$O(n_o)$
Inference	$O(d-1)n^2)$	$O((3n_h^2+6n_h)n_s)$	$O(4n_h^2+7n_h)n_s$	$O(n_h^2 n_s)$	$O(n_o)$

5. Advanced Sequential ML Methods

In Section 4, we introduced traditional sequential models, such as RNN, LSTM, GRU, and CNN. The key question we aim to answer in this section is how to effectively incorporate the unique characteristics of time-series data into the modeling process and leverage the temporal convolution model to mitigate channel distortion. Compared to other DL models like transformers and Fourier-based neural networks, convolutional models exhibit better generalization performance. Convolutional models are also more robust to changes in their parameter values when applied to new datasets, unlike the other models that require careful parameter initialization and hyperparameter tuning when used on new data [61].

This section starts with channel modeling, encompassing four distinct noise models. We derive the characteristics and capabilities required for the algorithm based on these models. Subsequently, we provide a detailed exposition of the architectures and fundamental assumptions underlying three models: Frequency-Calibrated Sampled-Interaction Neural Network (FC-SCINet) [62], Light Time-Series (LightTS) [63], and DLinear [64].

5.1. Distortion Model

The main limiting factor for the equalization task in a short-reach/PON system is ISI as a result of CD, sampling error (jitter), frequency shift (chirp), and Kerr-induced nonlinearity [47]. In this section, we will focus on the effects of CD, jitter, and chirp, as these are the dominant distortion mechanisms in short-reach PAM-based systems. The impact of Kerr-nonlinearities is limited in single-channel PONs due to the relatively short fiber lengths and low optical powers involved.

CD in an optical communication system is caused by different phase velocities with respect to frequency. It fundamentally constitutes a linear transformation, and its mathematical representation involves a differential equation that considers spatial position and time, which can be presented as

$$\frac{\partial A}{\partial z} = -j\frac{\beta_2}{2}\frac{\partial^2 A}{\partial t^2} \tag{1}$$

where A denotes the amplitude of the complex signal; t denotes time; z is the spatial position along the fiber, where the pulse pattern propagates [47]; and β_2 is the dispersion coefficient. Following the Fourier transformation, we have

$$D(z,w) = \exp(j\frac{\beta_2}{2}\omega^2 z) \tag{2}$$

w is the angular frequency. In the time domain, it is primarily manifested by significant attenuation in the high-frequency components and rapidly changing components.

Jitter is caused by fluctuations in sampling time. It presents itself as signal distortion, exhibiting a comparable impact to superimposed interference signals that adhere to the Gaussian distribution. Timing jitter can be described as

$$\begin{aligned} y_*(t) &= y(t) \sum_{n=-\infty}^{+\infty} \delta(t - n * t_A - \tau) \\ &= y(n * t_A + \tau) \end{aligned} \tag{3}$$

where τ is the timing sampling error, where the correctly sampled value is $y(n * t_A)$. This sampling error can be quantitively estimated as follows:

$$|y(n * t_A + \tau) - y(n * t_A)| \leq M_1|\tau| \tag{4}$$

where M_1 is the first moment of the band-limited spectrum of Fourier transformation of original signal $Y(f)$, can be simply written as:

$$|\frac{\partial y(t)}{\partial t}| \leq \sum_{-f_g}^{f_g} |2\pi f||Y(f)| = M_1 \tag{5}$$

Jitter refers to high-frequency fluctuations in the amplitude of a signal. This high-frequency perturbation can have a significant impact on neural networks that rely on low-frequency signals.

In conclusion, the error can be estimated as $|y(t_n) - y(n\tau_a)| \leq M_1|t_n - n\tau_a| = M_1|\tau_n|$. The error $|e_n|$ is bounded by $M_1 \cdot |\tau_n|$ for a given n. The value of e_n depends solely on τ_n. Assuming that the timing error τ_n follows a statistical nature with $E\{\tau_n\} = 0$ and $E\{\tau_n^2\} = \sigma_\tau^2$, it follows that the amplitude errors e_n are statistically independent. Consequently, the error variance is then given by $E\{e_n^2\} \leq M_1^2\sigma_\tau^2$. For more details, please refer to [65].

Chirp is a signal whose frequency varies with time. Mathematically, it can be described as follows,

$$s(t) = a(t) \cdot \exp[j(\omega_0 \cdot t + \theta(t))] \tag{6}$$

The frequency spectrum of this waveform is obtained as

$$S(\omega) = \int_{-\infty}^{\infty} a(t) \cdot \exp[j(\omega_0 t + \theta(t))] \cdot \exp(-j\omega t)\, dt \tag{7}$$

Simplifying further:

$$S(\omega) = \int_{-\infty}^{\infty} a(t) \cdot \exp[j\{(\omega_0 - \omega)t + \theta(t)\}]\, dt \tag{8}$$

In summary, all types of effects encountered in equalization issues, except for jitter, involve concurrent alterations in both the time and frequency domains. It is noteworthy that such changes are not statistically independent. Consequently, no single effect can be eliminated through straightforward nonlinear operations in a single domain.

5.2. Temporal Convolution Neural Network

DL-based equalizers fundamentally capture domain-specific nonlinear disturbances by employing linear transformations and nonlinear activation functions. Within the temporal CNN, the core modules comprise interval, continuous, and interaction sampling modules, alongside convolutional neurons and linear layers, as depicted in Figure 1. Each of these modules offers practical flexibility for hardware implementation, due to their computational efficiency. Subsequently, we delve into three of the most efficient convolution-based sampling networks.

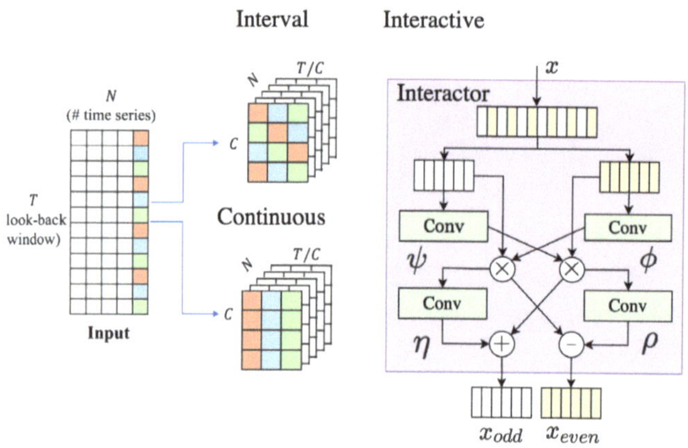

Figure 1. The overview of all sampling modules in temporal convolution networks is modified from [62,63], namely, interval sampling and continuous sampling in LightTS [63], and interactive sampling in [61,62].

FC-SCINet: This novel approach introduces an improved series decomposition technique as a spectrum correction module. In conjunction with the interaction sampling module, it has proven to be a robust tool for mitigating CD and addressing various real-world channel effects [62].

Decomp: In the case of FC-SCINet, it utilizes a moving averaging filter with kernel size w_1 to extract low-frequency signals from the input. Additionally, high-frequency signals are obtained by calculating the residuals between the original and low-frequency signals. The final output signal is generated through a weighted linear combination of these two components, which is $\hat{x}$ defined as Equation (9).

$$\hat{x} = W_s^T x_s + W_f^T x_f \tag{9}$$

The complexity of this module is $O(k)$, where k is the size of the kernel and is independent of the input sequence length.

However, as demonstrated in the empirical study in [62], the performance of FC-SCINet in mitigating CD remains strong. Moreover, the plug-and-play nature, low complexity, and interpretability of FC-SCINet make it highly flexible for seamless integration with various other algorithms. The DLinear architecture is another impressively low-complexity yet high-performance design.

SCIBlock: The SCIBlock, is a key component of FC-SCINet, because it can iteratively decompose a signal into sub-sequences at various scales while incorporating nonlinear transformations between adjacent layers. In contrast, the decomp block is restricted to enhancing fixed signal components and is limited to a single scale. From a mathematical perspective, the SCIBlock applies a hierarchical structure by systematically downsampling the input sequence into even-positioned and odd-positioned samples, denoted as x_{even} and x_{odd}. Following the convolutional layer, the sub-sequences in adjacent layers are iteratively multiplied together utilizing exponential and multiplication operations, as shown in Equations (10) and (11).

$$x^s_{even} = x_{even} \odot \exp(\psi(x_{odd})), \quad x^s_{odd} = x_{odd} \odot \exp(\phi(x_{even})) \tag{10}$$

$$x'_{odd} = x^s_{even} + \exp(\eta(x^s_{odd})) \quad x'_{even} = x^s_{odd} - \exp(\rho(x^s_{even})) \tag{11}$$

Here, $\odot$ represents an element-wise product, and ψ, ϕ, η, and ρ are independent 1D convolutional layers. The intermediate outputs can be presented as x^s_{even}, x^s_{odd}, x'_{even}, and x'_{odd}. Upon completion of the processing, the resulting sub-sequences are then reassembled and aligned back to their original positions within the original signal. Ultimately, all the sub-series are concatenated based on their original index in the raw sequence, as illustrated in Figure 1. To sum up, FC-SCINet is a framework capable of efficiently learning local-dependent patterns. Its distinctive feature lies in performing interactive learning on sub-sequences with odd-even positions after odd-even sampling, allowing for a larger receptive field under the premise of using the same convolutional kernel.

DLinear: As previously mentioned in FC-SCINet, while the concatenation of the decomp module may not offer optimal equalization, it has exhibited great performance in real-world datasets. Therefore, we will provide a brief introduction to this module: It first decomposes a raw data input into a low $\mathbf{x_s}$ and high-frequency $\mathbf{x_f}$ signal. $\mathbf{x_s}$ is extracted by a moving average kernel. It is equivalent to filtering the signal using a *sinc* function in the frequency domain. These two components are added in a linear combination form, expressed by W_s, W_f. The operation above is presented in Equation (12).

$$\mathbf{x_s} = \text{AvgPool}(\mathbf{x}) \quad \mathbf{x_f} = \mathbf{x} - \mathbf{x_s} \quad \mathbf{x}' = W_s\mathbf{x} + W_f\mathbf{x_f} \tag{12}$$

By iterative decomposition with different kernel sizes, DLinear can be extended to a deeper network. The complexity is $O(kn_s)$, where k is the number of the layer, and n_s is the length of the model input. To simplify the complexity, the weight matrix W could be replaced by the convolutional kernel.

LightTS: Both FC-SCINet and DLinear utilize only convolution and different sampling modules to capture the local and global dependencies. The LightTS architecture, detailed in [63], employs a multi-layer perceptron (MLP) structure to enhance predictive abilities.

Sampling: In contrast to SCIBlock, which samples the raw sequence using odd and even indices, LightTS introduces two generic sampling strategies: Interval Sampling and Continuous Sampling, as shown in Figure 1. Interval sampling partitions time-series data into non-overlapping sub-sequences based on fixed time intervals, as shown in Figure 1. This approach helps identify periodic patterns or regularities within the data while minimizing information loss. On the other hand, continuous sampling divides sequences into corresponding sub-sequences, extracting data points continuously throughout the time series and preserving temporal continuity. This sampling method enables the capture of patterns within the period, ensuring a more comprehensive representation of the underlying dynamics. The subsequent section presents an MLP-based architecture to extract useful features from both the downsampled sub-sequences and continuously sampled sub-sequences.

Information Exchange Block (IEBlock): The IEBlock serves as the central module in LightTS, designed to effectively process the 2D matrix resulting from continuous sampling and interval sampling. This block comprises three essential components: (1) temporal projection, which identifies temporal features following continuous sampling; (2) channel projection, which captures inter-channel information following interval sampling; and (3) the exchange block, which integrates the information from the outputs mentioned above, facilitating information fusion. All of them utilize MLP as the nonlinear behavior learning module. The design of LightTS is notably concise, employing only two sampling modules and an MLP. On certain datasets, it surpasses the performance of FC-SCINet [64].

Compared to the models mentioned earlier, the FC-SCINet model requires less structural adaptation and pre-processing when applied in practical PON applications. The FC-SCINet has been successful in recent PON-related work [62]. Different from LSTM, which offers the advantage of ensuring information flow strictly from past to future, temporal CNN goes beyond this by modeling the global dependency between input and output, while also leveraging stacked causal convolution layers. Additionally, the FC-SCINet introduces interaction modeling, enabling the explicit capture of interactions between elements within a sequence, making it a more advanced alternative to LSTM. In addition to these benefits, CNNs and SCINet offer several advantages over LSTM:

- CNNs can identify patterns regardless of their position within the input sequence. This property makes them well-suited for tasks where the position or timing of relevant features is not fixed, providing greater flexibility compared to LSTM.
- CNNs excel at capturing local patterns and extracting relevant features from the input sequence. This ability is particularly useful for tasks that require identifying and recognizing specific patterns or motifs within the data.
- Both CNNs and SCINet architectures typically have fewer parameters compared to LSTM models. This reduced parameter count can make training and inference more efficient, especially when working with limited computational resources or when dealing with large datasets.

Recent Progress: CNNs play a crucial role in current time-series prediction research and applications. This is due to their high parameter efficiency, model stability, and strong theoretical foundation (Multiscale Decomposition). The complexity of these convolutional networks mostly depends on the number of layers and the size of the convolutional kernels. Nowadays, more advanced designs like dilated convolution and inception are often combined with other modules to create complex DL models, but they are rarely used on their own. Even so, temporal CNNs still have a distinct advantage in terms of the performance-to-complexity ratio. They are also straightforward to implement in hardware.

5.3. *Transformer-Based Network*

Attention: The scaled dot-product attention mechanism is the key component aiming to aggregate information across different parts of the input sequence. Each input vector is transformed into three distinct vectors: Queries (Q), Keys (K), and Values (V). The process involves calculating the dot products of the queries with all keys, scaling them by the square root of the dimension, and applying a softmax function to obtain weights on the values [66].

$$\text{Attention}(Q, K, V) = \text{softmax}\left(\frac{QK^T}{\sqrt{d_k}}\right)V \tag{13}$$

The resulting matrix of outputs is obtained through a weighted sum of the values, where the weights are determined by the softmax-processed dot products of queries and keys. This attention module allows the model to focus on relevant parts of the input sequence, capturing local dependencies during the training process. A residual connection is applied around the two sub-layers, followed by layer normalization to maintain the information flow.

Transformers use multiple attention heads to look at the input sequence from different perspectives. This allows the model to simultaneously learn and consider various views of the input data. Equations (14) and (15) represent the functionality of the multi-head attention step. $head_i$ represents the single attention head. The final result of the multi-head attention is concatenating all the attention heads.

$$\text{MultiHead}(Q, K, V) = \text{Concat}(\text{head}_1, \ldots, \text{head}_h)W^o \tag{14}$$

$$\text{head}_i = \text{Attention}(QW_i^Q, KW_i^K, VW_i^V) \tag{15}$$

While the standard ("vanilla") transformer model has shown great performance on time-series data, the computational complexity of its attention mechanism makes it struggle to handle long sequences effectively. To overcome this limitation, researchers have developed various attention mechanism variants. An example is the locality-sensitive gashing (LSH) attention mechanism, which was introduced as part of the Reformer architecture [67]. The LSH attention mechanism utilizes specialized hash functions to transform queries and keys, thereby facilitating the categorization of similar items into shared hash buckets. Through sorting tokens based on their hash codes, items with similarities are grouped together, enabling the aggregation of relevant information. To enable parallel processing, the sorted sequence is divided into chunks. Subsequently, attention mechanisms are selectively applied to these chunks and their neighboring segments, allowing for focused examination of localized portions. The LSH attention mechanism uses hash coding to greatly improve the computational efficiency of the transformer compared to the original version. This helps address the challenge of processing long sequences by reducing complexity without sacrificing performance.

Decoder: The decoder consists of several stacked sub-decoders. The ground truth follows a process similar to that of the encoder, being transformed into Query Q, Key K, and Value V representations. The attention weights are calculated by comparing the Queries Q from the decoder with every value V in the encoder. This process is repeated in parallel across N sub-decoders, resulting in a final attention matrix. The attention matrix then undergoes a softmax operation, yielding probabilities for each value. In addition to the two sub-layers in each encoder layer, the decoder introduces a third sub-layer, performing multi-head attention over the encoder stack's output.

During the decoding process, the model is auto-regressive, using the previously generated outputs as additional input to generate the next output. Residual connections and layer normalization are used around each sub-layer. The self-attention sub-layer is modified to prevent positions from attending to future positions. This, along with the offset output embedding, ensures that the predictions for a position only depend on the known outputs at earlier positions in the sequence.

Attention Variants: The purpose of the attention layer is to identify connections and dependencies among the various input embeddings. This allows the model to evaluate the importance of each element in relation to the others. The attention mechanism explicitly computes the relationships between different elements in the sequence, providing insights into how information flows through the model. However, except for the computational complexity issue, the mechanism in the vanilla transformer [66] needs to be improved in terms of processing inter-dependencies and periodicity of signal data. The Autoformer model [68] introduces a new type of encoder that replaces the original encoder. This new encoder applies series decomposition and autocorrelation to detect dependencies between different parts of the input sequence, and then combines the representations of the sub-series. The series decomposition component divides the original signal into two distinct parts: the seasonal component, which captures short-term patterns, and the trend component, which captures long-term behavior. This partitioning allows for identifying and representing both the short-term and long-term characteristics present in the time-series data. Additionally, the auto-correlation mechanism utilizes the fast Fourier transform (FFT) to compute correlations between the time series and its delayed version, providing insights into how the series relates to its past values at different time lags. The combination of series decomposition and autocorrelation effectively captures and represents the underlying trend and seasonality in the time-series data.

Recent Progress: In this section, we provide a comprehensive overview of the vanilla transformer and its architecture, particularly within the domain of time series and traffic prediction. Over the years, substantial improvements have been made to enhance the transformer for accurate time-series prediction. Notably, advancements have been achieved in reducing computational complexity while improving the effectiveness of the attention mechanism [69,70]. However, recent research has introduced compact models based on multi-scale transformation [71], which surprisingly outperforms benchmark-designed models. This new development has sparked an important debate on the fundamental structure of sequence models. In the following sections, we summarize and explore this particular model in depth, providing insights into its implications. For the latest work, please refer to Table 3.

Table 3. Benchmark models.

<table>
<tr><th colspan="2">Models</th><th>Efficient Techniques</th><th>Literature</th></tr>
<tr><td rowspan="12">Transformer</td><td rowspan="7">Attention</td><td>Sparsity inductive bias</td><td>Ref. [69] LogTrans leverages convolutional self-attention for improved accuracy with $O(L(\log L)^2)$ lower memory costs.</td></tr>
<tr><td>Low-rank property</td><td>Ref. [70] Informer selects dominant queries based on queries and key similarities.</td></tr>
<tr><td>Learned rotate attention (LRA)</td><td>Ref. [72] Quatformer introduces learnable period and phase information to depict intricate periodical patterns.</td></tr>
<tr><td>Hierarchical pyramidal attention</td><td>Ref. [73] Pyraformer proposed one hierarchial attention mechanism with a binary tree following the path with linear time and memory complexity</td></tr>
<tr><td>Frequency attention</td><td>Ref. [74] FEDformer: proposed the attention operation with Fourier transform and wavelet transform.</td></tr>
<tr><td>Correlation attention</td><td>Ref. [68] Autoformer: the Auto-Correlation mechanism to capture sub-series similarity based on auto-correlation and seires decomposition</td></tr>
<tr><td>Cross-dimension dependency</td><td>Ref. [75] Crossformer utilizes multiple attention matrices to capture cross-dimension dependency</td></tr>
<tr><td rowspan="2">Architecture</td><td>triangular patch attention</td><td>Ref. [76] Triformer features a triangular, variable-specific patch attention with a lightweight and linear complexity</td></tr>
<tr><td>Multi-scale framework</td><td>Ref. [71] Scaleformer iteratively refines a forecasted time series at multiple scales with shared weights</td></tr>
<tr><td rowspan="3">Positional rncoding</td><td>Vallina Position</td><td>Ref. [66] cos and sin functions with a sampling rate-relevant period.</td></tr>
<tr><td>Relative positional encoding</td><td>Ref. [77] Introduces an embedding layer that learns embedding vectors for each position index.</td></tr>
<tr><td>Model-based learned</td><td>Ref. [69] LogSparse utilize one LSTM to learn relative position between series tokens</td></tr>
<tr><td rowspan="3">Fourier-NN</td><td rowspan="2">Time Domain</td><td>Series Decomposition</td><td>Ref. [64] DLinear performs one linear series decomposition with multiple layers</td></tr>
<tr><td>Frequency Attention</td><td>Ref. [78] TimesNet proposes the attention mechanism related to the amplitude of the signal</td></tr>
<tr><td>Frequency Domain</td><td>Frequency MLP</td><td>Ref. [79] FreqMLP performs MLP in frequency domain by leveraging the global view and energy compaction characteristic</td></tr>
<tr><td rowspan="7">TConv-NN</td><td rowspan="5">Sampling</td><td>Continous</td><td>Refs. [63,78] both utilize continous sampling to split original signal into windowed subseries similar to short time transformation</td></tr>
<tr><td>Interval</td><td>Ref. [63] Interval sampling with fixed step to extract periodic feature</td></tr>
<tr><td>Even-Odd/Multiscale</td><td>Ref. [62] proposes one iterative multiscale framework where even and odd series are interacted between layers</td></tr>
<tr><td>Frequency Continous</td><td>Ref. [64] leverages series decomposition module in a iterative manner to decompose signal in frequency domain with *sinc* function.</td></tr>
<tr><td>Negative sampling</td><td>Ref. [80] custom loss function is employed in an unsupervised manner, wherein distant or non-stationary subseries maximize the loss, while similar subseries minimize the loss.</td></tr>
<tr><td rowspan="2">Feature module</td><td>MLP</td><td>Ref. [63] applies an MLP-based structure to both interval sampling and continuous sampling for extracting trend and detail information.</td></tr>
<tr><td>Dilated convolutions</td><td>Ref. [81] leverages stacked dilated casual convolutions to handle spatial-temporal graph data with long-range temporal sequences</td></tr>
</table>

5.4. Fourier Convolution Neural Network

In the previous section, we introduced models based on convolutional kernels and subsampling as fundamental modules. Their core principle involves decomposing signals into different scales in the time domain and subsequently applying nonlinear transformations to learn salient features. However, for the majority of real-world signals, transforming them into the Fourier domain is often more efficient. This efficiency is attributed to the following factors: (1) The majority of real-world signals are bandpass or lowpass, and in the Fourier domain, their dynamic range decreases from n to $exp(-n)$; (2) The Fourier transformation is a bijective (one-to-one) transformation, which ensures energy conservation and controllable error in both the forward and inverse transformations; (3) The computational complexity of existing (fast) Fourier transform algorithms, after improvements, is $O(nlog(n))$, making it convenient for hardware implementation.

In this section, we introduce TimesNet [78], which utilizes a frequency-attention mechanism, and FreTS [79], which explicitly performs non-linear transformations in the frequency domain. For extensive models please refer to Table 3. The fundamental concept of TimesNet involves transforming the initial signal into k distinct 2D tensors instead of directly processing the original sequence. This approach empowers the model to effectively capture both intra-periodic and inter-periodic variations within these fixed windows. A variant of this model has been recently reported in [82].

TimesNet: An attention mechanism based on spectral amplitude is employed to determine the significance of signal segments at various frequencies. Simultaneously, across different temporal resolution scales, a shared convolution module is utilized to reconstruct nonlinear distortions introduced by the channel. It does not explicitly perform nonlinear transformations in the frequency domain; instead, it combines reconstructed signals at different window lengths through a linear combination. The FConvNet primarily comprises four key blocks: Component Detection, Alignment, ConvNet, and Reconstruction.

Component Detection: The identification of the k most crucial frequencies is based on the amplitude of the Fourier coefficients. Then, using only the selected components within the k frequency range, the signal is sampled using a continuous sampling method, and these sub-series are arranged into a two-dimensional tensor.

Alignment: The aligned sub-series are then fed into a convolution-based module, specifically an Inception network, to mitigate distortion caused by channel effects. The Fourier coefficients pass through a softmax function to generate attention weights, which are then multiplied by the output of each convolution module to produce the final output.

Fourier Attention: The Fourier transformation is a global operation, meaning that any changes in the signal's amplitude will cause periodic oscillations throughout the entire signal. Significant variations in the amplitude of the primary components lead to substantial fluctuations. TimesNet leverages this characteristic by using the Fourier spectrum of the nonlinearly transformed signal to determine the attention value for each component.

Reconstruction: Finally, employing a residual form, we obtain the reconstructed individual sub-components multiplied by their respective attention values, denoted as y', and add them to the input signal x to yield y.

FreTS: Time-domain-based processing is limited by information bottlenecks, as the local characteristics vary. FreTS explicitly uses frequency-domain features in its model architecture to directly mitigate distortion without manipulating the time-domain. FreTS is essentially an MLP-based network that is able to effectively learn patterns of time-series data in the frequency domain. As presented in [79], FreTS consists of two learners: a Frequency Channel Learner and a Frequency Temporal Learner. In the equalization problem, there is no actual channel dimension, but rather a stack of independent experiments. Therefore, FreTS only introduces a frequency-domain MLP.

Frequency MLP: The frequency temporal learner aims to capture temporal patterns in the frequency domain. Specifically, for a complex number input $\mathcal{H} \in \mathbb{C}^{m \times d}$, the MLP aims to optimize the weight matrix $\mathcal{W} \in \mathbb{C}^{d \times d}$ and bias $\mathcal{B} \in \mathbb{C}^{d}$ so that the final output $\mathcal{Y} \in \mathbb{C}^{m \times d}$ could approximately reconstruct the ground truth.

$$\mathcal{Y}_\ell = \sigma(\mathcal{Y}_{\ell-1}\mathcal{W}_\ell + \mathcal{B}_\ell) \tag{16}$$

$$\mathcal{Y}_0 = \mathcal{H} \tag{17}$$

The MLP in the frequency domain is equivalent to global convolutions in the time domain as detailed in [79]. An increasing number of studies have demonstrated the feasibility of DL models operating in the frequency domain. Simultaneously, the corresponding computational complexity of frequency-domain processing has decreased from $O(n)$ to $O(nlogn)$ due to the reduction in the signal's dynamic range. However, the advantages and disadvantages of networks in both the frequency and time domains remain inadequately explored. Due to space limitations, we offer a detailed categorization, along with corresponding references and keywords, in Table 3 for researchers with specific interests.

6. Model Compression

In recent years, the proliferation of large-scale ML models has significantly advanced state-of-the-art technology across various domains, ranging from natural language processing to computer vision. The surge in model complexity, often characterized by sophisticated architectures and many parameters, has driven the need for efficient hardware implementations to harness their full potential. The advent of single graphics processing units (GPUs) as a critical computational resource has been pivotal, offering a parallelized architecture suitable for accelerating the training and inference processes [83]. The significance of deploying large ML models on a single GPU lies in optimizing computational efficiency and reducing latency. Single GPU implementations facilitate parallel processing, enabling the simultaneous execution of multiple tasks and handling extensive model parameters. This enhances the speed of model training and facilitates real-time inference, a critical requirement in applications such as autonomous systems and edge computing. However, the hardware implementation of large ML models on a single GPU is challenging. The complexity of these models often exceeds the computational capacity and memory constraints of a single GPU, necessitating innovative solutions for efficient utilization [84]. Techniques such as model pruning, quantization, vector quantization, and knowledge distillation have emerged as strategies to mitigate these challenges, ensuring that even formidable models can be accommodated within the limitations of a single GPU without compromising performance. The authors in [84] examine how to use a single GPU effectively for implementing large ML models. They discuss methods that balance complexity and computational efficiency to maximize hardware utilization [84].

In addition, conducting a comprehensive performance-versus-complexity analysis is necessary to evaluate the suitability of various ANNs in short-reach optical communication systems. DL models, including CNNs, RNNs, and LSTMs, find applications in critical tasks such as equalization, fault detection, subcarrier allocation, nonlinearity compensation, and bandwidth request and allocation. The complexity of these models is a significant factor affecting their feasibility. For instance, CNNs may introduce convolutional and pooling layers, increasing model complexity. Similarly, RNNs and LSTMs, designed for sequential data, introduce recurrent connections that enhance their ability to capture temporal dependencies and contribute to increased complexity [85]. Analyzing the neural network architectures in detail, including their depth, the number of parameters, and computational demands, is crucial for understanding the trade-offs between performance and complexity. DL models often exhibit enhanced capabilities in capturing complex patterns and relationships in optical communication data. Still, their complexity may pose challenges regarding training time, computational resources, and practical implementation [86]. A thorough examination of these complexities is essential for identifying optimal models, such as choosing between a CNN for image-based tasks or an LSTM for sequential data, that balance high performance and manageable complexity, facilitating their efficient integration into short-reach optical communication systems [85]. Four prominent types, namely, the feed-forward neural network (FFNN), the radial basis function neural network (RBF-NNs), the auto-regressive RNN (AR-RNN), and the layer-RNN (L-RNN), offer distinct trade-offs in complexity and performance. Among nonlinear neural-network-based equalizers with equivalent numbers of inputs and hidden neurons, FFNN-based equalizers have the lowest computational complexity; however, AR-RNN demonstrates superior transmission performance in 50 Gb/s PAM4 systems [87].

Distillation model: Knowledge distillation, a model compression technique, transfers knowledge from complex, large-scale models or groups to more compact, feasible models suitable for real-world applications. Pioneered by Bucilua et al. in 2006 [88], knowledge distillation primarily operates on neural network architectures characterized by multifaceted structures comprising multiple layers and parameters. Knowledge distillation has been recently considered an important technique for practical DL applications such as speech recognition, image recognition, and natural language processing [89]. Deploying large deep neural network models can be especially challenging for edge devices, which are limited in memory and computational power. To address this challenge, an innovative model compression method was developed in [89], allowing transferring knowledge from larger, more complex models to train smaller, more efficient models without significant performance loss. This process, where a smaller model learns from a larger one, was formalized into the "Knowledge Distillation" framework by Hinton et al. [90]. This framework has become crucial for deploying the essential knowledge from sophisticated, large-scale models on computationally constrained edge devices.

Optimizing DL models through knowledge distillation shows great potential for advancing short-range optical communication systems. RNNs have been particularly effective at addressing nonlinear distortions [57,85]. However, the feedback loop inherent in RNN structures makes it difficult to parallelize them, preventing their use in low-complexity hardware designed for high-speed processing in optical networks [91]. Using knowledge distillation is a promising approach to enable parallelization of RNNs [85,92]. This application of knowledge distillation is set to revolutionize the implementation of RNNs, ensuring compatibility with low-complexity hardware and meeting the stringent processing requirements of high-speed optical networks [93].

Beyond just RNNs, knowledge distillation can be applied to many different ML models important for optical communications, such as CNNs, LSTMs, FFNNs, RBF-NNs, AR-RNNs, and L-RNNs [92]. These models each have their own challenges regarding complexity, adaptability, and real-time implementation. For example, using knowledge distillation in LSTMs for optical communication systems, can reduce model complexity without losing the ability to handle time-dependent patterns [92].

Another promising application of knowledge distillation is when facing challenges with limited time-series data. As "big data" impacts various fields, the scarcity of target events or high data acquisition costs can hinder ML in certain scenarios. A proposed method uses "privileged information" from partial time-series data during training to enhance long-term predictions for small datasets. Applied to optical communications, this distillation approach offers a solution to data constraints, demonstrating effectiveness on both synthetic and real-world data [94].

Vector quantization: Vector quantization (VQ) is a model compression technique targeting large-scale ML models. VQ represents complex data with a small set of prototype vectors, significantly cutting the computational load during inference. This makes VQ useful for applications that require balanced model efficiency and performance, such as when resources are limited. The VQ process involves partitioning the input space into regions, each with a representative prototype vector. During encoding, input vectors are assigned to the nearest prototype, quantizing the data. In the decoding or reconstruction phase, these prototype vectors are used to rebuild the original data.

The effectiveness of VQ relies on carefully selecting and updating the prototype vectors. The goal is to optimize the prototypes so they can effectively capture the essential information in the dataset [95]. By clustering and quantizing input vectors into a representative codebook, VQ enables encoding information in a more compact form. This is particularly beneficial in scenarios with limited data availability or high computational demands [96]. For example, VQ can be useful when applying knowledge distillation to RNNs. RNNs face challenges with parallelization due to their feedback loop structure. Using VQ in the distillation process for RNNs can help address the parallelization issue. VQ can represent the essential information from the RNN using a smaller set of prototype vectors [87,97,98]. This compression not only aids in overcoming hardware complexity but also contributes to faster processing in high-speed optical networks.

VQ uses an iterative process to improve the prototype vectors and enhance their ability to represent the data. Commonly, algorithms like k-means clustering are used for this. The prototypes are adjusted to minimize the difference between the original data and the quantized representation. This iterative refinement allows VQ to adapt to the patterns and structures in the data. This optimizes the compression capabilities of VQ while still preserving the critical information needed for training tasks [95].

Finally, VQ can be beneficial in optimizing other DL models, such as CNNs or LSTMs, by efficiently capturing essential features with a minimal set of representative vectors [99]. Exploring the use of VQ together with these models provides a promising way to improve the performance and scalability of ML applications in short-reach optical communication systems.

7. Conclusions

In this survey, we have undertaken a comprehensive examination of powerful machine learning models that exhibit the potential to achieve robust equalization in cost-sensitive short-reach optical systems, with a particular focus on PONs. Our objective has been to explore these models' capacity to operate efficiently and deliver effective computational performance. For the first time, we have classified the current models into three distinct types and conducted an extensive analysis of their core concepts, highlighting their differences, similarities, and the underlying insights they provide. Additionally, we have presented a simplified complexity analysis considering various input sizes. In the final stages of our survey, we have also investigated the potential of machine learning solutions in addressing the challenges associated with hardware implementation and complexity. We firmly believe that this survey will serve as a valuable resource, inspiring future research endeavors to develop efficient models explicitly tailored for short-reach and PON systems.

Author Contributions: Conceptualization, C.S.; data curation, J.L., C.S. and E.G.; formal analysis, M.F.; funding acquisition, T.K. and A.R.; investigation, M.F.; methodology, C.S. and E.G.; project administration, T.K.; resources, M.F. and T.K.; software, C.S. and J.L.; supervision, M.F. and T.K.; validation, C.S. and J.L.; visualization, J.L. and C.S.; writing—original draft preparation, C.S., J.L., P.S. and S.M.B.; writing—review and editing, S.L., M.F. and E.G. All authors have read and agreed to the published version of the manuscript.

Funding: This research was funded by Federal Ministry of Education and Research grant number 16KIS1228, and 16KIS1230.

Conflicts of Interest: Elias Giacoumidis, Shi Li, and Andre Richter were employed by the VPIphotonics (Germany) company. The remaining authors declare that the research was conducted in the absence of any commercial or financial relationships that could be construed as a potential conflict of interest.

References

1. Durisi, G.; Koch, T.; Popovski, P. Toward massive, ultrareliable, and low-latency wireless communication with short packets. *Proc. IEEE* **2016**, *104*, 1711–1726. [CrossRef]
2. Kapoor, R.; Porter, G.; Tewari, M.; Voelker, G.M.; Vahdat, A. Chronos: Predictable low latency for data center applications. In Proceedings of the Third ACM Symposium on Cloud Computing 2012, San Jose, CA, USA, 14–17 October 2012; pp. 1–14. [CrossRef]

3. Xie, Y.; Wang, Y.; Kandeepan, S.; Wang, K. Machine learning applications for short reach optical communication. *Photonics* **2022**, *9*, 30. [CrossRef]
4. Wu, Q.; Xu, Z.; Zhu, Y.; Zhang, Y.; Ji, H.; Yang, Y.; Qiao, G.; Liu, L.; Wang, S.; Liang, J.; et al. Machine Learning for Self-Coherent Detection Short-Reach Optical Communications. *Photonics* **2023**, *10*, 1001. [CrossRef]
5. Ranzini, S.M.; Da, R.F.; Bülow, H.; Zibar, D. Tunable Optoelectronic Chromatic Dispersion Compensation Based on Machine Learning for Short-Reach Transmission. *Appl. Sci.* **2019**, *9*, 4332. [CrossRef]
6. Che, D.; Hu, Q.; Shieh, W. Linearization of Direct Detection Optical Channels Using Self-Coherent Subsystems. *J. Light. Technol.* **2015**, *34*, 516–524. [CrossRef]
7. Li, G.; Li, Z.; Ha, Y.; Hu, F.; Zhang, J.; Chi, N. Performance assessments of joint linear and nonlinear pre-equalization schemes in next generation IM/DD PON. *J. Light. Technol.* **2022**, *40*, 5478–5489. [CrossRef]
8. Seb, S.J.; Gavioli, G.; Killey, R.I.; Bayvel, P. Electronic compensation of chromatic dispersion using a digital coherent receiver. *Opt. Express* **2007**, *15*, 2120–2126. [CrossRef]
9. Fludger, C.R.; Duthel, T.; Van den Borne, D.; Schulien, C.; Schmidt, E.; Wuth, T.; Geyer, J.; De Man, E.; Khoe, G.; de Waardt, H. Coherent Equalization and POLMUX-RZ-DQPSK for Robust 100-GE Transmission. *J. Light. Technol.* **2008**, *26*, 64–72. [CrossRef]
10. DeLange, O.E. Optical heterodyne detection. *IEEE Spectrum* **1968**, *5*, 77–85. [CrossRef]
11. Kahn, J.M.; Barry, J.R. Wireless infrared communications. *Proc. IEEE* **1997**, *85*, 265–298. [CrossRef]
12. Huang, L.; Xu, Y.; Jiang, W.; Xue, L.; Hu, W.; Yi, L. Performance and complexity analysis of conventional and deep learning equalizers for the high-speed IMDD PON. *J. Light. Technol.* **2022**, *40*, 4528–4538. [CrossRef]
13. Kartalopoulos, S.V. *Free Space Optical Networks for Ultra-Broad Band Services*; IEEE: New York, NY, USA, 2011; Volume 256, ISBN 978-111-810-423-1.
14. Tsiatmas, A.; Willems, F.M.J.; Baggen, C.P.M.J. Square root approximation to the Poisson channel. In Proceedings of the 2013 IEEE International Symposium on Information Theory (ISIT), Istanbul, Turkey, 7–13 July 2013; pp. 1695–1699. [CrossRef]
15. Moser, S.M. Capacity results of an optical intensity channel with input-dependent Gaussian noise. *IEEE Trans. Inf. Theory* **2012**, *58*, 207–223. [CrossRef]
16. Safari, M. Efficient optical wireless communication in the presence of signal-dependent noise. In Proceedings of the ICCW, London, UK, 8–12 June 2015; pp. 1387–1391. [CrossRef]
17. Fadlullah, Z.M.; Fouda, M.M.; Kato, N.; Takeuchi, A.; Iwasaki, N.; Nozaki, Y. Toward intelligent machine-to-machine communications in smart grid. *IEEE Commun. Mag.* **2011**, *49*, 60–65. [CrossRef]
18. Yi, L.; Li, P.; Liao, T.; Hu, W. 100 Gb/s/λ IM-DD PON Using 20G-Class Optical Devices by Machine Learning Based Equalization. In Proceedings of the 44th European Conference on Optical Communication, Roma, Italy, 23–27 September 2018; pp. 1–3. [CrossRef]
19. Kaur, J.; Khan, M.A.; Iftikhar, M.; Imran, M.; Haq, Q.E.U. Machine learning techniques for 5G and beyond. *IEEE Access* **2021**, *9*, 23472–23488. [CrossRef]
20. Simeone, O. A very brief introduction to machine learning with applications to communication systems. *IEEE Trans. Cogn. Commun. Netw.* **2018**, *4*, 648–664. [CrossRef]
21. Rodrigues, T.K.; Suto, K.; Nishiyama, H.; Liu, J.; Kato, N. Machine learning meets computation and communication control in evolving edge and cloud: Challenges and future perspective. *IEEE Commun. Surv. Tutor.* **2019**, *22*, 38–67. [CrossRef]
22. Zhang, Q.; Li, B.; Wu, R. A dynamic bandwidth allocation scheme for GPON based on traffic prediction. In Proceedings of the FSKD 2012: 9th International Conference on Fuzzy Systems and Knowledge Discovery, Chongqing, China, 29–31 May 2012; pp. 2043–2046. [CrossRef]
23. Sarigiannidis, P.; Pliatsios, D.; Zygiridis, T.; Kantartzis, N. DAMA: A data mining forecasting DBA scheme for XG-PONs. In Proceedings of the 2016 5th International Conference on Modern Circuits and Systems Technologies (MOCAST), Thessaloniki, Greece, 12–14 May 2016; pp. 1–4. [CrossRef]
24. Ruan, L.; Wong, E. Machine intelligence in allocating bandwidth to achieve low-latency performance. In Proceedings of the 2018 International Conference on Optical Network Design and Modeling (ONDM), Dublin, Ireland, 14–17 May 2018; pp. 226–229. [CrossRef]
25. Yi, L.; Liao, T.; Huang, L.; Xue, L.; Li, P.; Hu, W. Machine Learning for 100 Gb/s/λ Passive Optical Network. *J. Light. Technol.* **2019**, *37*, 1621–1630. [CrossRef]
26. Mikaeil, A.M.; Hu, W.; Hussain, S.B.; Sultan, A. Traffic-Estimation-Based Low-Latency XGS-PON Mobile Front-Haul for Small-Cell C-RAN Based on an Adaptive Learning Neural Network. *Appl. Sci.* **2018**, *8*, 1097. [CrossRef]
27. Chen, T.; Guestrin, C. XGBoost, A Scalable Tree Boosting System. In Proceedings of the 22nd ACM SIGKDD (2016), International Conference on Knowledge Discovery and Data Mining, San Francisco, CA, USA, 13–17 August 2016; pp. 1–3. [CrossRef]
28. Ye, C.; Zhang, D.; Hu, X.; Huang, X.; Feng, H.; Zhang, K. Recurrent Neural Network (RNN) Based End-to-End Nonlinear Management for Symmetrical 50Gbps NRZ PON with 29dB+ Loss Budget. In Proceedings of the 44th European Conference on Optical Communication, Roma, Italy, 23–27 September 2018; pp. 1–3. [CrossRef]
29. Kanonakis, K.; Giacoumidis, E.; Tomkos, I. Physical-Layer-Aware MAC Schemes for Dynamic Subcarrier Assignment in OFDMA-PON Networks. *J. Light. Technol.* **2012**, *30*, 1915–1923. [CrossRef]
30. Lim, W.; Kourtessis, P.; Senior, J.M.; Na, Y.; Allawi, Y.; Jeon, S.B.; Chung, H. Dynamic Bandwidth Allocation for OFDMA-PONs Using Hidden Markov Model. *IEEE Access* **2017**, *5*, 21016–21019. [CrossRef]

31. Bi, M.; Xiao, S.; Wang, L. Joint subcarrier channel and time slots allocation algorithm in OFDMA passive optical networks. *Opt. Commun.* **2013**, *287*, 90–95. [CrossRef]
32. Zhu, M.; Gu, J.; Chen, B.; Gu, P. Dynamic Subcarrier Assignment in OFDMA-PONs Based on Deep Reinforcement Learning. *IEEE Photonics J.* **2022**, *14*, 1–11. [CrossRef]
33. Senoo, Y.; Asaka, K.; Kanai, T.; Sugawa, J.; Saito, H.; Tamai, H.; Minato, N.; Suzuki, K.I.; Otaka, A. Fairness-Aware Dynamic Sub-Carrier Allocation in Distance-Adaptive Modulation OFDMA-PON for Elastic Lambda Aggregation Networks. *J. Opt. Commun. Netw.* **2017**, *9*, 616–624. [CrossRef]
34. Nakayama, Y.; Onodera, Y.; Nguyen, A.H.N.; Hara-Azumi, Y. Real-Time Resource Allocation in Passive Optical Network for Energy-Efficient Inference at GPU-Based Network Edge. *IEEE Internet Things J.* **2022**, *9*, 17348–17358. [CrossRef]
35. Cabrera, A.; Almeida, F.; Castellanos-Nieves, D.; Oleksiak, A.; Blanco, V. Energy efficient power cap configurations through Pareto front analysis and machine learning categorization. *Clust. Comput.* **2023**, *27*, 3433–3449. [CrossRef]
36. Chen, Z.; Wang, W.; Zou, D.; Ni, W.; Luo, D.; Li, F. Real-Valued Neural Network Nonlinear Equalization for Long-Reach PONs Based on SSB Modulation. *IEEE Photonics Technol. Lett.* **2022**, *35*, 167–170. [CrossRef]
37. Xue, L.; Yi, L.; Lin, R.; Huang, L.; Chen, J. SOA pattern effect mitigation by neural network based pre-equalizer for 50G PON Opt. *Opt. Express* **2021**, *29*, 24714–24722. [CrossRef]
38. Abdelli, K.; Tropschug, C.; Griesser, H.; Pachnicke, S. Fault Monitoring in Passive Optical Networks using Machine Learning Techniques. In Proceedings of the 23rd International Conference on Transparent Optical Networks, Bucharest, Romania, 2–6 July 2023; pp. 1–5. [CrossRef]
39. Vela, A.P.; Shariati, B.; Ruiz, M.; Cugini, F.; Castro, A.; Lu, H.; Proietti, R.; Comellas, J.; Castoldi, P.; Yoo, S.J.B.; et al. Soft Failure Localization During Commissioning Testing and Lightpath Operation. *J. Opt. Commun. Netw.* **2018**, *10*, A27–A36. [CrossRef]
40. Wang, Z.; Zhang, M.; Wang, D.; Song, C.; Liu, M.; Li, J.; Lou, L.; Liu, Z. Failure prediction using machine learning and time series in optical network. *Opt. Express* **2017**, *25*, 18553–18565. [CrossRef]
41. Tufts, D.W. Nyquist's Problem—The Joint Optimization of Transmitter and Receiver in Pulse Amplitude Modulation. *Proc. IEEE* **1965**, *53*, 248–259. [CrossRef]
42. Munagala, R.L.; Vijay, U.K. A novel 3-tap adaptive feed forward equalizer for high speed wireline receivers. In Proceedings of the 2017 IEEE International Symposium on Circuits and Systems (ISCAS), Baltimore, MD, USA, 28–31 May 2017; pp. 1–4. [CrossRef]
43. Williamson, D.; Kennedy, R.A.; Pulford, G.W. Block decision feedback equalization. *IEEE Trans. Commun.* **1992**, *40*, 255–264. [CrossRef]
44. Forney, G.D. The Viterbi Algorithm: A Personal History. *arXiv* **2005**, arXiv:cs/0504020.
45. Du, L.B.; Lowery, A.J. Digital Signal Processing for Coherent Optical Communication Systems. *IEEE J. Light. Technol.* **2013**, *31*, 1547–1556.
46. Malik, G.; Sappal, A.S. Adaptive Equalization Algorithms: An Overview. *Int. J. Adv. Comput. Sci. Appl.* **2011**, *2*. [CrossRef]
47. Karanov, B.; Chagnon, M.; Thouin, F.; Eriksson, T.A.; Bülow, H.; Lavery, D.; Bayvel, P.; Schmalen, L. End-to-End Deep Learning of Optical Fiber Communications. *J. Light. Technol.* **2018**, *36*, 4843–4855. [CrossRef]
48. Graves, A. Long Short-Term Memory. In *Supervised Sequence Labelling with Recurrent Neural Networks*; Springer: Berlin/Heidelberg, Germany, 2012; pp. 37–45. [CrossRef]
49. Pascanu, R.; Gulcehre, C.; Cho, K.; Bengio, Y. How to Construct Deep Recurrent Neural Networks. *arXiv* **2013**, arXiv:1312.6026.
50. Jing, L.; Gulcehre, C.; Peurifoy, J.; Shen, Y.; Tegmark, M.; Soljacic, M.; Bengio, Y. Gated Orthogonal Recurrent Units: On Learning to Forget. *Neural Comput.* **2019**, *31*, 765–783. [CrossRef] [PubMed]
51. Qin, X.; Yang, C.; Guo, H.; Gao, Y.; Zhou, Q.; Wang, X.; Wang, Z. Recurrent Neural Network Based Joint Equalization and Decoding Method for Trellis Coded Modulated Optical Communication System. *J. Light. Technol.* **2023**, *41*, 1734–1741. [CrossRef]
52. Pascanu, R.; Mikolov, T.; Bengio, Y. On the difficulty of training recurrent neural networks. In Proceedings of the 2013 International Conference on Machine Learning (ICML), Atlanta, GA, USA, 16–21 June 2013; pp. 1310–1318. [CrossRef]
53. Bengio, Y.; Simard, P.; Frasconi, P. Learning long-term dependencies with gradient descent is difficult. *IEEE Trans. Neural Netw.* **1994**, *5*, 157–166. [CrossRef]
54. Hochreiter, S.; Schmidhuber, J. Long Short-Term Memory. *Neural Comput.* **1997**, *9*, 1735–1780. [CrossRef]
55. Ling, P.; Li, M.; Guan, W. Channel-Attention-Enhanced LSTM Neural Network Decoder and Equalizer for RSE-Based Optical Camera Communications. *Electronics* **2022**, *11*, 1272. [CrossRef]
56. Dey, R.; Salem, F.M. Gate-variants of Gated Recurrent Unit (GRU) neural networks. In Proceedings of the IEEE 60th International Midwest Symposium on Circuits and Systems (MWSCAS 2017), Boston, MA, USA, 6–9 August 2017; pp. 1597–1600. [CrossRef]
57. Deligiannidis, S.; Mesaritakis, C.; Bogris, A. Performance and Complexity Analysis of Bi-Directional Recurrent Neural Network Models versus Volterra Nonlinear Equalizers in Digital Coherent Systems. *J. Light. Technol.* **2021**, *39*, 5791–5798. [CrossRef]
58. Liu, X.; Wang, Y.; Wang, X.; Xu, H.; Li, C.; Xin, X. Bi-directional gated recurrent unit neural network based nonlinear equalizer for coherent optical communication system. *Opt. Express* **2021**, *29*, 5923–5933. [CrossRef] [PubMed]
59. Wu, N.; Wang, X.; Lin, B.; Zhang, K. A CNN-Based End-to-End Learning Framework Toward Intelligent Communication Systems. *IEEE Access* **2019**, *7*, 110197–110204. [CrossRef]
60. Szegedy, C.; Liu, W.; Jia, Y.; Sermanet, P.; Reed, S.E.; Anguelov, D.; Erhan, D.; Vanhoucke, V.; Rabinovich, A. Going deeper with convolutions. In Proceedings of the 2015 IEEE Conference on Computer Vision and Pattern Recognition (CVPR), Columbus, OH, USA, 24–27 June 2014; pp. 1–9. [CrossRef]

61. Liu, M.; Zeng, A.; Chen, M.; Xu, Z.; Lai, Q.; Ma, L.; Xu, Q. Scinet: Time series modeling and forecasting with sample convolution and interaction. In Proceedings of the 36th Conference on Neural Information Processing Systems (NeurIPS 2022), New Orleans, LA, USA, 28 November–9 December 2022; pp. 5816–5828. [CrossRef]
62. Shao, C.; Giacoumidis, E.; Li, S.; Li, J.; Farber, M.; Kafer, T.; Richter, A. A Novel Machine Learning-based Equalizer for a Downstream 100G PAM-4 PON. In Proceedings of the 2024 Optical Fiber Communication Conference and Exhibition (OFC'24), San Diego, CA, USA, 24–28 March 2024. Available online: https://opg.optica.org/abstract.cfm?uri=ofc-2024-W1H.1&origin=search (accessed on 1 March 2024).
63. Zhang, T.; Zhang, Y.; Cao, W.; Bian, J.; Yi, X.; Zheng, S.; Li, J. Less Is More: Fast Multivariate Time Series Forecasting with Light Sampling-oriented MLP Structures. In Proceedings of the Conference on Robot Learning (CoRR 2022), Auckland, New Zealand, 14 December 2022. [CrossRef]
64. Zeng, A.; Chen, M.; Zhang, L.; Xu, Q. Are Transformers Effective for Time Series Forecasting? In Proceedings of the Conference on Artificial Intelligence (AAAI 2022), Arlington, VA, USA, 17–19 November 2022. [CrossRef]
65. Kiencke, U.; Eger, R. *Messtechnik: Systemtheorie für Elektrotechniker*; Springer: Berlin/Heidelberg, Germany, 2008; XII 341, ISBN 978-3-540-78429-6.
66. Vaswani, A.; Shazeer, N.M.; Parmar, N.; Uszkoreit, J.; Jones, L.; Gomez, A.N.; Kaiser, L.; Polosukhin, I. Attention is All you Need. In Proceedings of the Neural Information Processing Systems (NIPS'17), Long Beach, CA, USA, 4–9 December 2017; pp. 6000–6010. [CrossRef]
67. Kitaev, N.; Kaiser, L.; Levskaya, A. Reformer: The Efficient Transformer. In Proceedings of the International Conference on Learning Representations (ICLR 2020), Addis Ababa, Ethiopia, 26–30 April 2020. [CrossRef]
68. Wu, H.; Xu, J.; Wang, J.; Long, M. Autoformer: Decomposition Transformers with Auto-Correlation for Long-Term Series Forecasting. In Proceedings of the Neural Information Processing Systems, (NIPS 2021), Virtual, 6–14 December 2021; Volume 34, pp. 22419–22430.
69. Li, S.; Jin, X.; Xuan, Y.; Zhou, X.; Chen, W.; Wang, Y.; Yan, X. Enhancing the Locality and Breaking the Memory Bottleneck of Transformer on Time Series Forecasting. In Proceedings of the 33rd Conference on Neural Information Processing Systems (NeurIPS 2019), Vancouver, WA, USA, 8–14 December 2019.
70. Zhou, H.; Zhang, S.; Peng, J.; Zhang, S.; Li, J.; Xiong, H.; Zhang, W. Informer: Beyond Efficient Transformer for Long Sequence Time-Series Forecasting. In Proceedings of the AAAI conference on artificial intelligence, Vancouver, BC, Canada, 2–9 February 2021; pp. 11106–11115. [CrossRef]
71. Shabani, A.; Abdi, A.; Meng, L.; Sylvain, T. Scaleformer: Iterative multi-scale refining transformers for time series forecasting. In Proceedings of the International Conference on Learning Representations (2022), Virtual, 25–29 April 2022. [CrossRef]
72. Chen, W.; Wang, W.; Peng, B.; Wen, Q.; Zhou, T.; Sun, L. Learning to Rotate: Quaternion Transformer for Complicated Periodical Time Series Forecasting. In Proceedings of the 28th ACM SIGKDD Conference on Knowledge Discovery and Data Mining (KDD'22), Association for Computing Machinery, Washington, DC, USA, 14–18 August 2022; pp. 146–156. [CrossRef]
73. Liu, S.; Yu, H.; Liao, C.; Li, J.; Lin, W.; Liu, A.X.; Dustdar, S. Pyraformer: Low-Complexity Pyramidal Attention for Long-Range Time Series Modeling and Forecasting. In Proceedings of the International Conference on Learning Representations (ICLR 2022), Virtual, 25–29 April 2022. [CrossRef]
74. Zhou, T.; Ma, Z.; Wen, Q.; Wang, X.; Sun, L.; Jin, R. FEDformer: Frequency Enhanced Decomposed Transformer for Long-term Series Forecasting. In Proceedings of the 39th International Conference on Machine Learning (PMLR 2022), Baltimore, MD, USA, 17–23 July 2022; p. 162.
75. Zhang, Y.; Yan, J. Crossformer: Transformer Utilizing Cross-Dimension Dependency for Multivariate Time Series Forecasting. In Proceedings of the International Conference on Learning Representations (2023), Kigali, Rwanda, 1–5 May 2023.
76. Cirstea, R.; Guo, C.; Yang, B.; Kieu, T.; Dong, X.; Pan, S. Triformer: Triangular, Variable-Specific Attentions for Long Sequence Multivariate Time Series Forecasting-Full Version. In Proceedings of the International Joint Conference on Artificial Intelligence (IJCAI 2022), Vienna, Austria, 23–29 July 2022. [CrossRef]
77. Zerveas, G.; Jayaraman, S.; Patel, D.; Bhamidipaty, A.; Eickhoff, C. A transformer-based framework for multivariate time series representation learning. In Proceedings of the 27th ACM SIGKDD Conference on Knowledge Discovery & Data Mining (KDD 2021), Singapore, 8–14 December 2021; pp. 2114–2124. [CrossRef]
78. Wu, H.; Hu, T.; Liu, Y.; Zhou, H.; Wang, J.; Long, M. TimesNet: Temporal 2D-Variation Modeling for General Time Series Analysis. In Proceedings of the International Conference on Learning Representations (ICLR 2022), Virtual, 25–29 April 2022. [CrossRef]
79. Yi, K.; Zhang, Q.; Fan, W.; Wang, S.; Wang, P.; He, H.; An, N.; Lian, D.; Cao, L.; Niu, Z. Frequency-domain MLPs are More Effective Learners in Time Series Forecasting. In Proceedings of the Thirty-Seventh Conference on Neural Information Processing Systems (NIPS 2023), New Orleans, LA, USA, 10–16 December 2023; arXiv:2311.06184. [CrossRef]
80. Franceschi, J.; Dieuleveut, A.; Jaggi, M. Unsupervised Scalable Representation Learning for Multivariate Time Series. In Proceedings of the Neural Information Processing Systems (NIPS 2019), Vancouver, WA, USA, 8–14 December 2019. [CrossRef]
81. Wu, Z.; Pan, S.; Long, G.; Jiang, J.; Zhang, C. Graph WaveNet for Deep Spatial-Temporal Graph Modeling. In Proceedings of the Twenty-Eighth International Joint Conference on Artificial Intelligence (IJCAI-19), Macao, China, 10–16 August 2019. [CrossRef]
82. Shao, C.; Giacoumidis, E.; Matalla, P.; Li, J.; Li, S.; Randel, S.; Richter, A.; Faerber, M.; Kaefer, T. Advanced Equalization in 112 Gb/s Upstream PON Using a Novel Fourier Convolution-Based Network, Submitted to the European Conference on Optical Communication (ECOC 2024). Available online: https://arxiv.org/pdf/2405.02609 (accessed on 1 March 2024).

83. Lew, J.; Shah, D.; Pati, S.; Cattell, S.; Zhang, M.; Sandhupatla, A.; Ng, C.; Goli, N.; Sinclair, M.D.; Rogers, T.G.; et al. Analyzing Machine Learning Workloads Using a Detailed GPU Simulator. In Proceedings of the IEEE International Symposium on Performance Analysis of Systems and Software (ISPASS 2019), Wisconsin, WI, USA, 24–26 March 2019; pp. 151–152. [CrossRef]
84. Marculescu, D.; Stamoulis, D.; Cai, E. Hardware-Aware Machine Learning: Modeling and Optimization. In Proceedings of the 2018 IEEE/ACM International Conference on Computer-Aided Design (ICCAD), New York, NY, USA, 5–8 November 2018; pp. 1–8. [CrossRef]
85. Freire, P.J.; Osadchuk, Y.; Spinnler, B.; Napoli, A.; Schairer, W.; Costa, N.; Prilepsky, J.E.; Turitsyn, S.K. Performance Versus Complexity Study of Neural Network Equalizers in Coherent Optical Systems. *J. Light. Technol.* **2021**, *39*, 6085–6096. [CrossRef]
86. Asghar, M.Z.; Abbas, M.; Zeeshan, K.; Kotilainen, P.; Hämäläinen, T. Assessment of Deep Learning Methodology for Self-Organizing 5G Networks. *Appl. Sci.* **2019**, *9*, 2975. [CrossRef]
87. Xu, Z.; Sun, C.; Ji, T.; Manton, J.H.; Shieh, W. Computational complexity comparison of feedforward/radial basis function/recurrent neural network-based equalizer for a 50-Gb/s PAM4 direct-detection optical link. *Opt. Express* **2019**, *27*, 36953–36964. [CrossRef] [PubMed]
88. Bucila, C.; Caruana, R.; Niculescu-Mizil, A. Model compression. In Proceedings of the 12th ACM SIGKDD International Conference on Knowledge Discovery and Data Mining, New York, NY, USA, 20–23 August 2006; pp. 535–541. [CrossRef]
89. Gou, J.; Yu, B.; Maybank, S.J.; Tao, D. Knowledge Distillation: A Survey. *Int. J. Comput. Vis.* **2020**, *129*, 1789–1819. [CrossRef]
90. Hinton, G.; Vinyals, O.; Dean, J. Distilling the Knowledge in a Neural Network. *arXiv* **2015**. [CrossRef].
91. Chang, A.X.; Culurciello, E. Hardware accelerators for recurrent neural networks on FPGA. In Proceedings of the 2017 IEEE International Symposium on Circuits and Systems (ISCAS 2017), Baltimore, MD, USA, 28–31 May 2017; pp. 1–4. [CrossRef]
92. Srivallapanondh, S.; Freire, P.J.; Spinnler, B.; Costa, N.; Napoli, A.; Turitsyn, S.K.; Prilepsky, J.E. Knowledge Distillation Applied to Optical Channel Equalization: Solving the Parallelization Problem of Recurrent Connection. In Proceedings of the 2023 Optical Fiber Communications Conference and Exhibition (OFC), San Diego, CA, USA, 5–9 March 2023; p. Th1F.7. [CrossRef]
93. Robert, R.; Yuliana, Z. *Parallel and High Performance Computing*; Manning Publications: Greenwich, CT, USA, 2021; Volume 704, ISBN 978-161-729-646-8.
94. Hayashi, S.; Tanimoto, A.; Kashima, H. Long-Term Prediction of Small Time-Series Data Using Generalized Distillation. In Proceedings of the International Joint Conference on Neural Networks (IJCNN 2019), Budapest, Hungary, 14–19 July 2019; pp. 1–8. [CrossRef]
95. Gray, R.M. Vector quantization. *IEEE Assp Mag.* **2019**, *1*, 4–29. [CrossRef]
96. Pourghasemi, H.R.; Gayen, A.; Lasaponara, R.; Tiefenbacher, J.P. Application of learning vector quantization and different machine learning techniques to assessing forest fire influence factors and spatial modelling. *Environ. Res.* **2020**, *184*, 109321. [CrossRef] [PubMed]
97. Wang, X.; Takaki, S.; Yamagishi, J.; King, S.; Tokuda, K. A Vector Quantized Variational Autoencoder (VQ-VAE) Autoregressive Neural F_0 Model for Statistical Parametric Speech Synthesis. *IEEE/ACM Trans. Audio Speech Lang. Process.* **2019**, *28*, 157–170. [CrossRef]
98. Rasul, K.; Park, Y.J.; Ramström, M.N.; Kim, K.M. VQ-AR: Vector Quantized Autoregressive Probabilistic Time Series Forecasting. In Proceedings of the Twelfth International Conference on Learning Representations (ICLR 2024), Vienna, Austria, 7–11 May 2024. [CrossRef]
99. Ozair, S.; Li, Y.; Razavi, A.; Antonoglou, I.; Van Den Oord, A.; Vinyals, O. Vector quantized models for planning. In Proceedings of the International Conference on Machine Learning (ICML 2021), Virtual, 18–24 July 2021; pp. 8302–8313. [CrossRef]

photonics

MDPI

Review

Machine Learning for Self-Coherent Detection Short-Reach Optical Communications

Qi Wu [1,2], Zhaopeng Xu [2,*], Yixiao Zhu [1], Yikun Zhang [1], Honglin Ji [2], Yu Yang [2], Gang Qiao [2], Lulu Liu [2], Shangcheng Wang [2], Junpeng Liang [2], Jinlong Wei [2], Jiali Li [2], Zhixue He [2], Qunbi Zhuge [1,2] and Weisheng Hu [1,2]

1 State Key Laboratory of Advanced Optical Communication System and Networks, Department of Electronic Engineering, Shanghai Jiao Tong University, Shanghai 200240, China; sjtuwuqi@sjtu.edu.cn (Q.W.); yixiaozhu@sjtu.edu.cn (Y.Z.); yyyk94@sjtu.edu.cn (Y.Z.); qunbi.zhuge@sjtu.edu.cn (Q.Z.); wshu@sjtu.edu.cn (W.H.)

2 Peng Cheng Laboratory, Shenzhen 518055, China; jihl@pcl.ac.cn (H.J.); yangy07@pcl.ac.cn (Y.Y.); qgstayfoolish@stu.pku.edu.cn (G.Q.); liull@pcl.ac.cn (L.L.); wangshch@pcl.ac.cn (S.W.); liangjp@pcl.ac.cn (J.L.); weijl01@pcl.ac.cn (J.W.); lijl01@pcl.ac.cn (J.L.); hezhx01@pcl.ac.cn (Z.H.)

* Correspondence: xuzhp@pcl.ac.cn

Abstract: Driven by emerging technologies such as the Internet of Things, 4K/8K video applications, virtual reality, and the metaverse, global internet protocol traffic has experienced an explosive growth in recent years. The surge in traffic imposes higher requirements for the data rate, spectral efficiency, cost, and power consumption of optical transceivers in short-reach optical networks, including data-center interconnects, passive optical networks, and 5G front-haul networks. Recently, a number of self-coherent detection (SCD) systems have been proposed and gained considerable attention due to their spectral efficiency and low cost. Compared with coherent detection, the narrow-linewidth and high-stable local oscillator can be saved at the receiver, significantly reducing the hardware complexity and cost of optical modules. At the same time, machine learning (ML) algorithms have demonstrated a remarkable performance in various types of optical communication applications, including channel equalization, constellation optimization, and optical performance monitoring. ML can also find its place in SCD systems in these scenarios. In this paper, we provide a comprehensive review of the recent progress in SCD systems designed for high-speed optical short- to medium-reach transmission links. We discuss the diverse applications and the future perspectives of ML for these SCD systems.

Keywords: optical fiber communication; self-coherent detection; machine learning; short-reach transmission

1. Introduction

With the advent of the 6G era, the Internet of Things, and the metaverse, there has been an explosive growth in data traffic in recent years, which poses higher requirements for current optical interconnects in terms of capacity and reliability. Coherent detection transceivers were introduced in 2006, which have been widely utilized in optical communication systems spanning thousands of kilometers, such as transoceanic, transcontinental, and metropolitan networks. In a coherent system, a local oscillator (LO) laser is employed to linearly map the received optical field into the electrical domain. Linear mapping allows for the obtainment of the amplitude, phase, and polarization information of the optical signal and for compensation against a number of transmission impairments, including fiber chromatic dispersion (CD), nonlinearity, random polarization rotation, and polarization mode dispersion using advanced digital signal processing (DSP) techniques [1–9]. Consequently, coherent detection enables large-capacity and high-spectral-efficiency (SE) long-haul optical communications.

On the other hand, short-to-medium distance optical networks mainly encompass data-center interconnects, passive optical networks, mobile front-haul, and industrial internet. These networks typically involve a great number of optical connections, making cost sensitivity a crucial factor for the deployed optical devices [10]. However, the utilization of LO in coherent detection necessitates temperature control circuits at the receiver to align with the frequency of the laser at the transmitter. This significantly increases the manufacturing cost of optical devices and hinders the deployment of coherent transceivers in cost-sensitive and large-scale short-to-medium distance optical links [11]. Furthermore, since the LO and the transmitter laser are different, phase noise and frequency offset estimation need to be performed in DSP, leading to the increased power consumption of the DSP chips. In contrast, direct detection systems have a natural structural advantage over coherent detection systems as they do not require a narrow-linewidth and high-stable LO at the receiver. This eliminates the need for complicated temperature control circuits, frequency offset estimation, and carrier phase recovery [11]. As a result, the manufacturing cost of direct detection transceivers is lower, making them promising for short-to-medium distance optical networks over the past decade.

The intensity modulation and direct detection (IMDD) scheme, as a classic direct detection system, encodes information directly onto the optical intensity. At the receiver, the optical intensity is converted into photocurrents through square-law detection using a single photodetector (PD), achieving the mapping from the optical domain to the electrical domain. While the IM-DD system is simple and practical, its transmission performance is limited by power fading caused by fiber CD [12]. The frequency-selective fading limits its applications for high data rates or long-distance transmission [12].

To address the issue of power fading, researchers have proposed to use vestigial sideband (VSB) modulation systems. One approach is to utilize an optical filter to eliminate one of the sidebands of the real-valued double-sideband (DSB) signal [13], reducing the influence of the fiber CD. While VSB modulation enhances the system's resistance to CD, it also introduces nonlinear impairments due to the presence of an incompletely suppressed sideband. As a result, single-sideband (SSB) modulation systems without vestigial components have been developed as an alternative [14–17], which can be achieved using IQ modulators or optical frequency shifters. To further improve the electrical SE and transmission capacity beyond the SSB systems, single-polarization phase retrieval (PR) receiver [18–21], carrier-assisted differential detection (CADD) receiver [10,22,23], and asymmetric self-coherent detection (ASCD) receiver [24] have been proposed to achieve linear detection of complex-valued DSB signals, effectively doubling the electrical SE with respect to SSB and IM-DD systems. Additionally, polarization-division-multiplexing can double the capacity and SE of single-polarization direct detection systems. However, the random birefringence of the fiber leads to polarization rotation, resulting in polarization fading [25–30] in direct detection systems with a co-propagating optical carrier. In order to deal with this effect, Stokes-vector receiver (SVR) [25] and Jones-space field recovery (JSFR) [30] schemes have been proposed. SVR performs polarization rotation in Stokes space, allowing for up to three-dimensional real-valued modulation. The JSFR scheme, however, first recovers the optical field and then performs polarization rotation in Jones space, enabling four-dimensional modulation including the amplitude and phase of two polarizations [30]. The above-mentioned schemes in which the optical carrier and the signal are transmitted together, allowing for phase- or polarization-diversity, are commonly known as self-coherent detection (SCD) systems. SCD systems recover the optical field in the receiver DSP, allowing compensation for the CD similar to coherent detection. The power fading effect induced by the traditional IMDD channel will no longer be a problem in SCD systems.

Although SCD has numerous advantages, there are still several issues in SCD systems that need to be addressed, such as signal-to-signal beating interference (SSBI) and optical field reconstruction. In the past decade, machine learning (ML) technology has rapidly advanced, and its applications have spread across various fields, including image recognition [31], natural language processing [32], medical diagnosis [33], and optical fiber communications [34–78]. ML techniques often achieve a higher accuracy or lower complexity compared to traditional approaches in many scenarios. In optical fiber communications, ML has been extensively studied and has shown a promising performance in optical performance monitoring [34,35], modulation format recognition [36,37], channel equalization [38–72], and constellation shaping [73,74]. In this paper, we provide a comprehensive overview of the application of ML techniques in SCD communication systems, with a particular focus on their applications in nonlinear impairment compensation, IQ imbalance correction, PR, polarization demultiplexing, and optical signal processing. In Chapter 2, we provide a brief introduction to the principles and challenges of various self-coherent systems. In Chapter 3, we provide extensive applications, as well as a detailed analysis of the performance of ML techniques in SCD systems. Finally, in Chapter 4, we summarize the findings and provide an outlook on the future development of ML technology in SCD systems. All the abbreviations used in this paper are listed in Appendix A.

2. SCD Systems

2.1. VSB System

To mitigate the impact of CD-induced power fading impairment, a VSB system is proposed, employing a simple receiver configuration as depicted in Figure 1a. This receiver setup requires only a single PD and an analog-to-digital converter (ADC). The modulated optical spectrum of the VSB signal is shown in Figure 1b. The most commonly-used method in VSB systems is to employ an optical filter to remove unwanted spectral components in the optical domain. By selectively filtering out specific frequency components, the spectral shape of the VSB signal can be modified [13], allowing for effective suppression of the vestigial sideband. In addition to optical filters, VSB modulation can also be achieved through dual-drive Mach–Zehnder modulators (MZMs) and radio frequency delays. In the VSB system, the dominant impairment originates from the unfiltered residual spectral components, which can be expressed as in [21]:

$$E(t) = C + S_s(t) + S_r(t) \tag{1}$$

$$|E(t)|^2 = |C|^2 + C^*(S_s(t) + S_r(t)) + C(S_s(t) + S_r(t))^* + |S_s(t) + S_r(t)|^2 \tag{2}$$

where C, $S_s(t)$, and $S_r(t)$ denote the co-propagating optical carrier, the desired SSB signal, and the residual sideband signal, respectively. The superscript * denotes conjugate operation. After square-law detection as shown in Equation (2), the mirror image of residual components can cause nonlinear distortion to the signal, which can be compensated for by using nonlinear equalizers such as neural networks (NNs).

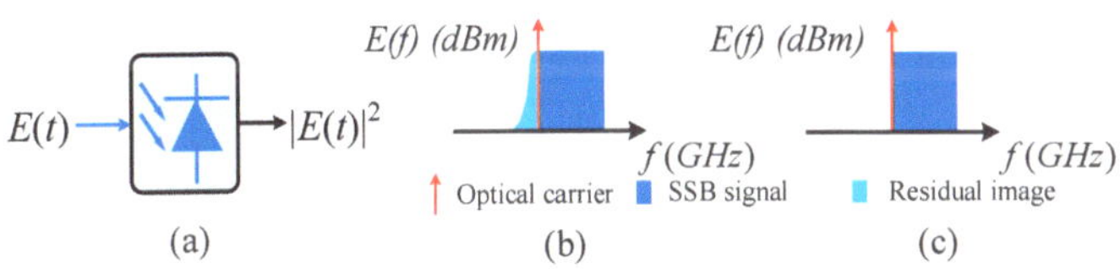

Figure 1. (**a**) Receiver structure for VSB and SSB systems. (**b**) Optical spectrum of a VSB signal. (**c**) Optical spectrum of an SSB signal.

2.2. SSB System

To suppress the residual signal $S_r(t)$ and nonlinear impairment, an SSB system is proposed based on the Hilbert transformation and IQ electrical-to-optical modulator [14,15]. The Hilbert transformation enables us to remove the unwanted sideband, generating a complex-valued electrical signal which drives the IQ modulator to convert into an optical signal, as shown in Figure 1c. The receiver structure of the SSB system is the same as the

VSB system. After optical-to-electronic conversion, the dominant distortion is called SSBI, denoted by $|S_s(t)|^2$. For the SSB system, the received signal can be expressed as in [21]:

$$|E(t)|^2 = |C|^2 + C^* S_s(t) + C S_s(t)^* + |S_s(t)|^2 \tag{3}$$

Fortunately, for the SSB signal, the impact of SSBI can be mitigated by employing phase recovery algorithms based on the minimum phase condition [79–81] or deep learning techniques. These methods help in recovering the phase information lost in optical-to-electronic conversion and enable the compensation of CD in the DSP, avoiding the impact of power fading.

2.3. PR Receiver

Although the resistance to CD is improved in VSB and SSB systems, the electrical SE of these systems is the same as the IM-DD system, defined as the achieved date rate divided by the electrical bandwidth of the receiver. To increase the SE, a PR receiver [18–21] is proposed to detect a complex-valued DSB signal, as shown in Figure 2a. The PR receiver consists of two PDs and one dispersive element, as shown in Figure 2b. The two detected photocurrents $i_1(t)$ and $i_2(t)$ are expressed as [82]

$$i_1(t) = |C + S_d(t)|^2, i_2(t) = |(C + S_d(t)) \otimes h_D(t)|^2, \tag{4}$$

where $S_d(t)$ and $h_D(t)$ are the DSB signal and the transfer function of the dispersive element. Using a fully-connected convolutional NN (CNN), or other nonlinear equalization algorithms, the optical field could be reconstructed in the receiver DSP [83,84]. Note that the PR receivers also enable to recover the phase of optical SSB signal, which will be introduced in Section 3.3.

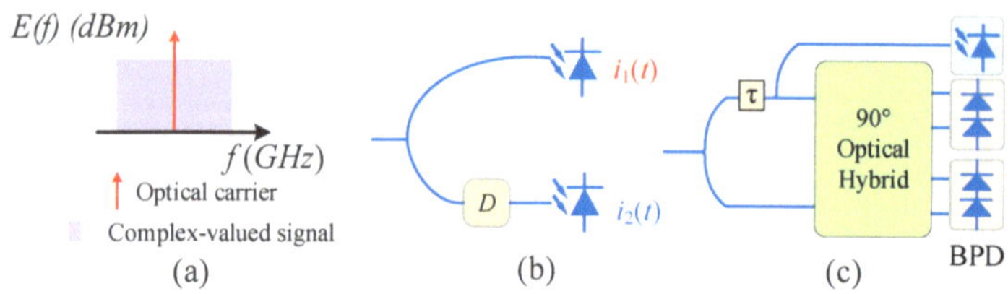

Figure 2. (**a**) Optical spectrum of a complex-valued DSB signal. (**b**) Phase retrieval receiver. *D* is the dispersive element. (**c**) Receiver of the CADD scheme.

2.4. CADD

Another kind of receiver used to detect a complex-valued DSB signal is the CADD receiver [10]. The receiver structure is shown in Figure 2c, consisting of one optical hybrid, one PD, and two balanced photodetectors (BPDs), which is more complex than the PR receiver. However, it can achieve a higher modulation bandwidth and electrical SE than the PR receiver. In the receiver DSP, certain SSBI iterative cancellation algorithms and deep NNs are also used for optical field reconstruction [10,85]. With the help of ML techniques, the channel parameters such as the optical delay values and the carrier-to-signal power ratio (CSPR) can be optimized accurately to achieve a better system performance than the SSBI iterative cancellation algorithm.

2.5. SVR

The direct detection system has been pursuing polarization division multiplexing because it can double its capacity and SE. However, for the optical field where the signal and carrier are transmitted together, the optical signal suffers from polarization fading after passing through a polarization beam splitter (PBS). Polarization fading can result in the failure of optical field recovery on random X- or Y-polarization, making it hard to achieve polarization demultiplexing using multi-input multi-output (MIMO) equalization. Thus, the famous SVR [25] was proposed to combat polarization fading in Stokes space. The receiver structure is shown in Figure 3a, where three received Stokes vectors S_1, S_2, and S_3 are used to address the polarization rotation. The transmitted Stokes vectors could be recovered using S_1, S_2, and S_3 and a de-rotation matrix. Thus, the polarization diversity of the DD system is successfully accomplished.

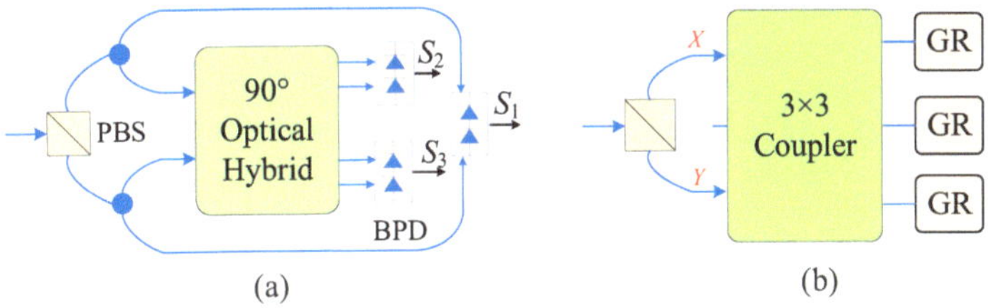

Figure 3. (**a**) Stokes-vector receiver. PBS: polarization beam splitter; BPD: balanced photodetectors. (**b**) Receiver structure of JSFR scheme. GR: generalized receiver, which could be a PD, PR receiver, CADD receiver, or ASCD receiver.

2.6. JSFR

Although the polarization fading issue is solved, at most three modulation dimensions are supported in the real-valued three-dimensional Stokes space. Great efforts are made to exploit the fourth modulation dimension, but these fail to compensate for CD. More recently, the JSFR scheme was proposed to realize polarization diversity for a direct detection system with a co-propagating optical carrier, as shown in Figure 3b. It utilizes the optical coupler to mix the two polarizations to eliminate the impact of the polarization fading effect. The generalized receiver (GR) in this scheme can be implemented using one PD, PR receiver, CADD, and ASCD, according to different modulation formats. Using JSFR, the amplitudes and phases of both X- and Y-polarizations can be recovered, which provides the potential of realizing high-SE and large-capacity optical interconnects for short-reach optical networks. For these polarization-diverse SCD systems, ML can be used to handle the coupling between the polarization modes, namely polarization tracking and polarization mode demultiplexing [82,85].

3. ML Techniques in SCD System

In this section, we will introduce the applications of ML techniques in SCD systems including nonlinearity compensation [86,87], IQ imbalance correction [88], PR in SSB [89–91], optical field recovery in PR receiver [83,84] and CADD schemes [82,92], and polarization tracking and demultiplexing in JSFR schemes [85]. In addition, the transfer learning [93–95] technique has been employed to realize fast remodeling in SSB system, which could be scalable to other DD systems. Finally, we briefly introduce the photonics neuromorphic computing [96] used in SCD systems to extract the phase information and demodulate the quadrature amplitude modulation (QAM) formats.

3.1. Nonlinear Compensation

3.1.1. Fiber Nonlinearity

In optical communication, the electrical field evolution of light in a single-mode fiber can be described by the well-known nonlinear Schrödinger equation (NLSE) [1], which takes the following form:

$$\frac{\partial A}{\partial z} + \frac{i\beta_2}{2}\frac{\partial^2 A}{\partial t^2} = -\frac{\alpha}{2}A + i\gamma|A|^2 A, \quad (5)$$

where z, α, β_2, and γ are, respectively, the propagation distance, the loss coefficient, the group-velocity dispersion (or second-order dispersion) coefficient, and the fiber nonlinear

Kerr coefficient. The NLSE is a nonlinear partial differential equation that does not have an analytical solution, and the nonlinear parameter γ describes the effects of self-phase modulation and cross-phase modulation. In the case of SCD systems, the transmitted optical field has a strong optical carrier, making it more susceptible to fiber nonlinear impairments. It is widely known that NNs have powerful nonlinear fitting capabilities. Therefore, researchers have proposed the use of NNs to compensate for fiber nonlinearity, including various types of NNs such as artificial neural networks (ANNs) [86], long short-term memory networks (LSTMs) [87], and others, showing a better performance compared to traditional digital back-propagation and perturbation algorithms. LSTMs are a specific type of recurrent NN (RNN) model designed to mitigate the vanishing gradient problem commonly encountered in traditional RNNs. LSTMs have proven to be effective tools for mitigating transmission impairments, including both linear and nonlinear distortions, making them valuable for various applications in signal processing and communication systems. In [87], a linear network-assisted LSTM is proposed to mitigate the fiber nonlinearity in the wavelength-division-multiplexing (WDM) SSB system. Figure 4 depicts the architecture of a linear network-assisted LSTM.

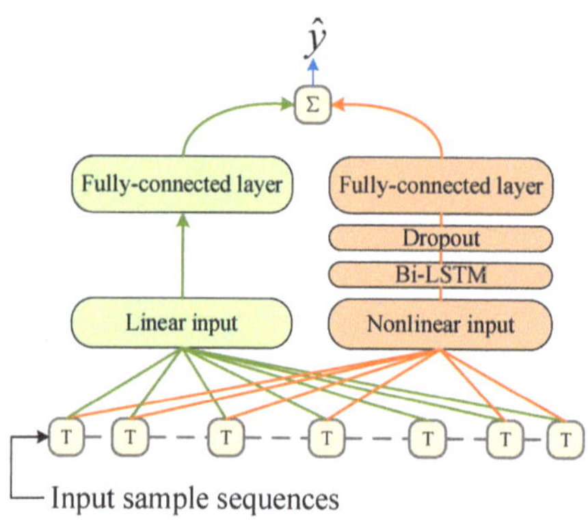

Figure 4. Conceptual illustration of a linear network-assisted LSTM.

The output $\hat{y}$ can be expressed as [87]:

$$\hat{y} = \mathbf{W}_L\mathbf{X} + \mathbf{W}_{NL}Bi - LSTM(\mathbf{X}_{NL}) + \mathbf{b}_{NL}, \tag{6}$$

where $\mathbf{X}$, $\mathbf{X}_{\mathrm{NL}}$, $\mathbf{W}_{\mathrm{L}}$, $\mathbf{W}_{\mathrm{NL}}$, *Bi-LSTM*, and $\mathbf{b}_{\mathrm{NL}}$ are, respectively, the linear input vector, nonlinear vector, the weight matrix for the fully-connected layer of the linear module, the weight matrix for the nonlinear module, the one-layer Bi-LSTM network operations, and the bias vector of the nonlinear modules. Compared to conventional Bi-LSTM, the linear network-assisted LSTM achieves a significant improvement in terms of the Q-factor while also significantly reducing computation complexity.

3.1.2. Device Nonlinearity

Apart from the fiber nonlinearity, another nonlinear impairment comes from the electro-optic modulation. When the dual-drive MZM or IQ modulator is used for complex-valued QAM formats, the modulation nonlinearity will be enhanced with an increase in the peak-to-peak voltage. Figure 5a shows the bias point of the MZM and the modulation nonlinearity induced by the function of $\sin(\cdot)$. Additionally, other device nonlinearity such as the responsibility curve of PD also deteriorates the system performance. In scenarios involving multiple nonlinear impairments, traditional methods face challenges in accurately estimating the channel parameters and compensating for the mixed nonlinear effects. However, ML demonstrates its excellent capability for parameter optimization in such complex channels. In [86], a sparsely connected ANN is proposed to address the fiber nonlinearity and modulation nonlinearity. The principle of ANN pruning is shown in Figure 5b. A weight threshold is set, and connections with weights below this threshold are pruned, thereby reducing the complexity of the NN. The pruned sparsely connected ANN is shown in Figure 5c. By implementing this method, the number of connections in the ANN is reduced by an order of magnitude, while maintaining the bit-error-rate (BER) performance without significant degradation.

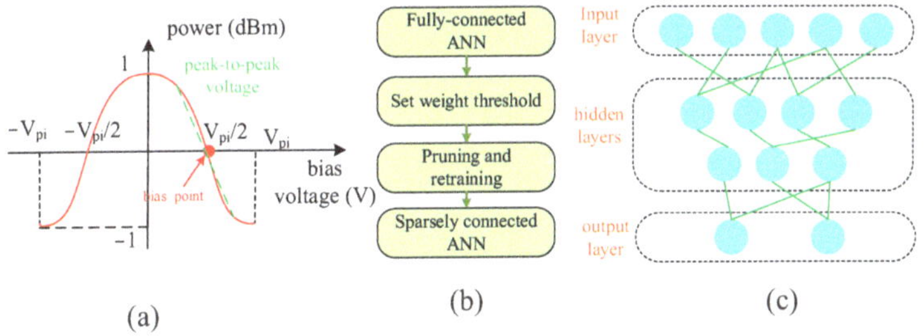

Figure 5. (**a**) Transfer function of MZM. (**b**) Principle for ANN pruning. (**c**) Sparsely connected ANN.

3.1.3. SSBI Cancellation

Unlike coherent detection, the direct detection system does not utilize LO and BPD to cancel the common-mode component inside the photocurrent, known as SSBI. Therefore, for direct detection systems, SSBI generated by the PD becomes the primary impairment limiting the system's transmission capacity. As observed from the fourth term in Equation (3), SSBI takes the form of a quadratic term of the original signal. The spectra of the signal and its SSBI are depicted in Figure 6a, illustrating that the bandwidth of SSBI is twice that of the original signal in the electrical domain. Consequently, SSBI distorts the signal, degrading system performance. Certain methods have been proposed to handle SSBI in direct detection systems such as the Volterra nonlinear equalization and SSBI iterative mitigation methods. Additionally, ML methods such as NNs can also play a significant role in SSBI cancellation. Compared to traditional algorithms, an NN-based equalizer offers tremendous improvements in SSBI elimination, improving the performance of the transmission system. In [81], a soft-combined ANN was proposed and its structure is shown in Figure 6b. The output of the soft-combined ANN is an average of the outputs of all of the ANNs. The results reveal that the soft-combined ANN exhibits a superior performance compared to a single ANN in compensating for both linear and nonlinear SSBI impairments in the signal. Remarkably, this improved performance is achieved while maintaining the same symbol length of the required training sequence.

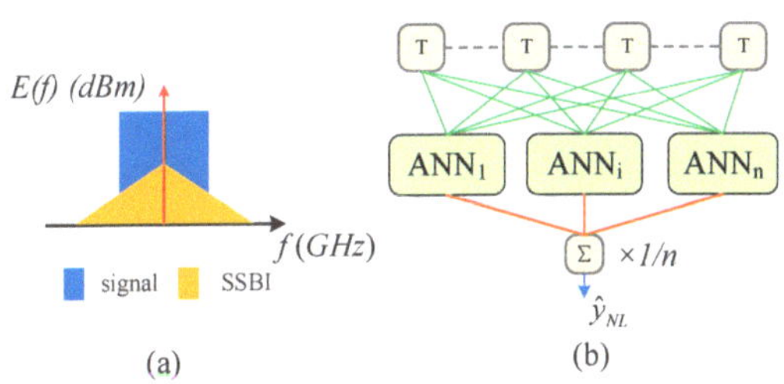

Figure 6. (**a**) Electrical spectra of a typical direct detection signal and its SSBI. (**b**) Structure of a soft-combined ANN.

3.2. *IQ Imbalance Correction*

For complex QAM modulation, IQ imbalance and crosstalk can lead to an incorrect signal decision resulting in a degraded BER performance. Therefore, in classical DSP steps,

IQ orthogonalization algorithms are commonly used for compensation. In optical communication systems where both nonlinear impairments and IQ imbalance exist, the traditional DSP algorithms used for compensation can be replaced by a MIMO-ANN. The joint compensation approach, which addresses both types of impairments simultaneously, generally yields superior results compared to using separate compensation for each impairment individually. In [88], a MIMO-ANN is proposed to compensate for the fiber nonlinearity, SSBI, and IQ imbalance simultaneously. Figure 7 displays its structure, consisting of two ANNs. The in-phase and quadrature components, X_I and X_Q, and their delay copies are fed into these two ANNs. Y_I and Y_Q are the outputs of the MIMO-ANN. In order to minimize the cost function, the back-propagation algorithm is employed to update the weights and biases of layers. After the training processes, the optimized ANNs are used to equalize the received data. The experimental results confirm the outstanding performance of the MIMO-ANN scheme in mitigating interference between two orthogonal signals.

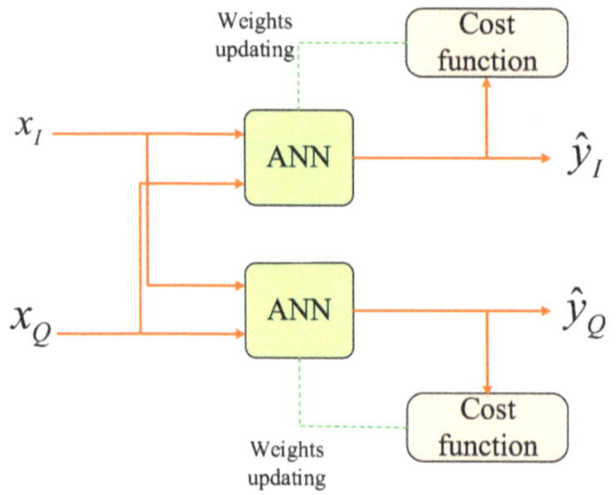

Figure 7. Block diagram of the MIMO-ANN.

3.3. PR and Optical Field Recovery

In a DD system, the phase information of the optical signal is lost during envelope detection by the PD while the intensity information is retained in the photocurrent. To recover the phase information, the Kramers–Kronig (KK) receiver algorithm was proposed for DD in 2016 [79], which relies on the minimum phase condition. If the SSB signal satisfies the minimum phase condition, the phase can be extracted from the intensity information using a Hilbert transformation. To successfully apply the KK receiver algorithm, a high CSPR is required. However, achieving a high CSPR comes with certain challenges. It introduces an additional sensitivity penalty and increases the impact of nonlinear fiber propagation effects. These factors need to be carefully considered when implementing the KK receiver algorithm in DD systems. To alleviate the CSPR requirement, a supervised learning CNN model was proposed in [89,90] to emulate the KK algorithm for the PR task. The architecture of the NN model is illustrated in Figure 8a. The input of the NN is the received photocurrent, namely $|E(t)|^2$ in Equation (2). The down-sampling blocks, labeled as D_i ($i = 1, 2, 3$), consist of a convolutional layer followed by the Rectified Linear Unit (ReLU) activation function. The up-sampling blocks, labeled as U_i ($i = 1, 2, 3$), incorporate a combination of convolutional layers, transposed convolutional layers, and ReLU activation functions. In this NN model, the target outputs are selected as the in-phase and quadrature components, rather than the amplitude and phase. Through simulations, it has been demonstrated that the ML-based PR scheme accurately reconstructs the phase of a modulation phase signal even at weak carrier power levels. This ML-based approach relaxes the CSPR requirement and improves the receiver sensitivity compared to the original KK algorithm. Overall, the proposed NN model provides a promising solution for PR, leveraging the power of deep learning techniques to enhance the performance of SCD systems.

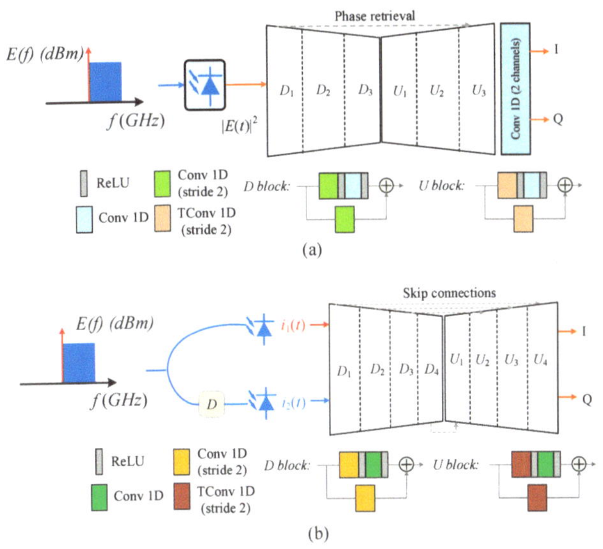

Figure 8. (**a**) Temporal CNN architecture with a detailed view of the down-sampling (D) and up-sampling (U) blocks. Conv/TConv 1D: 1D convolutional/transposed convolutional layer. (**b**) NN architecture for phase retrieval receiver.

In addition to constructing SSB signals that do not satisfy the minimum phase condition, PR receivers and their corresponding algorithms can be utilized to restore the phase of the optical field. It can be applied to phase recovery of SSB signals or complex-valued DSB signals. Gerchberg–Saxton algorithm is most commonly employed for PR [18–21], but it requires multiple iterations to converge. For SSB signals, the received two optical photocurrents ($i_1(t)$ and $i_2(t)$) can be fed as inputs to an NN [91], as shown in Figure 8b. This NN consists of eight convolutional blocks aimed at down-sampling and up-sampling. The outputs of the NN are the real and imaginary parts of the optical field. By implementing this NN to achieve PR, the required dispersion value of the dispersive element is significantly decreased and the computational complexity is also reduced by 30%. Most importantly, the SSB signal no longer requires a strong optical carrier to satisfy the minimum phase condition. With the same Erbium-doped fiber amplifier launch power, it is possible to increase the number of WDM channels and reduce nonlinear fiber impairments, which potentially provides a larger capacity.

On the other hand, the optical and electrical SE could be improved if the PR receiver is utilized to detect the complex-valued DSB optical signal. The deep-learning-enabled direct detection scheme [83,84] was proposed to recover the optical field at a low CSPR, which is shown in Figure 9. Similarly, the inputs are two samples of photocurrents. The NN based on deep residual learning blocks consists of two convolutional layers and several residual modules. Its output is the desired phase information of the optical field. Residual learning is a technique that introduces shortcut connections into the traditional CNN structure, providing benefits in terms of training speed and network performance. The deep residual network architecture is built around stacked residual blocks, with each block consisting of two convolutional layers and a shortcut connection. The shortcut connections enable the direct propagation of information from one layer to another, bypassing intermediate layers. The integration of shortcut connections and stacked residual blocks improves the training efficiency and enables the effective learning of deep CNN models. This architecture has been proven highly effective in various computer vision tasks, enabling the construction of deeper networks without the issues of vanishing or exploding gradients. In [83], the residual learning technique is applied to accurately recover the transmitted signal in the presence of a large SSBI under the low CSPR condition. Compared with the conventional SSBI cancellation scheme, the deep-learning-enabled DD receiver shows a significant reduction of 8 dB in the optimum CSPR when detecting a complex-valued DSB signal.

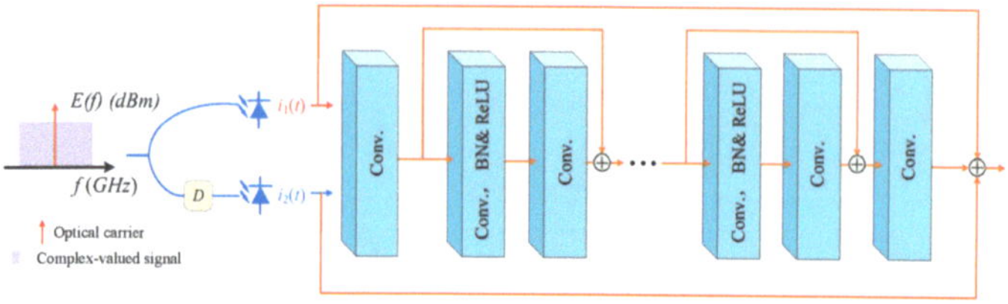

Figure 9. Time-domain data channels in the deep CNN.

3.4. Polarization Demultiplexing

For polarization-multiplexed optical communication systems, random birefringence in optical fibers can lead to random coupling between polarization modes. Therefore, at the receiver end, DSP algorithms are required to accomplish polarization demultiplexing. Additionally, the coupling of polarization states varies over time, necessitating algorithms with the ability to track polarization changes. In the phase- and polarization-diverse JSFR scheme, a MIMO-NN was proposed [81,85] to simultaneously achieve linear optical field recovery, polarization demultiplexing, and non-linear SSBI mitigation. The receiver structure, along with the MIMO-NN, is depicted in Figure 10. The MIMO-NN consists of four layers and takes the six digital waveforms as inputs. It first extracts the in-phase and quadrature components of the dual-polarization optical field. Then, the MIMO-NN performs polarization mode demultiplexing by utilizing the inverse matrix of the polarization rotation unitary matrix. This integrated scheme enables the reconstruction of the optical field, the demultiplexing of the polarization modes, and the mitigation of nonlinear SSBI effects. By harnessing the capabilities of the MIMO-NN, the receiver achieves the detection of four-dimensional modulated signals, encompassing the amplitudes and phases of both polarizations. This advanced technique significantly enhances the SE of the DD system, approaching the performance levels of coherent detection systems.

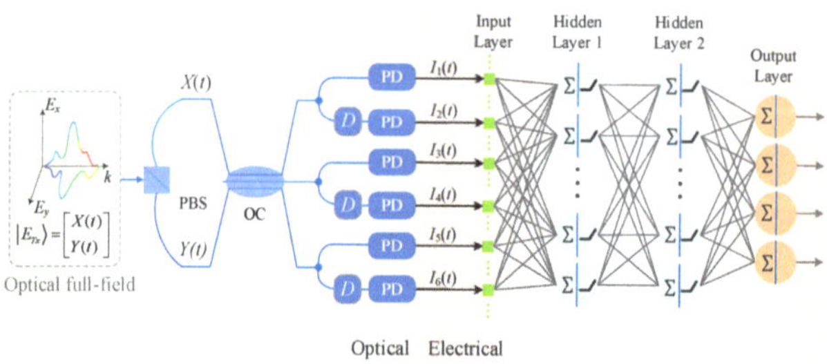

Figure 10. Receiver of the JSFR scheme concatenated with a four-layer NN used for polarization demultiplexing. PBS: polarization beam splitter; OC: 3 × 3 optical coupler; *D*: dispersive element; PD: photodetector.

3.5. Fast Remodeling

Transfer learning (TL) refers to the process of leveraging knowledge and experience gained from previous tasks to improve performance on new target tasks. In TL, the source task and the target task may not be consistent, meaning that they may differ in terms of data distribution, input/output spaces, or even objectives. In optical fiber communications, to reduce the number of training symbols and epochs, TL has been introduced and proven to enable fast remodeling [93,94], nonlinear equalization, and optical signal-to-noise ratio estimation [95]. In [93], a TL-assisted ANN was proposed for a multi-channel nonlinearity mitigation scheme in an SSB system. In the case of multi-channel transmission where multiple channels co-propagate in the same fiber, there exists a correlation of nonlinear distortion. This means that the nonlinear effects introduced by one channel can impact the other channels. Understanding and accounting for this correlation is crucial in designing and optimizing multi-channel transmission systems. By considering the correlation of nonlinear distortion, more accurate modeling and compensation techniques can be developed to mitigate the impact of nonlinearities and improve the overall system performance.

The principle of TL-ANN for multiple channels is shown in Figure 11. At the initial stage, an ANN is trained using labeled training data that have been collected. Once the initial training is complete, the prior distribution of parameters from the trained source model is transferred to accelerate the remodeling process. This parameter transfer avoids the need for re-initialization in the retraining method. By leveraging the learned knowledge from the source model, the remodeling process can be accelerated and potentially achieve a better performance. Subsequently, a few samples are used to train the parameters of the TL-ANN so that it can converge and accurately compensate for the impairments in the current channel. The experimental results show that the required training epochs can be reduced by 80% without BER performance degradation, saving considerable computational complexity.

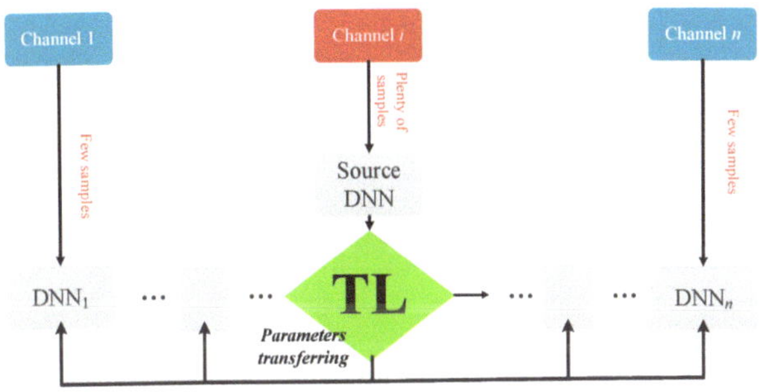

Figure 11. Schematic diagram of TL-assisted nonlinear compensation in a multichannel scheme.

3.6. Optical Signal Processing

Photonic NNs, also known as optical NNs, are a class of NNs that utilize the principles of photonics to perform signal processing in the optical domain. Instead of relying on traditional electronic components, these networks employ optical elements for computation and communication. One specific implementation of photonic NNs is photonic reservoir computing (RC), which is an ML framework that utilizes a fixed, random dynamical system called the reservoir to process data. In the case of a photonic RC, the reservoir is implemented using photonic components and principles. In [96], a recurrent optical spectrum slicing (ROSS) neuromorphic accelerator was proposed to realize an SCD receiver. This network aimed to extract phase information and demodulate QAM formats while simultaneously mitigating CD. The structure of the neuromorphic receiver based on ROSS is illustrated in Figure 12.

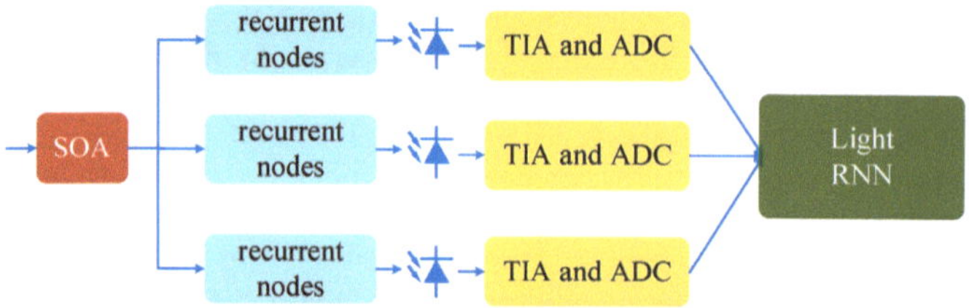

Figure 12. The neuromorphic receiver based on ROSS concatenated with a light RNN. SOA: semiconductor optical amplifier; TIA: transimpedance amplifier; ADC: analog-to-digital converter. Three recurrent nodes comprised from an MZDI in a loop equipped with a variable optical attenuator, phase shifters, and optical delays.

At the receiver side, a semiconductor optical amplifier compensates for the transmission and insertion losses of the integrated chip. The structure includes three recurrent nodes, each consisting of a Mach–Zehnder delay interferometer (MZDI) in a loop equipped with variable optical attenuators, phase shifters, and optical delays. PDs follow these nodes and are then connected to transimpedance amplifiers and ADCs. The subsequent DSP includes a light-based RNN for each quadrature. This configuration enables the extraction of phase information, demodulation of QAM formats, and effective mitigation of CD using the photonic components and principles employed in the ROSS structure. The photonic RC contributes to reducing the power consumption associated with high-bandwidth PDs/ADCs and heavy digital equalization algorithms.

4. Conclusions

This paper introduced the challenges and advancements in SCD systems and reviewed the application of ML techniques in addressing these challenges. The utilization of ML algorithms has exhibited promising results in compensating for various impairments such as fiber nonlinearity, IQ imbalance, SSBI, PR, polarization demultiplexing, and fast channel remodeling. CNNs, LSTMs, sparsely connected ANNs, and MIMO-NNs have been successfully employed to achieve accurate nonlinear impairment compensation and efficient signal processing. Furthermore, transfer learning has been utilized to reduce training time and improve modeling in multi-channel scenarios, while the residual learning method combined with a CNN has been proven effective for optical field recovery. Additionally, the emergence of photonic NNs, such as photonic reservoir computing, harnesses the advantages of photonics for information processing in SCD systems. The integration of ML techniques into SCD systems has resulted in significant enhancements in modulation dimensions, SE, transmission performance, and capacity. Integrating machine learning into direct detection systems may also raise costs due to specialized hardware needs for efficient computation. The actual impact varies with the technology advancements and performance benefits gained. However, further research is necessary to optimize ML models, explore novel network architectures, and address practical implementation challenges to fully leverage the potential of ML in SCD systems. In the context of SCD systems, machine-learning techniques are increasingly favored for tasks such as optical field recovery or phase retrieval: tasks that traditional nonlinear equalization algorithms struggle to achieve. Regarding challenges linked to applying ML in SCD systems, these involve concerns about computational complexity and hardware requirements, especially for ASIC chips. Consequently, it is essential to focus future endeavors on exploring and resolving the intricacies of ML algorithms to facilitate their practical implementation.

In summary, the combination of SCD systems and ML techniques holds tremendous promise for enabling high-capacity, cost-effective, and reliable optical communication networks in the 6G era and beyond. The advancements in ML algorithms offer new avenues for overcoming the challenges and improving the overall performance of SCD systems.

Author Contributions: Conceptualization, Q.W., Z.X., Y.Z. (Yixiao Zhu), Q.Z. and W.H.; methodology, Q.W., Z.X. and Y.Z. (Yixiao Zhu); investigation, Q.W., Y.Z. (Yikun Zhang), H.J., Y.Y., G.Q., L.L., S.W., J.L. (Junpeng Liang), J.W., J.L. (Jiali Li) and Z.H.; writing—original draft preparation, Q.W.; writing—review and editing, Q.W., Z.X. and Y.Z. (Yixiao Zhu). All authors have read and agreed to the published version of the manuscript.

Funding: This work was supported by National Natural Science Foundation of China (62271305).

Institutional Review Board Statement: Not applicable.

Informed Consent Statement: Not applicable.

Data Availability Statement: Data underlying the results presented in this paper are not publicly available at this time but may be obtained from the authors upon reasonable request.

Conflicts of Interest: The authors declare no conflict of interest.

Appendix A

Table A1. This table gives the abbreviations and their definitions used in the paper.

Abbreviation	Definition	Abbreviation	Definition
LO	Local oscillator	CD	Chromatic dispersion
DSP	Digital signal processing	IM-DD	Intensity modulation and direct detection
SE	Spectral efficiency	PD	Photodetector
VSB	Vestigial sideband	DSB	Double-sideband
PR	Phase retrieval	SSB	Single-sideband
CADD	Carrier-assisted differential detection	ASCD	Asymmetric self-coherent detection
SVR	Stokes-vector receiver	JSFR	Jones-space field recovery
SCD	Self-coherent detection	SSBI	Signal-to-signal beating interference
ML	Machine learning	ADC	Analog-to-digital converter
MZM	Mach–Zehnder modulator	NN	Neural network
CNN	Convolutional neural network	BPD	Balanced photodetector
CSPR	Carrier-to-signal power ratio	PBS	Polarization beam splitter
MIMO	Multi-input multi-output	GR	Generalized receiver
QAM	Quadrature amplitude modulation	NLSE	Nonlinear Schrödinger equation
ANN	Artificial neural network	LSTM	Long short-term memory network
RNN	Recurrent neural network	WDM	Wavelength-division-multiplexing
BER	Bit-error-rate	KK	Kramers–Kronig
ReLU	Rectified Linear Unit	TL	Transfer learning
RC	Reservoir computing	ROSS	Recurrent optical spectrum slicing
MZDI	Mach–Zehnder delay interferometer		

References

1. Kikuchi, K. Fundamentals of Coherent Optical Fiber Communications. *J. Light. Technol.* **2016**, *34*, 157–179. [CrossRef]
2. Cvijetic, M. *Coherent and Nonlinear Lightwave Communications*; Artech House: Boston, MA, USA, 2016; ISBN 13-978-0890065907.
3. Batagelj, B.; Janyani, V.; Tomazic, S. Research challenges in optical communications towards 2020 and beyond. *Inf. MIDEM* **2014**, *44*, 177–184.
4. Ip, E.M.; Kahn, J.M. Fiber Impairment Compensation Using Coherent Detection and Digital Signal Processing. *J. Light. Technol.* **2009**, *28*, 502–519. [CrossRef]
5. Uzunidis, D.; Logothetis, M.; Stavdas, A.; Hillerkuss, D.; Tomkos, I. Fifty Years of Fixed Optical Networks Evolution: A Survey of Architectural and Technological Developments in a Layered Approach. *Telecom* **2022**, *3*, 619–674. [CrossRef]

6. Savory, S.J.; Gavioli, G.; Killey, R.I.; Bayvel, P. Electronic compensation of chromatic dispersion using a digital coherent receiver. *Opt. Express* **2007**, *15*, 2120–2126. [CrossRef]
7. Winzer, J.; Essiambr, R. Advanced Optical Modulation Formats. *Proc. IEEE* **2006**, *94*, 952–985. [CrossRef]
8. Zhong, K.; Zhou, X.; Huo, J.; Yu, C.; Lu, C.; Lau, A.P.T. Digital Signal Processing for Short-Reach Optical Communications: A Review of Current Technologies and Future Trends. *J. Light. Technol.* **2018**, *36*, 377–400. [CrossRef]
9. Wu, Q.; Zhu, Y.; Cheng, Z.; Yin, L.; Hu, W. Spectrally sliced heterodyne coherent receiver with halved electrical bandwidth. *Chin. Opt. Lett.* **2022**, *20*, 090601-1-090601-6. [CrossRef]
10. Shieh, W.; Sun, C.; Ji, H. Carrier-assisted differential detection. *Light. Sci. Appl.* **2020**, *9*, 1–9. [CrossRef]
11. Che, D.; Sun, C.; Shieh, W. Maximizing the spectral efficiency of Stokes vector receiver with optical field recovery. *Opt. Express* **2018**, *26*, 28976–28981. [CrossRef]
12. Che, D.; Hu, Q.; Shieh, W. Linearization of Direct Detection Optical Channels Using Self-Coherent Subsystems. *J. Light. Technol.* **2016**, *34*, 516–524. [CrossRef]
13. Zhang, J.; Yu, J.; Li, X.; Wei, Y.; Wang, K.; Zhao, L.; Zhou, W.; Kong, M.; Pan, X.; Liu, B.; et al. 100 Gbit/s VSB-PAM-n IM/DD transmission system based on 10 GHz DML with optical filtering and joint nonlinear equalization. *Opt. Express* **2019**, *27*, 6098–6105. [CrossRef] [PubMed]
14. Li, Z.; Erkilinc, M.S.; Shi, K.; Sillekens, E.; Galdino, L.; Thomsen, B.C.; Bayvel, P.; Killey, R.I. SSBI Mitigation and the Kramers–Kronig Scheme in Single-Sideband Direct-Detection Transmission with Receiver-Based Electronic Dispersion Compensation. *J. Light Technol.* **2017**, *35*, 1887–1893. [CrossRef]
15. Wu, Q.; Zhu, Y.; Hu, W. Low-complexity and Non-iterative SSBI Decomposition and Cancellation Algorithm for SSB Direct Detection System. In Proceedings of the Optical Fiber Communication Conference, San Diego, CA, USA, 6–10 March 2022; p. M2H.6.
16. Bo, T.; Kim, B.; Yu, Y.; Kim, D.; Kim, H. Generation of Broadband Optical SSB Signal Using Dual Modulation of DML and EAM. *J. Light Technol.* **2021**, *39*, 3064–3071. [CrossRef]
17. Bo, T.; Kim, H. Generalized model of optical single sideband generation using dual modulation of DML and EAM. *Opt. Express* **2020**, *28*, 28491–28501. [CrossRef]
18. Chen, H.; Fontaine, N.K.; Gene, J.M.; Ryf, R.; Neilson, D.T.; Raybon, G. Dual Polarization Full-Field Signal Waveform Reconstruction Using Intensity Only Measurements for Coherent Communications. *J. Light. Technol.* **2020**, *38*, 2587–2597. [CrossRef]
19. Gerchberg, R. A practical algorithm for the determination of phase from image and diffraction plane pictures. *Optik* **1972**, *35*, 237–246.
20. Karar, A.; Falou, A.; Barakat, J.; Gürkan, Z.; Zhong, K. Recent Advances in Coherent Optical Communications for Short-Reach: Phase Retrieval Methods. *Photonics* **2023**, *10*, 308. [CrossRef]
21. Wu, Q.; Zhu, Y.; Zhuge, Q.; Hu, W. Carrier-Assisted Phase Retrieval with Multiple Projections. *J. Light. Technol.* **2022**, *36*, 465–468. [CrossRef]
22. Qin, P.; Bai, C.; Wang, Z.; Xu, H.; Yang, L.; Zhang, Y.; Ge, P.; Luo, X. Offset double sideband carrier assisted differential de-tection with field recovery at low carrier-to-signal power ratio. *Opt. Express* **2022**, *30*, 48112–48132. [CrossRef]
23. Qin, P.; Bai, C.; Xu, H.; Yang, L.; Zhang, Y.; Luo, X.; Chi, X.; Zhang, R.; Li, P.; Wu, Y. Offset double-sideband signal field re-covery at low CSPR using filter-assisted direct detection. *Opt. Express* **2022**, *30*, 48112–48132. [CrossRef]
24. Li, X.; O'sullivan, M.; Xing, Z.; Alam, S.; Mousa-Pasandi, M.E.; Plant, D.V. Asymmetric self-coherent detection. *Opt. Express* **2021**, *29*, 25412–25427. [CrossRef] [PubMed]
25. Che, D.; Li, A.; Chen, X.; Hu, Q.; Wang, Y.; Shieh, W. Stokes Vector Direct Detection for Linear Complex Optical Channels. *J. Light. Technol.* **2015**, *33*, 678–684. [CrossRef]
26. Kikuchi, K. Quantum Theory of Noise in Stokes Vector Receivers and Application to Bit Error Rate Analysis. *J. Light. Technol.* **2020**, *38*, 3164–3172. [CrossRef]
27. Qian, D.; Cvijetic, N.; Hu, J.; Wang, T. 108 Gb/s OFDMA-PON with Polarization Multiplexing and Direct Detection. *J. Light. Technol.* **2009**, *28*, 484–493. [CrossRef]
28. Ji, H.; Li, J.; Li, X.; Dong, S.; Xu, Z.; Su, Y.; Shieh, W. Complementary Polarization-Diversity Coherent Receiver for Self-Coherent Homodyne Detection with Rapid Polarization Tracking. *J. Light. Technol.* **2022**, *40*, 2773–2779. [CrossRef]
29. Wu, Q.; Zhu, Y.; Zhuge, Q.; Hu, W. Dual-Carrier-Assisted Phase Retrieval for Polarization Division Multiplexing. *J. Light. Technol.* **2022**, *40*, 7297–7306. [CrossRef]
30. Wu, Q.; Zhu, Y.; Jiang, H.; Zhuge, Q.; Hu, W. Optical Field Recovery in Jones Space. *J. Light. Technol.* **2022**, *41*, 66–74. [CrossRef]
31. Krizhevsky, A.; Sutskever, I.; Hinton, G. ImageNet Classification with Deep Convolutional Neural Networks. In Proceedings of the Advances in Neural Information Processing Systems, Lake Tahoe, NV, USA, 3–16 December 2012; pp. 1097–1105.
32. Otter, D.W.; Medina, J.R.; Kalita, J.K. A Survey of the Usages of Deep Learning for Natural Language Processing. *IEEE Trans. Neural Netw. Learn. Syst.* **2021**, *32*, 604–624. [CrossRef]
33. Richens, J.; Lee, C.; Johri, S. Improving the accuracy of medical diagnosis with causal machine learning. *Nat. Commun.* **2022**, *11*, 3923. [CrossRef]
34. Musumeci, F.; Rottondi, C.; Nag, A.; Macaluso, I.; Zibar, D.; Ruffini, M.; Tornatore, M. An Overview on Application of Machine Learning Techniques in Optical Networks. *IEEE Commun. Surv. Tutor.* **2018**, *21*, 1383–1408. [CrossRef]

35. Rottondi, C.; Barletta, L.; Giusti, A.; Tornatore, M. Machine-learning method for quality of transmission prediction of unestablished lightpaths. *J. Opt. Commun. Netw.* **2018**, *10*, A286. [CrossRef]
36. Khan, F.N.; Zhou, Y.; Lau, A.P.T.; Lu, C. Modulation format identification in heterogeneous fiber-optic networks using artificial neural networks. *Opt. Express* **2012**, *20*, 12422–12431. [CrossRef] [PubMed]
37. Saif, W.S.; Esmail, M.A.; Ragheb, A.M.; Alshawi, T.A.; Alshebeili, S.A. Machine Learning Techniques for Optical Performance Monitoring and Modulation Format Identification: A Survey. *IEEE Commun. Surv. Tutor.* **2020**, *22*, 2839–2882. [CrossRef]
38. Eriksson, T.A.; Bulow, H.; Leven, A. Applying Neural Networks in Optical Communication Systems: Possible Pitfalls. *IEEE Photon-Technol. Lett.* **2017**, *29*, 2091–2094. [CrossRef]
39. Chen, Z.; Wang, W.; Zou, D.; Ni, W.; Luo, D.; Li, F. Real-Valued Neural Network Nonlinear Equalization for Long-Reach PONs Based on SSB Modulation. *IEEE Photon-Technol. Lett.* **2022**, *35*, 167–170. [CrossRef]
40. Xu, Z.; Ji, H.; Yang, Y.; Qiao, G.; Wu, Q.; Li, J.; Lu, W.; Liu, L.; Wang, S.; Wei, J.; et al. Mixed-Precision Integer-Arithmetic-Only Neural Network-Based Equalizers for DML-Based Short-Reach IM/DD Systems. In Proceedings of the 2023 European Conference on Optical Communication (ECOC), Glasgow, Scotland, 1–5 October 2023; p. Th.A.3.4.
41. Xu, Z.; Dong, S.; Manton, J.H.; Shieh, W. Low-Complexity Multi-Task Learning Aided Neural Networks for Equalization in Short-Reach Optical Interconnects. *J. Light. Technol.* **2021**, *40*, 45–54. [CrossRef]
42. Giacoumidis, E. Blind Nonlinearity Equalization by Machine Learning based Clustering for Single-and Multi-Channel Coherent Optical OFDM. *J. Light. Technol.* **2018**, *36*, 721–727. [CrossRef]
43. Zibar, D.; Piels, M.; Jones, R.; Schaeffer, C.G. Machine Learning Techniques in Optical Communication. *J. Light. Technol.* **2016**, *34*, 1442–1452. [CrossRef]
44. Lu, X.; Qiao, L.; Zhou, Y.; Yu, W.; Chi, N. An I-Q-Time 3-dimensional post-equalization algorithm based on DBSCAN of machine learning in CAP VLC system. *Opt. Commun.* **2019**, *430*, 299–303. [CrossRef]
45. Nguyen, T.; Mhatli, S.; Giacoumidis, E.; Van Compernolle, L.; Wuilpart, M.; Mégret, P. Fiber Nonlinearity Equalizer Based on Support Vector Classification for Coherent Optical OFDM. *IEEE Photon-J.* **2016**, *8*, 2. [CrossRef]
46. Giacoumidis, E.; Mhatli, S.; Stephens, M.F.; Tsokanos, A.; Wei, J.; McCarthy, M.E.; Doran, N.J.; Ellis, A.D. Reduction of Non-linear Inter-Subcarrier Intermixing in Coherent Optical OFDM by a Fast Newton-based Support Vector Machine Nonlinear Equalizer. *J. Light. Technol.* **2017**, *35*, 2391–2397. [CrossRef]
47. Chagnon, M.; Karanov, B.; Schmalen, L. Experimental demonstration of a dispersion tolerant end-to-end deep learning-based im-dd transmission system. In Proceedings of the European Conference on Optical Communication (ECOC), Roma, Italy, 23–27 September 2018; pp. 1–3.
48. Katumba, A.; Yin, X.; Dambre, J.; Bienstman, P. A Neuromorphic Silicon Photonics Nonlinear Equalizer for Optical Communications with Intensity Modulation and Direct Detection. *J. Light. Technol.* **2019**, *37*, 2232–2239. [CrossRef]
49. Haigh, P.A.; Ghassemlooy, Z.; Rajbhandari, S.; Papakonstantinou, I.; Popoola, W. Visible Light Communications: 170 Mb/s Using an Artificial Neural Network Equalizer in a Low Bandwidth White Light Configuration. *J. Light. Technol.* **2014**, *32*, 1807–1813. [CrossRef]
50. Tanimura, T.; Hoshida, T.; Kato, T.; Watanabe, S.; Morikawa, H. Data-analytics-based optical performance monitoring technique for optical transport networks. In Proceedings of the Optical Fiber Communication Conference, San Diego, CA, USA, 11–15 March 2018; p. Tu3E.3.
51. Xu, Z.; Sun, C.; Manton, J.H.; Shieh, W. Joint Equalization of Linear and Nonlinear Impairments for PAM4 Short-Reach Direct Detection Systems. *IEEE Photon-Technol. Lett.* **2021**, *33*, 425–428. [CrossRef]
52. You, Y.; Jiang, Z.; Janz, C. Machine Learning-Based EDFA Gain Model. In Proceedings of the 2018 European Conference on Optical Communication (ECOC), Rome, Italy, 23–27 September 2018; pp. 1–3.
53. Morais, R.M.; Pedro, J. Machine Learning Models for Estimating Quality of Transmission in DWDM Networks. *J. Opt. Commun. Netw.* **2018**, *10*, D84–D99. [CrossRef]
54. Yankov, M.; Ros, F.; Moura, U.; Carena, A.; Zibar, D. Flexible Raman Amplifier Optimization Based on Machine Learn-ing-Aided Physical Stimulated Raman Scattering Model. *J. Light. Technol.* **2017**, *35*, 1887–1893.
55. Panayiotou, T.; Savva, G.; Shariati, B.; Tomkos, I.; Ellinas, G. Machine learning for QoT estimation of unseen optical network states. In Proceedings of the Optical Fiber Communication Conference 2019, San Diego, CA, USA, 3–7 March 2019; p. Tu2E.2.
56. Jimenez, T.; Aguado, J.C.; de Miguel, I.; Duran, R.J.; Angelou, M.; Merayo, N.; Fernandez, P.; Lorenzo, R.M.; Tomkos, I.; Abril, E.J. A Cognitive Quality of Transmission Estimator for Core Optical Networks. *J. Light. Technol.* **2013**, *31*, 942–951. [CrossRef]
57. Xu, Z.; Sun, C.; Ji, T.; Ji, H.; Manton, J.; Shieh, W. Cascade Recurrent Neural Network Enabled 100-Gb/s PAM4 Short-Reach Optical Link Based on DML. In Proceedings of the Optical Fiber Communication Conference, San Diego, CA, USA, 8–12 March 2020; p. W2A.45.
58. Sambo, N.; Pointurier, Y.; Cugini, F.; Valcarenghi, L.; Castoldi, P.; Tomkos, I. Light path Establishment Assisted by Offline QoT Estimation in Transparent Optical Networks. *J. Opt. Commun. Netw.* **2010**, *2*, 928–937. [CrossRef]
59. Seve, E.; Pesic, J.; Delezoide, C.; Bigo, S.; Pointurier, Y. Learning process for reducing uncertainties on network parameters and design margins. *J. Opt. Commun. Netw.* **2018**, *10*, A298–A306. [CrossRef]
60. Wang, D.; Zhang, M.; Li, J.; Li, Z.; Li, J.; Song, C.; Chen, X. Intelligent constellation diagram analyzer using convolutional neural network-based deep learning. *Opt. Express* **2017**, *25*, 17150–17166. [CrossRef] [PubMed]

61. Musumeci, F.; Rottondi, C.E.M.; Corani, G.; Shahkarami, S.; Cugini, F.; Tornatore, M. A Tutorial on Machine Learning for Failure Management in Optical Networks. *J. Light. Technol.* **2019**, *37*, 4125–4139. [CrossRef]
62. Varughese, S.; Lippiatt, D.; Richter, T.; Tibuleac, S.; Ralph, S.E. Identification of Soft Failures in Optical Links Using Low Complexity Anomaly Detection. In Proceedings of the Optical Fiber Communication Conference (OFC) 2019, San Diego, CA, USA, 3–7 March 2019; p. W2A.46.
63. Shahkarami, S.; Musumeci, F.; Cugini, F.; Tornatore, M. Machine-Learning-Based Soft-Failure Detection and Identification in Optical Networks. In Proceedings of the Optical Fiber Communications Conference and Exposition (OFC 2018), San Diego, CA, USA, 11–15 March 2018; pp. 1–3. [CrossRef]
64. Khan, F.N.; Fan, Q.; Lu, C.; Lau, A.P.T. An Optical Communication's Perspective on Machine Learning and Its Applications. *J. Light. Technol.* **2019**, *37*, 493–516. [CrossRef]
65. Vela, A.P.; Ruiz, M.; Fresi, F.; Sambo, N.; Cugini, F.; Meloni, G.; Poti, L.; Velasco, L.; Castoldi, P. BER Degradation Detection and Failure Identification in Elastic Optical Networks. *J. Light. Technol.* **2017**, *35*, 4595–4604. [CrossRef]
66. Qin, P.; Bai, C.; Qi, Q.; Xu, H.; Yang, F.; Li, P.; Zhang, Y. Analytical Study and Noise Mitigation for Complex-Valued Optical DSB Transmission with Carrier-Assisted Differential Detection Receiver. *IEEE Access* **2023**, *11*, 31443–31455. [CrossRef]
67. Zhang, K.; Fan, Y.; Ye, T.; Tao, Z.; Oda, S.; Tanimura, T.; Akiyama, Y.; Hoshida, T. Fiber nonlinear noise-to-signal ratio estimation by machine learning. In Proceedings of the Optical Fiber Communication Conference (OFC) 2019, San Diego, CA, USA, 3–7 March 2019; p. Th2A.45.
68. Schaedler, M.; Kuschnerov, M.; Pachnicke, S.; Bluemm, C.; Pittala, F.; Changsong, X. Subcarrier Power Loading for Coherent Optical OFDM optiMized by Machine Learning. In Proceedings of the Optical Fiber Communication Conference, San Diego, CA, USA, 3–7 March 2019.
69. Wang, C.; Fu, S.; Xiao, Z.; Tang, M.; Liu, D. Long Short-Term Memory Neural Network (LSTM-NN) Enabled Accurate Optical Signal-to-Noise Ratio (OSNR) Monitoring. *J. Light. Technol.* **2019**, *37*, 4140–4146. [CrossRef]
70. Kamalov, V.; Jovanovski, L.; Vusirikala, V.; Zhang, S.; Yaman, F.; Nakamura, K.; Inoue, T.; Mateo, E.; Inada, Y. Evolution from 8QAM live Traffic to PS 64-QAM with Neural-Network Based Nonlinearity Compensation on 11000 km Open Subsea Cable. In Proceedings of the Optical Fiber Communication Conference, San Diego, CA, USA, 11–15 March 2018.
71. Wang, Z.; Yang, A.; Guo, P.; He, P. OSNR and nonlinear noise power estimation for optical fiber communication systems using LSTM based deep learning technique. *Opt. Express* **2018**, *26*, 21346–21357. [CrossRef]
72. Jones, R.; Eriksson, T.; Yankov, M.; Zibar, D. Deep Learning of Geometric Constellation Shaping Including Fiber Nonlinearities. In Proceedings of the 2018 European Conference on Optical Communication (ECOC), Rome, Italy, 23–27 September 2018; pp. 1–3.
73. Jones, R.; Yankov, M.; Zibar, D. End-to-end learning for GMI optimized geometric constellation shape. In Proceedings of the 2018 European Conference on Optical Communication (ECOC), Dublin, Ireland, 22–26 September 2019; pp. 1–3.
74. Khan, F.N.; Zhong, K.; Zhou, X.; Al-Arashi, W.H.; Yu, C.; Lu, C.; Lau, A.P.T. Joint OSNR monitoring and modulation format identification in digital coherent receivers using deep neural networks. *Opt. Express* **2017**, *25*, 17767–17776. [CrossRef]
75. Jarajreh, M.A.; Giacoumidis, E.; Aldaya, I.; Le, S.T.; Tsokanos, A.; Ghassemlooy, Z.; Doran, N.J. Artificial Neural Network Nonlinear Equalizer for Coherent Optical OFDM. *IEEE Photon-Technol. Lett.* **2015**, *27*, 387–390. [CrossRef]
76. Khan, F.N.; Yu, Y.; Tan, M.C.; Yu, C.; Lau, A.P.T.; Lu, C. Simultaneous OSNR monitoring and modulation format identification using asynchronous single channel sampling. In Proceedings of the Asia Communications and Photonics Conference 2015, Hong Kong, China, 19–23 November 2015; p. AS4F.6.
77. Tanimura, T.; Hoshida, T.; Kato, T.; Watanabe, S.; Morikawa, H. Convolutional Neural Network-Based Optical Performance Monitoring for Optical Transport Networks. *J. Opt. Commun. Netw.* **2019**, *11*, A52–A59. [CrossRef]
78. Antonelli, C.; Mecozzi, A.; Shtaif, M.; Chen, X.; Chandrasekhar, S.; Winzer, P.J. Kramers-Kronig coherent receiver. *Optica* **2016**, *3*, 1220–1227. [CrossRef]
79. Bo, T.; Kim, H. Kramers-Kronig receiver operable without digital upsampling. *Opt. Express* **2018**, *26*, 13810–13818. [CrossRef] [PubMed]
80. Bo, T.; Kim, H. Toward Practical Kramers-Kronig Receiver: Resampling, Performance, and Implementation. *J. Light. Technol.* **2018**, *37*, 461–469. [CrossRef]
81. Ming, H.; Zhang, L.; Yang, F.; Ruan, X.; Zhang, F. High-speed SSB transmission with a silicon MZM and a soft combined artificial neural network-based equalization. *Opt. Lett.* **2020**, *45*, 2066–2069. [CrossRef] [PubMed]
82. Wu, Q.; Li, X.; Zhu, Y.; Chen, C.; Xu, Z.; Li, J.; He, Z.; Hu, W. Single-Wavelength Net 1 Tb/s Transmission in SMF and 6.4 Tb/s in Weakly-Coupled 7-Core MCF Using a Phase-and Polarization-Diverse Direct Detection Receiver in Jones spacey. In Proceedings of the 2023 OptoElectronics and Communications Conference (OECC), Shanghai, China, 2–6 July 2023.
83. Li, X.; An, S.; Ji, H.; Li, J.; Shieh, W.; Su, Y. Deep-learning-enabled high-performance full-field direct detection with dispersion diversity. *Opt. Express* **2022**, *30*, 11767–11788. [CrossRef]
84. Li, X.; Li, J.; An, S.; Liu, H.; Shieh, W.; Su, Y. Deep-learning-enabled Direct Detection with Reduced Computational Complexity and High Electrical-spectral-efficiency. *J. Light. Technol.* **2023**, *41*, 5495–5502. [CrossRef]
85. Wu, Q.; Zhu, Y.; Jiang, H.; Fu, M.; Zhang, Y.; Zhuge, Q.; Hu, W. Four-dimensional direct detection with Jones space optical full field recovery. *arXiv* **2022**, arXiv:2212.14587.
86. Wan, Z.; Li, J.; Shu, L.; Luo, M.; Li, X.; Fu, S.; Xu, K. Nonlinear equalization based on pruned artificial neural networks for 112-Gb/s SSB-PAM4 transmission over 80-km SSMF. *Opt. Express* **2018**, *26*, 10631–10642. [CrossRef]

87. Fang, X.; Chen, X.; Yang, F.; Zhang, L.; Zhang, F. 6.4Tb/s SSB WDM Transmission Over 320km SSMF with Linear Network-Assisted LSTM. *Photonic. Tech. Lett.* **2021**, *33*, 1407–1410. [CrossRef]
88. An, S.; Zhu, Q.; Li, J.; Ling, Y.; Su, Y. 112-Gb/s SSB 16-QAM signal transmission over 120-km SMF with direct detection using a MIMO-ANN nonlinear equalizer. *Opt. Express* **2019**, *27*, 12794–12805. [CrossRef]
89. Orsuti, D.; Antonelli, C.; Chiuso, A.; Santagiustina, M.; Mecozzi, A.; Galtarossa, A.; Palmieri, L. Phase Retrieval Receiver Based on Deep Learning for Minimum-Phase Signal Recovery. In Proceedings of the 2018 European Conference on Optical Com-munication (ECOC), Basel, Switzerland, 18–22 September 2022; pp. 1–3.
90. Orsuti, D.; Antonelli, C.; Chiuso, A.; Santagiustina, M.; Mecozzi, A.; Galtarossa, A.; Palmieri, L. Deep Learning-Based Phase Retrieval Scheme for Minimum-Phase Signal Recovery. *J. Light. Technol.* **2023**, *41*, 578–592. [CrossRef]
91. Orsuti, D.; Cappelletti, M.; Santagiustina, M.; Galtarossa, A.; Palmieri, L. Edge-carrier-assisted Phase-Retrieval Based on Deep Learning Enabling low CSPR and low Applied Dispersion Values. In Proceedings of the Optical Fiber Communication Con-ference, San Diego, CA, USA, 5–9 March 2023; p. M2F.3.
92. Zhu, Y.; Li, L.; Fu, Y.; Hu, W. Symmetric carrier assisted differential detection receiver with low-complexity signal-signal beating interference mitigation. *Opt. Express* **2020**, *28*, 19008–19022. [CrossRef] [PubMed]
93. Zhang, J.; Xia, L.; Zhu, M.; Hu, S.; Xu, B.; Qiu, K. Fast remodeling for nonlinear distortion mitigation based on transfer learning. *Opt. Lett.* **2019**, *44*, 4243–4246. [CrossRef] [PubMed]
94. Xu, Z.; Sun, C.; Ji, T.; Ji, H.; Manton, J.; Shieh, W. Transfer learning aided neural networks for nonlinear equalization in short-reach direct detection systems. In Proceedings of the Optical Fiber Communication Conference, San Diego, CA, USA, 8–12 March 2020; p. T4D.4.
95. Xia, L.; Zhang, J.; Hu, S.; Zhu, M.; Song, Y.; Qiu, K. Transfer learning assisted deep neural network for OSNR estimation. *Opt. Express* **2019**, *27*, 19398–19406. [CrossRef]
96. Sozos, K.; Deligiannidis, S.; Mesaritakis, C.; Bogris, A. Self-Coherent Receiver Based on a Recurrent Optical Spectrum Slicing Neuromorphic Accelerator. *J. Light. Technol.* **2023**, *41*, 2666–2674. [CrossRef]

MDPI AG
Grosspeteranlage 5
4052 Basel
Switzerland
Tel.: +41 61 683 77 34

Photonics Editorial Office
E-mail: photonics@mdpi.com
www.mdpi.com/journal/photonics

www.ingramcontent.com/pod-product-compliance
Lightning Source LLC
Chambersburg PA
CBHW041410210926
51850CB00013B/3235
* 9 7 8 3 7 2 5 8 4 0 9 6 0 *